P9-DNA-516

FROM COLD WAR TO HOT PEACE

FROM COLD WAR TO HOT PEACE

AN AMERICAN AMBASSADOR IN PUTIN'S RUSSIA

Michael McFaul

Houghton Mifflin Harcourt
Boston • New York
2018

hmhco.com

Library of Congress Cataloging-in-Publication Data is available.
ISBN 978-0-544-71624-7 (hardcover)

Book design by Chloe Foster

Printed in the United States of America
DOC 10 9 8 7 6 5 4 3 2 1

CONTENTS

PROLOGUE

As *Air Force One* began its initial descent into Prague on a clear, sunny day in April 2010, President Obama asked Gary Samore and me to join him in his office at the front of the plane to run over the final talking points for his meeting with Russian president Dmitry Medvedev later that day. The president was in an ebullient mood. In the same city a year earlier, Obama had given what may have been his most important foreign policy speech to date, calling for a nuclear-free world. Now, just a year later, he was delivering a major piece of business toward his Prague Agenda, as we called it. With his Russian partner, he was about to sign the New Strategic Arms Reduction Treaty (New START), reducing by 30 percent the number of nuclear weapons allowed in the two countries. Samore, our special assistant to the president for weapons of mass destruction at the National Security Council, was our lead at the White House in getting this treaty done; I was his wingman. So this quick trip to Prague was a day of celebration for the two of us as well.

On the tarmac, Obama asked Gary and me to ride with him in his limousine — "the Beast" — to the majestic Prague Castle, where the signing ceremony would take place. For a while, we discussed our game plan for pushing Medvedev to support sanctions on Iran, but mostly the drive was a victory lap. Obama waved to the crowds on the streets, flashing his broad smile through the tinted bulletproof glass. We all felt good about getting something concrete done, always a challenge in government work. We also allowed ourselves that day to imagine even deeper cooperation between the United States and Russia on issues beyond arms control. Maybe we truly had come to a turning point in the bumpy road of U.S.-Russia relations since the collapse of the Soviet Union, a turn toward genuine strategic cooperation that had eluded previous American and Russian leaders.

With new young presidents in the White House and in the Kremlin, the Cold War felt distant. We were signing the first major arms control treaty in decades, working together to stop Iran from acquiring a nuclear weapon, joined in efforts to fight the Taliban and al-Qaeda in Afghanistan, and increasing trade and investment between our two countries. We seemed to have put contentious issues such as NATO expansion and the Iraq War behind us, and were now digging deeper into areas of mutual interest. That year, for instance, Russian and American paratroopers were jumping out of airplanes together in Colorado, conducting joint counterterrorist-training operations, at the same time that American and Russian entrepreneurs were working together to develop Skolkovo, Russia's aspirational Silicon Valley. We were even discussing the possibilities of cooperating on missile defense. These were breakthroughs unimaginable just two years earlier. Medvedev seemed like a pro-Western modernizer, albeit a cautious one. On that celebratory drive through the cobblestone streets of Prague, the Reset — the bumper sticker for Obama's Russia policy — appeared to be working.

Later that afternoon, Medvedev and Obama signed the New START treaty, drank champagne together, and spoke glowingly about possibilities for further cooperation. At the time, solid majorities in both countries were convinced of such possibilities. Russia was popular in America, and America was popular in Russia. I flew home the next day in a great mood, convinced that we were making history.

Only two years later, on a cold, dark day in January 2012, I arrived in Moscow as the new U.S. ambassador to the Russian Federation, charged with continuing the Reset. I had thought about, written about, and worked toward closer relations with the Soviet Union and then Russia since my high school debating days, so this new mission should have been a crowning achievement of my career: an opportunity of a lifetime to further my ideas about American-Russian relations. It was not. On my first day of work at the embassy, the Russian state-controlled media accused President Obama of sending me to Russia to foment revolution. On his evening commentary show, *Odnako,* broadcast on the most popular television network in Russia, Mikhail Leontiev warned his viewers that I was neither a Russia expert nor a traditional diplomat, but a professional revolutionary whose assignment was to finance and organize Russia's political opposition as it plotted to overthrow the Russian government; to finish *Russia's Unfinished Revolution,* the title of one of my books written a decade earlier. This portrayal of my mission to Moscow would haunt me for the rest of my days as ambassador.

A few months later, in May 2012, I accompanied my former boss at the White House, National Security Advisor Tom Donilon, to his meeting with President-elect Putin. This was the first meeting between a senior Obama official and Putin since Putin's reelection in March 2012. We met at Novo-Ogaryovo, Putin's country estate, the same place where Obama had enjoyed a cordial, constructive, three-hour breakfast with the then prime minister four years earlier. Putin listened politely to Tom's arguments for continued cooperation. At some point in their dialogue, however, he turned away from Tom to stare intensely at me with his steely blue eyes and stern scowl to accuse me of purposely seeking to ruin U.S.-Russia relations. Putin seemed genuinely angry with me; I was genuinely alarmed. The hair on the back of my neck stood on end and sweat covered my brow as I endured this tongue-lashing from one of the most powerful people in the world.

In Prague, I had been the author of the Reset, the driver of closer relations with Russia. In Moscow, I was now a revolutionary, a usurper, and Vladimir Putin's personal foe.

What happened? How did we go from toasting the Reset in Prague in 2010 to lamenting its end in Moscow just two years later? Nothing fundamental had changed in our policy toward Russia. Nor had Russia done anything abroad that might trigger new animosity — that would come later. The one obvious change between these two meetings was Russia's leadership. Medvedev was president when we were in Prague to sign New START; Putin was elected president soon after my arrival in Moscow as U.S. ambassador. But that seemed too simple an explanation. After all, Putin was prime minister during the heyday of the Reset, and most people thought he was calling the shots during that time.

And then things became even worse. Two years into his third term as Russian president, in February 2014, Putin invaded Ukraine, annexing Crimea and supporting separatist militias in the eastern part of the country. Not since

World War II had a European country violated the sovereignty of another country in this way. Putin's outrageous and shocking actions reaffirmed the end of the Reset, and with it, three decades of American and Russian leaders' efforts to build a more cooperative relationship after the end of the Cold War. The project of Russian integration into the West — started by Ronald Reagan and sustained to varying degrees by all post–Cold War presidents — was over.

Sometime between the Obama-Medvedev summit in Prague in 2010 and Putin's invasion of Ukraine in 2014, public opinion in both countries also flipped: solid majorities in both Russia and the United States now perceived each other as enemies.

This new era of confrontation did not mark a return to the Cold War, exactly, but most certainly could be described as a hot peace. Unlike during the Cold War, the Kremlin today no longer promotes an ideology with worldwide appeal. Russian economic and military power is growing, but Russia has not obtained superpower status like the Soviet Union. And the entire globe is not divided between red and blue states — the communist bloc versus the free world — reinforced by opposing alliance structures. Russia today has very few allies. And yet, our new era of hot peace has resurrected some features eerily reminiscent of the Cold War, while also adding new dimensions of confrontation. A new ideological struggle has emerged between Russia and the West, not between communism and capitalism but between democracy and autocracy. Putin also has championed a new set of populist, nationalist, conservative ideas antithetical to the liberal, international order anchored by the United States. Putinism now has admirers in Western democracies, just as communism did during the Cold War. Europe is divided again between East and West; the line has just moved farther east. Russia's military, economic, cyber, and informational capabilities to project power and ideas have also grown considerably in the last several years. And sometimes, this hot peace has morphed into hot wars, not directly between the United States and Russia but between proxies in Ukraine and Syria. And then there are some acts of Russian aggression that Soviet leaders during the Cold War never dared to attempt, including annexing territory in Ukraine and intervening in the 2016 U.S. presidential election. Khrushchev and Brezhnev confronted the United States around the world, even at times using military force as an instrument to fight the Cold War, but never annexed land or audaciously violated American sovereignty. As Putin's Russia has amassed new capabilities and greater intent to challenge the United States and the West more generally, America's standing as the leader of the free world and anchor of the liberal international order has waned. Compared to the Cold War, that's new too. Today's hot peace is not as dangerous as the worst moments of the Cold War, but most certainly is tenser than some of the more cooperative periods of the Cold War.

What went wrong? Why did the end of the Cold War not produce closer relations between our two countries? Could this new, tragic era of confrontation have been avoided?

And what had I gotten wrong? From my days as a high school debater in Bozeman, Montana, in 1979 to my years as ambassador to Russia ending in 2014, I had argued that closer relations with Moscow served American national interests. As a college student, I'd been so eager to deepen engagement with the Russians — or the Soviets, back then — that I enrolled in first-year Russian in the fall quarter of my freshman year at Stanford University and then traveled to what Ronald Reagan called "the evil empire," against the wishes of my mother, on my first trip abroad, to study in Leningrad in the summer of 1983. When the Soviet Union collapsed in 1991, I again packed my bags and moved to Russia to help support market and democratic reforms there, believing that those changes would help bring our two countries closer together. Beginning on January 21, 2009, I worked for President Obama at the National Security Council, and in 2012, I became his ambassador, animated by the belief that a more cooperative relationship served American national interests. I have spent the greater part of my life trying to deepen relations between Russia and the United States. But in 2014 all these efforts seemed for naught. Heightening my sense of frustration, I was banned by Putin from traveling to Russia — the first U.S. ambassador since George Kennan, in 1952, to be barred entry to the country. What had I personally done wrong? Had I pursued the wrong strategies? Or had I embraced the wrong goals in the first place?

This book seeks to answer these big, hard questions about Russian-American relations over the last thirty years, from my perspective as both analyst and participant. The story begins with the first reset in U.S.-Soviet relations, in the Reagan-Gorbachev era, and ends with the last, failed attempt initiated by President Obama. It's a complex story, shaped by Cold War legacies, punctuated by economic depression and civil wars, and interrupted by popular uprisings and foreign interventions. I will address all of these factors as potential causes of cooperation and confrontation. However, in my account, individuals — their ideas and their decisions — drive the narrative of U.S.-Russia relations over the last three decades. Real people made decisions that sometimes produced cooperation and other times led to confrontation. Relations between the United States and Russia were not determined by the balance of power between our two countries, or by innate forces such as economic development, culture, history, or geography. Where others see destiny, determinism, and inevitability, I see choices, contingency, and opportunities, realized and missed.[1] In this book, I evaluate both American and Russian leaders — their ideas, decisions, and behavior — by how well they succeeded in achieving what in my view were desirable and attainable objectives. I try to evaluate my own work through the same critical lens.

U.S. leaders have certainly pursued policies that by turns nurtured and undermined cooperation with Russia. I will discuss how and why. I personally made some mistakes in both analysis and actions. Those are not glossed over here. But in my account, decisions and actions by Russian leaders — shaped in large part by their domestic politics — drive most of the drama in American-Russian relations, sometimes in a positive direction, other times in a negative direction.

This book is one participant's recollections, not a dispassionate history of U.S.-Russian relations over the last thirty years. As I have learned from writing other scholarly books, memories are imperfect sources for the writing of history. It is too soon to draft the definitive account of these events, especially those that occurred in the Obama era, because almost none of the important documents — including ones that I wrote — have been declassified. Few of the central decision makers, in Russia or the United States, have been interviewed. Only a handful of the central policymakers in the United States have written memoirs. The memoir literature from Russian decision makers is even thinner. On some big policy issues during the Obama era, such as Iran and Syria, I also recognize that I participated in only a portion of the policymaking — that which had to do with Russia. Others in our government will have to write their accounts to complete the picture. And as this book goes to press, the role of Russia in the 2016 U.S. presidential election is a story still unfolding. This book is an early, interpretative take — my take — on this era in U.S.-Russia relations.[2] I hope it will stimulate and inform other scholars in the writing of more definitive studies in the future.

By design, this book is a mix of abstract analysis, historical narrative, and personal anecdote. I hope that President Obama, my fellow policy wonks in government, my colleagues at Stanford, and my mother in Montana will all read it. That's a tall order. My mom will tire when reading about telemetry, and my academic colleagues may not be that intrigued by the two-step. Therefore, I give all readers license to skip those passages not of interest to them, which I trust can be done without losing the arc of the story. Scholars will quickly identify the social science theories that shaped both the analysis in this book and my thinking about policy while in government. But I deliberately hid, rather than highlighted, the academic scaffolding and academic references to allow for better storytelling. I also wrote this book from the sometimes competing perspectives of scholar, policymaker, diplomat, and Montanan. Rather than try to maintain one voice, I deliberately deployed these multiple standpoints because I am all of these things, and therefore my understanding of and engagement in Russian-American relations was shaped by these multiple dimensions. These tensions in the book accurately reflect those tensions in my life.

FROM COLD WAR TO HOT PEACE

1

THE FIRST RESET

I was born during the height of the Cold War, a year after Kennedy and Khrushchev nearly blew up the world. Growing up in Montana, where ICBMs are located, nuclear-attack drills in the classroom were the only connection I had to Russia, the USSR, and the Cold War. I first became interested in the Soviet Union through a high school debate class, a course I'd selected only because of its reputation as an easy English credit. That class, however, set the course for my professional life. The debate topic that year was how to improve U.S. trade policy, and so my partner, Steve Daines (now a U.S. senator for Montana), and I tried to come up with an obscure case — we called them "squirrels" — that would be unfamiliar to other debaters in Montana. We settled on the repeal of Jackson-Vanik, an amendment to the 1974 Trade Act that denied the Soviet Union normal trade relations with the United States until the Kremlin allowed Jews to emigrate freely. I would not have predicted back then that this gimmicky debate strategy would spark an enduring interest in the Soviet Union and then Russia, or that I'd actually be part of the government team that lifted Jackson-Vanik three decades later.

In high school I started to develop my ideas about reset with Russia. If we could just engage with the Soviets more directly, maybe we could reduce tensions in our bilateral relations, or so I believed at the time. I developed this theory of great power relations more systematically as a college student. My first year at Stanford was Ronald Reagan's first year in the White House, when he was advocating for the need to challenge the evil empire, fight global communism, and restore America to its former "greatness." As a freshman, I enrolled in first-year Russian and Intro to International Relations — because I thought Reagan was dangerous. I worried that his confrontational policies toward Moscow would lead to nuclear war. I thought he exaggerated the Soviet threat, spooking us into thinking that the Soviets wanted to take over the world. And what was so wrong with the idea of greater equality, an idea that Soviet communists allegedly advocated, anyway? Or so I speculated back then with sympathetic housemates at Stanford's hippie co-op, Synergy. We needed to be reaching out to the Russians, not threatening them. If we could just get to know each other better, we could reduce tensions.

I was not the only American worried about world war back then. In 1983 a hundred million viewers tuned in to ABC to watch *The Day After,* a made-for-TV movie that graphically depicted nuclear war between the Soviet Union and the United States. It was a scary time. The Cold War seemed like it could become hot. I wanted to do something about it.

The way to test my hypotheses was to go to the Soviet Union and get to know the people that I thought we should be engaging. So I traveled to Leningrad (as St. Petersburg was called back then) in the summer of 1983 to participate in a language program for Americans. When my mother sent me off to college, she worried that I would go to California and come back with long hair and radical ideas. Imagine her horror when I called her to report that my first trip abroad would be not to London or Paris but the evil empire.

The Soviet Union I encountered that summer seemed neither scary nor oppressive. Yes, the sausage had tons of fat in it. Yes, we avoided using ice cubes to cool our drinks, for fear of giardia in the water. And yes, you had to stand in line for an hour for vanilla ice cream sprinkled with a few chocolate flakes; usually there were no other choices. But Russians did not seem poor or deprived. The paint may have been peeling on some of the buildings, the stores may have been empty, but Leningrad was a grand, impressive city, made all the more magical by the White Nights phenomenon in June. The Soviet government discouraged its citizens from talking to foreigners, and especially Americans, back then. But to my surprise, the Russians who risked talking to me and my compatriots (admittedly not a representative sample) were friendly and curious. Many were even Led Zeppelin fans, just like me! We seemed more alike than different. Our governments might have been rivals, but our peoples were not. If we just made an effort to get to know them, to show respect, to find common ground based on mutual interests, we could get along with the Soviets, or so I believed with even greater conviction after that summer.

If my first trip to the USSR affirmed my theory of peace through engagement, my second trip, in 1985, challenged it. This time I was in dreary Moscow, not imperial Leningrad, and in January, not June. With somewhat better language skills and more connections, I dug deeper into the complex, contradictory fabric of Soviet society. I met refuseniks,[1] black marketeers, the spoiled children of Communist Party leaders, and otherwise intelligent university students who nonetheless repeated inane slogans about the virtues of communism and the evils of capitalism. I rushed to stand in long, snaking lines for toilet paper and bananas. I hid in the closet one night when my terrified Russian hosts answered the door and reported to police that no foreigners were present in their apartment. I watched brave believers defy barricades to attend midnight Easter services. I shopped at the mall at the International Hotel, where only foreigners and hard currency were allowed. I lost twenty pounds, because it was genuinely difficult to find food, and would have lost even more without the stews that my kind Ghanaian dorm room friend, Adam, shared with me. Life was hard in this communist paradise. I sold books on the black market to pay the seventy-two rubles (the equivalent of eighty-three dollars in 1985) to make a six-minute phone call to my girlfriend — now my wife — Donna. This system was corrupt. This system was repressive. This system didn't work. This time around, I witnessed and experienced communist dictatorship more closely and didn't like it.

That semester in Moscow hardened my convictions about the importance of liberty, freedom, and democracy. I no longer believed that engagement between our two countries was enough. I started to consider an alternative theory: only democratic change inside the Soviet Union would allow our two governments and our two societies to come closer together. That possibility seemed very remote in the winter of 1985, but a little less remote after Mikhail Gorbachev emerged on the scene.

I was living in Moscow the day Gorbachev became general secretary of the Communist Party of the Soviet Union. News of the appointment didn't come until a few days later, however, as the Soviet media delayed announcing the death of General Secretary Chernenko. That's how totalitarian the regime was:

it could even postpone death and freeze time. My first glimpses on television of the new leader inspired hope. He was not a comrade of the original Bolsheviks, or a hero of the Great Patriotic War (what the Russians call World War II) — he was too young. My Russian friends saw these biographical and generational characteristics as reasons to dream of a better future — not democracy or markets, but a reformed communism.

Some U.S. leaders also sensed early on the difference in Gorbachev. Vice President George H. W. Bush and Secretary of State George Shultz led the American delegation to Chernenko's funeral in Moscow and squeezed in a meet and greet at the embassy, to which a friendly diplomat invited the few dozen American students in the city. It was my first chance to listen to genuine Cold Warriors up close. But Shultz, who would later become a mentor of mine at Stanford, didn't sound like a Cold War hawk that day: he hinted at fresh possibilities in U.S.-Soviet relations with a new leader in the Kremlin. That night, I left Uncle Sam's — the embassy bar where we'd gathered — feeling upbeat about my decision to study in the USSR. The food was awful and the winter was long, but maybe this was a place where big history was going to be made.

In the spring of 1985 Gorbachev assumed control of a political and economic system that was not performing well. Economic growth rates had declined for decades. The Soviet leadership allocated huge resources to the military-industrial complex, creating shortages of consumer goods. Inefficiencies in distribution opened opportunities for black market traders. Above all, the command economy of state-controlled prices and wages as well as the lack of private property suppressed incentives for work and innovation. In the winter of 1985 I learned from Russian friends a perfect saying to describe the system: "We pretend to work, they pretend to pay."

And yet Gorbachev was not pressured to reform this economic system.[2] State control of news and information kept societal expectations low, and Gorbachev and others in the Politburo could have muddled through comfortably for years, if not decades, on the profits of energy exports. A handful of dissidents challenged the regime episodically, but no organized political opposition ever emerged. Reagan's military buildup worried the Kremlin and strained the Soviet reach around the globe, but the system could have survived without major economic reforms well into the twenty-first century.

A different Soviet leader might have stayed the course. Gorbachev made conscious decisions to depart from business as usual. The new Soviet leader was an idealist. He did not accept the world as it was; rather, he sought to remake it as he thought it should be. He believed in socialism but wanted to make it work better. Those beliefs motivated him to initiate a series of economic reforms aimed at making the existing economic system more efficient. His first idea was *uskorenie,* or the acceleration of economic development. It was a very old-school Soviet concept: if people worked harder, the economy would produce more.[3] To encourage greater labor productivity, Gorbachev also tried to crack down on alcoholism. The new general secretary also pushed the idea of *khozraschet*—self-financing—a policy that compelled Soviet enterprises to balance their books, but also gave them more autonomy. While the goal was to increase the efficiency of enterprises, *khozraschet* fit nicely into the conventional Soviet philosophy of greater discipline and accountability within the economy. If enterprise directors were held responsible for their expenditures and income, they would have fewer opportunities to shirk and steal.

Gorbachev's early efforts produced modest results, prompting him to pursue more far-reaching reforms in 1987. Those initiatives—collectively known as *perestroika,* or "rebuilding"—began with the bold Law on Cooperatives, which legalized private economic activity for the first time. Employers could now contract with employees and retain some of the profits of their economic activities. Restaurants opened with real prices and attentive service. Small shoots of capitalism were sprouting up within Soviet communism.

Gorbachev, however, never clearly defined the desired end point of perestroika. He claimed to want to improve socialism, yet introduced market reforms to achieve that goal. He sought to give economic actors more autonomy, but kept the system of national planning in place. He seemed to be looking for a halfway point between socialism and capitalism. There was none.

To push his economic-reform agenda, Gorbachev needed new political allies. He purged the ruling Communist Party to make room for more reformers like himself. To stimulate supporters of perestroika outside the government, he loosened controls on society by means of *glasnost,* or "openness," which relaxed limits on free speech. Stale publications came alive, publishing provocative articles on everything from Stalinism to consumer-goods shortages. Gorbachev then amended the Soviet criminal code to legalize political assembly and demonstrations. He also released hundreds of political prisoners, even permitting the return from exile of Nobel Peace laureate Andrei Sakharov, who had been expelled from Moscow for his participation in dissident circles. The famous physicist quickly became the moral leader of Russia's opposition movement. For the first time since the Bolshevik Revolution in 1917, civil society began to emerge as an autonomous force.

Gorbachev also believed that better relations with the West could further his reforms at home. A more benign international environment would make it easier to focus on domestic issues and would weaken the Soviet hardliners who pushed for greater military resources to defend the fatherland from American imperialism. Gorbachev called his novel foreign policy approach "new thinking." In particular, Gorbachev sought improved relations with the United States.

Gorbachev and Reagan met for the first time in November 1985 in Geneva, Switzerland. Already, at that summit, Gorbachev championed dynamic ideas. Reagan was listening, not afraid to consider the possibility that this new Soviet leader might be different from his predecessors. Only a few years before, Reagan had labeled the USSR an evil empire. Yet Reagan, like Gorbachev, was also an idealist: not trapped by the status quo, but willing to believe in the possibility of a new relationship between the United States and the Soviet Union. More quickly than other world leaders or members of his administration, Reagan revised his assessment of his Kremlin counterpart. Animated in part by his most utopian of beliefs — the desire to abolish the global store of nuclear weapons — Reagan embraced the challenge of resetting relations with the Soviet Union.

Individuals matter. Leadership matters. A different American president might have adopted a more cautious approach to engagement with this communist leader. Reagan, however, dared to be bold.

As Gorbachev loosened political control to foster economic reform, citizens started to form independent groups and associations. These groups — called *neformaly,* or "informals," to distinguish them from the formal organizations affiliated with the Communist Party — initially organized around nonpolitical subjects, such as football or the humane treatment of animals. But focus eventually shifted to edgier concerns, like reforming socialism and supporting competitive elections. Thousands of these informal associations sprang up, unleashing pent-up demand for civic engagement. With time, they became more radical. When I first started interacting with them, some even advocated for a truly revolutionary idea: democracy.

My first encounter with the informals happened by chance. As the most radical of them began advocating revolution, I was studying revolution, but in Africa, not the USSR. I was writing my DPhil dissertation at Oxford about revolutionary movements in southern Africa, and took a research trip to the Soviet Union in the spring of 1988 to interview Soviet officials about their methods for supporting those movements. To my interviewees, I was obviously a CIA agent with a bad cover. After all, I was an American in the Reagan era trying to interview Communist Party officials about their tactics for spreading communism in the developing world. At one memorable event at the Institute of African Studies, I tried to explain my theory about revolutions in an international context, complete with elaborate descriptions of dependent and independent variables. My halting Russian, heavy accent, and poli-sci jargon combined to make me completely incomprehensible. At the end of my remarks, the audience merely stared. Years later, I learned that nearly everyone in attendance worked for the KGB.

Tanya Krasnopevtsova, however, did not. She wasn't even a member of the Communist Party — an amazing achievement for the Soviet Union's leading expert on Zimbabwe at the time. As I departed the institute, she followed me down the corridor and out the door. We walked around Patriarch's Pond in one of Moscow's oldest and most charming neighborhoods, at first chatting cautiously, and only about Zimbabwe. We kept moving; Tanya was worried that we might be followed. But as she became more comfortable, both with our surroundings — did the KGB hang listening devices from the trees? we joked — and with me, she started to tell me what her government was really doing in Africa. It wasn't just supporting equality and showing solidarity, but running guns, sending money, and plotting a Communist Party coup in South Africa. The African National Congress (ANC), she told me, was infiltrated by Communist Party members loyal to Moscow, who had plans to nationalize all private property, abolish markets, establish one-party rule, and run South Africa just like their comrades ran the Soviet Union. I pushed back. I didn't believe it. But Tanya spoke passionately and convincingly. She hated the communists, and despised the Soviet regime. During my last stint in Moscow in 1985, I had met critics of the regime, and some people who desperately wanted to leave the USSR. But Tanya had a different spark to her. She wanted to change the country, not flee.

When eventually she concluded that I truly was an academic, not a CIA agent, and had a genuine academic interest in theories of revolution, Tanya urged me to switch my research focus from Angola and Zimbabwe to the Soviet Union. She described how nongovernmental organizations were sprouting up everywhere in the USSR. They would probably fail, she warned, and their naive, idealistic leaders would most likely be arrested. Khrushchev's thaw, after all, had ended with Brezhnev's stagnation, during which most of Khrushchev's liberalizing reforms were rolled back. But as a subject of academic inquiry for a scholar of revolutions, she said, Moscow, not Harare, was the place to be.

At the time, I was a toiling graduate student, steeped in the dense academic literature on revolution and democratization and trying to plow forward on a dissertation that might land me a university teaching job. But I was increasingly becoming a committed democracy promoter as well. I studied southern Africa not just to explain nationalist liberation movements, but also to stay engaged in the international antiapartheid struggle. This balancing of my academic pursuits and political interests — analyst versus activist — would remain a struggle for me for decades to come. Now, Tanya was offering me a window into understanding political mobilization inside the USSR, an academic subject that I'd never conceived as possible, but also a historic event to which I was normatively attracted. I had to learn more.

A few days later, Tanya introduced me to Yuri Skubko, her young colleague at the institute. Skubko was a genuine revolutionary — ideologically committed, comfortable working outside the system, and a little kooky. He was active in Memorial, a new NGO dedicated to preserving the memory of Stalin's victims, and also a founding member of the Democratic Union, the first noncommunist political party in the Soviet Union (or so its manifesto claimed). The Democratic Union included liberals, social democrats, anarchists, and even communists, a mix that made for some unwieldy meetings. But the party unified behind two goals that were outrageously radical for the Soviet Union at the time: dismantling the cult of Leninism and creating a multiparty democracy. Unlike most other informal groups at the time, the Democratic Union did not seek to reform the Soviet Union, but to destroy it. In the spring of 1988 these were crazy ideas, as likely to happen, I thought at the time, as a socialist revolution in America.

I hung around with Yuri and his Democratic Union friends as they debated democratic principles, organized illegal public demonstrations, and moved in and out of jail. I admired their convictions, no matter how quixotic their mission seemed. And I found myself pivoting my intellectual focus from revolution in Africa to revolution in the USSR.

By the summer of 1988, Tanya and Yuri were not alone in criticizing the Communist Party of the Soviet Union (CPSU). The general secretary himself was losing faith in the Party as an instrument for reform. Out of frustration, Gorbachev launched his most audacious idea ever — partially competitive elections for the USSR Congress of People's Deputies to be held in the spring of 1989. Gorbachev hoped to empower this parliament-like body to replace the Communist Party as the main governing institution in the country. The newly elected Congress would go on to select Gorbachev as president. Over time, Gorbachev hoped his title as president would become more important and confer more power than his position as general secretary.

The elections of 1989 triggered an even bigger explosion of nongovernmental political activity. Thousands of nominating committees formed overnight to support independent candidates. The media, including those outlets still controlled by the government, dared to report on topics that had been taboo only a year earlier. All kinds of civil society organizations now felt safe to organize and advocate.

The 1989 national elections were followed in the spring of 1990 by parliamentary elections in the fifteen Soviet republics. At the same time, oblasts (roughly equivalent to American states), cities, and districts within cities held elections for soviets, or councils. After decades with no meaningful elections, it seemed like everyone was a candidate for something in 1990, in contests in which the outcomes were uncertain — that is, truly competitive contests. Most consequentially, citizens in every republic from Tajikistan to Estonia had the opportunity to elect representatives to their respective parliaments, an opportunity that produced majorities in favor of independence in legislative bodies in the Baltics, Ukraine, Georgia, and eventually even in Russia.

It's understandable that majorities in support of independence might form in those republics colonized by Moscow. It took some conceptual imagination to create a coalition in Russia in support of independence from the Soviet Union.

The metropole was declaring independence from the empire. A new electoral bloc, Democratic Russia, won hundreds of seats in the Russian Congress of People's Deputies and controlled majorities in the Moscow and Leningrad city councils. Boris Yeltsin, at the time the unquestioned leader of the democratic opposition, was elected chairman of the Russian Congress. Gorbachev had hoped that these newly elected deputies would ally with him against the mid-level bureaucrats in the Communist Party blocking his reforms, but Yeltsin and his coalition quickly turned against Gorbachev, instead advocating for much more radical change. The standoff between the Soviet and Russian governments became especially obvious after the Russian Congress of People's Deputies voted in favor of independence — a Declaration of State Sovereignty — on June 12, 1990.

As I observed these events, both from afar and up close, I became increasingly convinced that revolution in the Soviet Union was a real possibility. At the time, I was back at Stanford on a predoctoral fellowship, trying to finish my dissertation and prepare for the academic job market. It killed me, however, to be so far away, both physically and intellectually, from what I sensed was the greatest historical event of my life. Back home, I also was frustrated with how few people agreed with my assessment of the enormity of the revolutionary changes unfolding inside the USSR. In 1989 President George H. W. Bush and his new team initially pumped the brakes on engagement with Gorbachev, but eventually preferred Gorbachev to Yeltsin as a partner. Most senior members of the Bush administration, including the president himself, considered Yeltsin an erratic, populist drunk — in part based on Yeltsin's ill-fated trip to the United States in 1989, when he made a scene upon his arrival at the White House after discovering that Bush had not scheduled a formal meeting with him (only a "drop by").[4] At the time, most Sovietologists in the academy also revered Gorbachev and despised Yeltsin.

Back then I too was suspicious of Yeltsin, but for different reasons. He was a former candidate member of the Politburo of the CPSU. How could a senior official and lifetime member of the repressive Communist Party be the leader of Russia's democratic movement? He was no Nelson Mandela. As the party boss in his home region, Sverdlovsk, and then the city of Moscow, he had developed a reputation as a maverick, but then clashed with Gorbachev and was demoted. He had some progressive credentials, but I worried that he was an opportunist rather than a committed democrat. He also flirted with nationalist groups from time to time. Yet there was no question that Yeltsin's populist rhetoric resonated with millions of Russians, well beyond the liberal intelligentsia. By ignoring him and his allies, my colleagues in academia and Washington were missing the dynamics of revolutionary change from below.

In the spring of 1990 I became convinced that we were witnessing one of the great social revolutions of the twentieth century. I wrote as much in my first essay on Soviet political change, published in the *San Jose Mercury News* on August 19, 1990 (exactly one year before the August 1991 coup attempt). In the article, titled "1789, 1917 Can Guide '90s Soviets," I deployed analogies from the French and Bolshevik Revolutions to help explain the revolutionary change under way in the Soviet Union at the time. I depicted Gorbachev as the interim moderate (like Kerensky) who would eventually be overthrown by radicals such as Yeltsin and his allies. In the same essay, I expressed my concern that a strongman like Bonaparte or Stalin would eventually come along after the radicals failed. It would take two decades, but Vladimir Putin eventually assumed that role.

Unsatisfied with merely watching this revolution from Oxford or Palo Alto, I applied and was accepted as a Fulbright Scholar to attend Moscow State University for the 1990–91 academic year. In my proposal to Fulbright, I outlined my intent to conduct interviews about Soviet policies for supporting revolution in southern Africa. As it happened, however, I spent most of my time that year observing and occasionally helping anticommunist revolutionaries in Moscow.

Soon after arriving at Moscow State, I scored an observer's pass to the founding congress of Democratic Russia in October 1990. In downtown Moscow's cavernous Rossiya Theater, delegates from around the Federation raucously and inefficiently—but ultimately successfully—transformed their electoral coalition from the previous spring into a new national political movement. Democratic Russia was Russia's approximate equivalent to Solidarity in Poland: a loose coalition of political party leaders, civil society activists, businesspeople, coal miners, and a few poets dedicated to the peaceful transformation of their country from a communist dictatorship into a democracy. I watched raw democracy in action over those two days. The delegates argued and compromised, but eventually agreed. The Russian opposition was united, and I was inspired.

A few weeks later I tagged along with Viktor Kuzin — once an extremist, marginal Democratic Union cofounder and now an elected member of the Moscow city council — to march in the parade on the anniversary of the Great October Socialist Revolution — not to celebrate communism but to denounce it. I shuffled in the cold for hours, but few of the tens of thousands assembled that day seemed bothered by the subzero temperatures. Already in my first month back in the USSR, Democratic Russia's founding congress and this massive demonstration against the regime convinced me that Soviet communism was winding down. What also struck me at these two events was the warmth and solidarity extended to me by Russians. We believed in the same things — freedom, democracy, respect for human rights — or so it seemed at the time. If these people took power, a new era of partnership in U.S.-Russia relations seemed not only possible but probable.

Over the course of that academic year, I engaged in a variety of activities concerning democratization in Russia. I helped organize academic seminars at Moscow State on the transition to democracy attended by Russian and American scholars. With a Russian partner, Sergey Markov, I interviewed opposition leaders for a book called *The Troubled Birth of Russian Democracy*. I also got to know the chairman of the Social Democratic Party, Oleg Rumyantsev. At the time, Rumyantsev was also serving as the secretary of the Constitutional Commission of the Supreme Soviet of the Russian Federation. Reporting in the *Washington Post,* David Remnick called him the James Madison of Russia, a title that he readily embraced. One evening that fall, a visiting American delegation from the National Endowment for Democracy and I met with Rumyantsev for some horrible food at the drab Intourist Hotel, but for a very bright, festive occasion — toasting the first complete draft of Russia's new constitution. All of us at that table, Russians and Americans, were democratic idealists. Only a few years earlier, any discussion of democracy would have led to prison sentences for our Russian colleagues and deportation for us. But in the fall of 1990 Americans and Russians were now engaging freely over the merits of federalism, the perils of presidentialism, and the pluses and minuses of various electoral systems. Oleg and his colleagues wanted to interact with us, to hear our ideas, and to accept our intellectual support and modest technical assistance. We were united in our fight for the same cause: democracy.

Later that year, another American democracy-promoting organization, the National Democratic Institute (NDI), connected with me in Moscow. In a time before email, a few members of this group hand-delivered a letter from a former student of mine, Amy Biehl, who then worked in NDI's D.C. office. After a thirty-minute chat, they asked me to join their efforts in Russia, and I did. NDI had a complex pedigree — an organization affiliated with the U.S. Democratic Party and supported by U.S. government funds, but officially nonpartisan and nongovernmental. NDI's mission was to promote democracy by transmitting ideas about party development through conferences and training programs. NDI's identity — partisan and nonpartisan, governmental and not — confused me and most Russians, but history was in the making, and I no longer wanted only to watch. I wanted to help. NDI gave me that chance.

Despite the fact that my Fulbright stipend did not keep up with runaway inflation — I usually ate canned tuna or ramen for dinner — I did not want to be paid by NDI, as I worried it might violate the terms of my fellowship. And since NDI was a shoestring operation at the time, I'm not sure financial compensation was even an option. But that year of volunteering for NDI turned out to be some of the most exhilarating and meaningful work of my professional career.

Of course, most of the tasks I performed for NDI were mundane — arranging meetings, organizing conferences, deciding between chicken or beef for the banquet. When former vice president Walter Mondale, NDI's chairman back then, came to Moscow for a conference, my greatest achievement was securing him a stash of Diet Coke — a true feat in this period between the collapse of the command economy and the introduction of market reforms. The excitement of the work, however, came from the feeling that I was no longer an observer of Russia's revolution, but a participant. I was now attending meetings of the Coordinating Committee of Democratic Russia and sat in on founding congresses of various political parties. When Democratic Russia organized demonstrations that attracted hundreds of thousands of people across the country, I was behind the podium as its guest. When campaign specialists from Democratic Russia plotted Yeltsin's successful campaign to become Russia's first democratically elected president in June 1991, I occasionally added my two cents about

American campaign practices, though my involvement was mostly to show solidarity. The mere presence of someone from the United States — a country that inspired these new democrats with our democratic ways — lent a sense of common struggle. For decades, the Communists had shouted, "Proletariats of the World, Unite!" We were now proclaiming, "Democrats of the World, Unite!" We collectively believed that democratic change inside Russia would bring our two societies closer together. Democracies, after all, have developed some of the closest and most enduring alliances in the world. If we could forge deep political, economic, and security relationships with a democratic Germany or democratic Japan after World War II, why couldn't we do the same with a democratic Russia at the end of the Cold War? The logic seemed compelling. Motivated by the pursuit of this goal, I saw my activities as distinctly pro-Russian in the same way that nurturing democracy in Japan and Germany after World War II was also pro-Japanese and pro-German. In this great drama, I wanted to be on the right side of history.

NDI's first major event in the USSR was a conference in Moscow in December 1990. At the closing dinner, I sat next to Mondale at the VIP table, working as a translator. (My Russian had improved dramatically since my visit to the Institute of African Studies in the spring of 1988.) Mondale was tired. Well before dessert, he stood up and announced he was ready for bed. But instead of going directly to his car, he grabbed my arm and veered off to shake hands with the conference participants — all four hundred or so of them — with me translating at his side. As we made the rounds, one former dissident, who had sat in jail during the Carter administration, was in tears as he expressed his gratitude for the former president's human rights advocacy. The feeling of solidarity, of sharing a common commitment to democracy, of being part of a righteous struggle, was palpable in the room that night.

The following year, however, was a turbulent one for Russia's democrats. At times their righteous struggle appeared to be floundering. In December 1990, Soviet foreign minister Eduard Shevardnadze resigned to protest the tightening political system, using his last speech to the Soviet Congress to warn of coming dictatorship.[5] Weeks after Shevardnadze stepped down, Soviet Special Forces killed two dozen people and injured hundreds more in raids on a television station in Vilnius, Lithuania, and a government ministry building in Riga, Latvia.[6] The counterrevolution had begun.

Shortly after these tragic events, I sat in a window alcove at the Hotel Moscow and listened to Democratic Russia leaders debate their next moves. They were nervous, but decided to organize another major demonstration to express solidarity with their democratic allies in the Baltics. I was shocked by the numbers; at least one hundred thousand people, and by some estimates multiples more, turned out to denounce Soviet violence in Latvia and Lithuania.[7] Flags from Estonia, Latvia, Lithuania, Ukraine, and Georgia were sprinkled throughout the crowd, reminding me that this democratic revolutionary movement was much bigger than just Russia. In response, Gorbachev opted courageously for peace. He did not double down with more repression, but backed away, signaling a clear break from previous Soviet dictators who had used force when faced with similar moments of dissent. In this battle, the Baltic democrats emerged victorious, and so too did Russia's democratic forces.

But the democrats' growing strength produced a powerful conservative backlash from nationalists, communists, and conservatives not controlled by the Kremlin. One rising star among the nationalists was Vladimir Zhirinovsky, the leader of the Liberal Democratic Party of the Soviet Union, a party which was neither liberal nor democratic. Instead, Zhirinovsky espoused a fiery brand of ethnic populism. He pushed for a stronger Soviet state that would use force to suppress ethnic groups seeking independence. Zhirinovsky also championed transnational ideas of racial solidarity—ideas that could include Americans. As he explained to me when we met in May 1991, we "Northern" people (that is, white people) had to unite against those from "the South" to protect "our culture."[8] Zhirinovsky wasn't the only one: Pamyat at the time was an equally troubling political organization that championed extreme right, anti-Semitic views. There was also Viktor Alksnis, a former Air Force colonel and now elected member of the Soviet Congress of People's Deputies who aimed to unite communists and nationalists to defend the Soviet Union against my Russian friends in the democratic movement. These men were angry; they despised Gorbachev, but hated Yeltsin and the democrats even more.

In March 1991 conservative deputies called an emergency session of the Russian Congress of People's Deputies in an attempt to oust Yeltsin as chairman.

To deter this vote, Democratic Russia decided to organize a major demonstration in support of Yeltsin. This time, Gorbachev sided with the conservatives and banned public demonstrations that spring. Defying Gorbachev's decree, Democratic Russia leaders decided to move forward with their demonstration. Gorbachev and his government — now much more conservative — upped the ante, ordering tens of thousands of soldiers and special police, equipped with riot gear and water cannons, to barricade the city's main boulevard. Democratic Russia faced a dilemma: if it canceled the demonstration it would look weak, but marching meant risking violence. Democratic Russia leaders agreed to a compromise: rather than proceed all the way to Manezh Square as planned, they would hold their demonstration before reaching the police barricade.

That morning, I managed to make it to Manezh Square near the Kremlin, the original location of the Democratic Russia protest. The square was completely occupied by military and police vehicles. Gorbachev's new government had pivoted decidedly toward more confrontational tactics. I then walked up Moscow's main boulevard, Tverskaya, to Pushkin Square, where water cannons, horses, and riot police stood waiting for the marchers. The giant crowd inched toward this blockade but then halted. Instead of trying to break through the police line, they built a makeshift stage right there on the street in order to deliver their speeches, thereby holding their meeting but avoiding violence. I thought the demonstrators' move was brilliant, but not everyone agreed. A Western reporter standing next to me, who had previously worked in South Africa, lamented that Russia's democrats had no courage: protesters in South Africa would have forced the regime to use violence against them, which in turn would have further undermined the legitimacy of the regime (and landed his story on page 1!). I was not convinced. How would violence have advanced the democratic cause? Who knows how it might have ended? Besides, for me it was now personal: The protest leaders were people I knew. I didn't want to see them clubbed, sprayed, or arrested.

On the metro home that night, I struggled with a dilemma that would haunt me for decades. Was I an activist or an academic? At the time, I was still trying to finish my dissertation on national liberation movements in southern Africa, which meant that my research and my political activities remained separate. But that afternoon, I'd found myself defending the decisions of my Democratic Russia friends before a group of Western journalists and academics who were observing and reporting, not participating. Where did I stand? Professionally, I still aspired to become an analyst. Emotionally, especially that evening, I sided with the democrats.

In my work with NDI, most events weren't nearly as exciting as Democratic Russia's mass demonstration in March 1991. In May of that year NDI organized a new round of "technical training seminars" — a title we used to sound apolitical — on Democratic Government and Municipal Finance. How boring does that sound? By now, we were working directly with city government officials in Moscow and Leningrad to aid in the development of democratic practices. As the chief liaison with our Russian partners, I was in charge of making sure everything ran smoothly.

It was while planning these seminars that I first met Vladimir Vladimirovich Putin, then the deputy mayor in Leningrad in charge of international contacts. I was an international contact, and one who must have intrigued a KGB officer. Our NDI delegation had come to meet Anatoly Sobchak, the new mayor of Leningrad, soon to be renamed St. Petersburg. Sobchak, a law professor turned politician, was one of the truly inspirational leaders of Russia's surging democratic movement at the time. Emphatically pro-Western, pro-European, and pro-American, he radiated hope about the possibilities of a democratic Russia and closer relations between our two countries. At that meeting, I stepped in to translate, and aside from mixing up "Abkhazia" as "Oklahoma," helped communicate our proposal for a conference that would bring city council members from Los Angeles and New York to the Russian city to share their experiences about formulating a city budget in a democratic, transparent way. Sobchak embraced our proposal, our ideas, and us. We were ideological allies. We loved the guy. The following year, NDI gave Sobchak the W. Averell Harriman Democracy Award, given to an individual to honor his or her commitment to democracy and human rights.

Putin made much less of an impression on me than his boss. He was careful, unenthusiastic, diminutive — an apparatchik. I could not tell from this meeting or our subsequent encounters with him if he supported or detested NDI's work. He said little, but promised that his team would organize an orderly, successful workshop — and delivered. He handed me off to his trusted deputy, Igor Sechin, who ran the logistics to perfection — much more efficiently than his counterparts in the Moscow city government. (As a reward for his years of loyalty to Putin, Sechin would later become the CEO of Russia's largest oil company, Rosneft.)

Why was Putin organizing democracy-promotion seminars at which his sworn American enemy could lecture his people on how to run their government? It was well known at that time that Putin used to work for the KGB, and maybe still did. After all, is there really such a thing as a *former* KGB officer? The question we chewed on in our NDI delegation later that day was this: had Putin and his comrades flipped to the side of the white hats like Sobchak, or had the KGB inserted Putin into Sobchak's office to keep tabs on this dangerous democrat?

I concluded at the time, wrongly, that it didn't matter. To paraphrase Martin Luther King Jr., the arc of Russian history was at last bending toward justice, freedom, and democracy. That put the KGB on the wrong side of history. Putin's only chance to survive was to join the democrats, or so I believed at the time.

In our humble self-assessment, our seminars that spring were a smashing success. The mood in Leningrad was especially electric. Los Angeles City Council member Zev Yaroslavsky was given the honor of speaking before the city's entire legislative council. In the same chamber where Lenin once rose to speak in favor of shutting down bourgeois democracy, an American politician (with ancestors from Russia) now orated to inspire Russia's democratic renewal. The hall went wild. We felt like heroes; we were contributing to advancing freedom, eroding seventy years of communist dictatorship, and ending the Cold War.

At the concluding dinner that night, Russians and Americans together toasted our cooperation and success. In the midst of our celebration, Sechin, Putin's deputy, made a jarring confession. As we congratulated each other on a successful conference, he revealed to me that he too had worked in intelligence, just like his boss. He spoke Portuguese, just like I did, and had worked in southern Africa, just like I had. Although I am sure that I had met dozens of Soviet intelligence officers by then, none of them had admitted it. I wondered if he was telling me this information, especially about our shared experiences in Lusophone Africa, to suggest that he believed that I also was an intelligence officer, a CIA agent. Or was he just trying to be friendly? I finally concluded that it didn't really matter, since we were all on the same side now.

• • •

While NDI was providing training and technical assistance to Russia's democrats, President George H. W. Bush and his administration still focused their work on Gorbachev and the Soviet government. They did not trust our Russian partners, and for understandable reasons. Our friends in Russia sought to undermine a Soviet leader who was producing tremendous dividends for American foreign policy: Gorbachev had not tried to stop the overthrow of communist regimes in the fall of 1989, the destruction of the Berlin Wall, or German unification. He pulled Soviet troops out of Afghanistan and other parts of the developing world, and did not try to hinder the American-led liberation of Kuwait and invasion of Iraq. Yeltsin's democratic challenge to Gorbachev was extremely inconvenient for the Bush administration.

Yeltsin and his supporters tried to curry favor with the White House, but Bush and many of his senior foreign policy advisors did not support revolutionary change in the USSR. They feared that Yeltsin's call for Soviet dissolution would trigger a civil war, similar to the one that had unfolded after the collapse of Yugoslavia but with one big difference: thousands of nuclear weapons would be floating around in this anarchy. Consequently, even as Yeltsin became more powerful, Bush maintained a firm policy of noninterference in the internal affairs of the Soviet Union. In his infamous "Chicken Kiev" speech on August 1, 1991, Bush warned advocates of Ukrainian independence, saying, "Americans will not support those who seek independence in order to replace a far-off tyranny with a local despotism. They will not aid those who promote a suicidal nationalism based upon ethnic hatred."[9]

Bush's speech underscored for me how out of touch Washington was with the revolution under way in Russia. But maybe that was inevitable. Revolutionary time always moves faster than outside observers can grasp. Moreover, maybe Bush had no choice. Gorbachev was his counterpart. The whole world still recognized the Soviet Union as the legitimate, sovereign government.

For me in Moscow, however, it seemed like it was just a matter of time before sovereignty devolved to the republics. The point of no return was Yeltsin's presidential campaign in the summer of 1991. One day in early June just days before Russia's first presidential election, I was sitting in the office of Mikhail (Misha) Shneider at the Moscow city council. At that time, Misha was one of Democratic Russia's most talented grassroots organizers, whose assignment now was to help Yeltsin win. I watched him shout into multiple telephones, coordinating various events for the Yeltsin campaign throughout the country. These were the days before email and cell phones, so the Soviet Union's dilapidated landline system presented a real impediment. But in between calls, he stopped to smile. It was over. Yeltsin was going to win, and the end of communism was imminent. I was not as confident in predicting Russia's future as Misha that day, but in the end agreed. Momentum for their side — for our side — seemed too great to stop. That shared sense of mission was what united us in those days: not money, not strategic interests — ideas. In the summer of 1991 I firmly believed that an alliance of American and Russian democracies was possible, as did Shneider and other leaders in the Democratic Russia movement. I think that's why they liked having me around. As the young, idealistic American, I offered them hope for Russia's return to the international community and the forging of a special relationship with the United States.

Yeltsin did win the June 1991 presidential election, by a landslide, becoming Russia's first elected president. He won over 60 percent of the popular vote, finishing with three times as many votes as the runner-up, Nikolai Ryzhkov, the former Soviet prime minister. Zhirinovsky, the radical nationalist, surprised most by finishing third, with 8 percent: a sign of more dangerous developments to come. But in June 1991 few were focused on Zhirinovsky. The democrats — my friends — had won. Formally, Gorbachev was still the chief executive of the USSR. But to many, including me at the time, his role was become increasingly honorific. Real political power was now located in the Russian Republic, not the Soviet Union.

Not everyone was ready to watch political power shift from Gorbachev to Yeltsin without a fight. Conservatives were plotting to strike back. Apollon Davidson, Russia's leading historian of South Africa at the time, warned me as much in June 1991. He had served as my advisor that academic year at Moscow State University — specifically, at the Institute of Asian and African Studies, which was known back then for training KGB agents, not scholars. In my exit meeting with him at the end of my Fulbright fellowship on my very last day in Moscow in June 1991, I thanked Apollon for his efforts on my behalf and said that I hoped to see him again in the fall upon my return. He shocked me by replying grimly that we would never see each other again. The KGB had taken a keen interest in my activities over the last academic year. Davidson predicted that the conservative hardliners would attempt to reassert control and end the chaos of Gorbachev's failed reforms. And when they did take over again, there would be no place for people like me in his country anymore.

I was shaking as I left Davidson's office. It was true that I had expanded my activities well beyond research on liberation movements in southern Africa. But I was living in Moscow during one of the most tumultuous, revolutionary moments in world history. No one in Russia that year was doing what they had planned to do a year earlier. Davidson's reference to the KGB, however, had me truly rattled. Those three letters — *KGB* — always instilled fear, but now they were being invoked in relation to my own personal activities. Maybe they would never allow me to return? I drove directly from Apollon's office to the airport, wondering if this would be my last time in Moscow.

Davidson's prediction turned out to be true: conservatives did try to seize power just a few months later. But in a departure from my professor's analysis, they failed. His guys lost; my guys won.

2

DEMOCRATS OF THE WORLD, UNITE!

It was two months after my Fulbright year in Moscow ended that the conservatives launched a coup to stop Yeltsin's quest for Russian sovereignty. On August 19, 1991, the State Committee for the State of Emergency (GKChP) announced that it had assumed responsibility for governing the country. While Gorbachev was held under house arrest in Crimea, the Red Army surrounded strategic buildings throughout Moscow and other major cities. But they did not seize or close these buildings immediately; the Russian "White House," which at the time housed the Russian Congress of People's Deputies and the Russian president, Boris Yeltsin, was left unoccupied. Yeltsin reached the White House, denounced the coup plotters, and called upon all Russians to resist their actions. In direct defiance of GKChP orders, Yeltsin issued decrees of his own. Democratic Russia also sprang into action, amassing tens of thousands of coup resisters who formed a human barricade between the tanks and the White House. By the end of the first day, some senior Soviet military officers defected to Yeltsin's side and urged their fellow soldiers to respect the legitimate, democratically elected commander in chief.

In the chaos of the 1917 revolution, Russia's Provisional Government had wrestled with the Soviets for authority, each claiming to be the sole sovereign of Russia. Now the country again found itself caught between two competing governments, poised between the return to the Soviet past and the promise of a new future. After three days, one side — the democratic side, the anti-Soviet side, my side — won. The coup plotters gave up. Gorbachev may have started the process of reform, but it was Russia's democratic movement that thwarted the coup attempt. Russian idealists had changed their country radically and relatively peacefully. The failed August 1991 coup attempt was a victory for Russian democrats, and a victory for every small-*d* democrat around the world.

The days that followed must rank as one of the more optimistic periods in Russian history. As during many other times in Russia's past, Kremlin rulers had issued orders to suppress protesters, be they striking workers in Novocherkassk in 1962, when dozens were killed, or dissidents protesting the Soviet invasion of Czechoslovakia in 1968, or even just a few years earlier, in 1988, when Democratic Union demonstrations were dispersed and activists arrested. But this time, the people resisted and prevailed. The victory created an atmosphere of unlimited potential. Tens of thousands of elated Russians paraded through Red Square carrying a giant Russian flag. A *Time* magazine headline proclaimed "Serfdom's End: A Thousand Years of Autocracy Are Reversed."

Four months later, Yeltsin met with the leaders of Belarus and Ukraine to sign the Belovezhskaya Accord, which peacefully dissolved the Soviet Union. With that entity's extinction, communism also seemed to disappear and the Cold War to end. Even the ever-cautious President George H. W. Bush celebrated the moment in an address to the nation, declaring, "Eastern Europe is free. The Soviet Union itself is no more. This is a victory for democracy and freedom. It's a victory for the moral force of our values."[1]

For those in charge of a newly independent Russia, however, euphoria faded fast. The empire's collapse stranded millions of ethnic Russians in untethered Soviet states, and prompted demands for independence or secession in Chechnya and Tatarstan within the Russian Federation, and in the autonomous republics located within the newly independent states of Georgia, Moldova, and Azerbaijan. Overnight, Soviet institutions such as the army and KGB were broken into pieces and scattered across fifteen countries. Large state-owned enterprises were also thrown into disarray, now located in two or more different countries with ambiguously defined borders and property rights. Analysts inside and outside Russia worried that ethnic conflicts, civil wars, and state breakdown would proliferate throughout the former Soviet empire. Some even worried that the Russian Federation itself would collapse.

Independence, coupled with weak policies and falling oil and gas prices, brought about complete economic chaos. In 1991 Soviet GNP declined 13 percent, real wages fell by double digits, inflation skyrocketed, and the government went broke — a terrible foundation on which to begin building a newly independent, democratic Russia.[2] Gold and hard currency reserves were depleted, the budget deficit exploded, money was abundant but goods were scarce, production had halted, and trade had all but collapsed.[3] Some experts predicted starvation in the first month of Russian independence.[4]

This economic chaos threatened democratic development, especially as nationalistic forces gained strength. As St. Petersburg mayor Anatoly Sobchak told my NDI colleagues and me earlier that year, "If democratic government in the republics, the local authorities and the leadership in Moscow cannot stop inflation and falling living standards, then disenchantment and disaffection with democracy will set in."[5] And once the "democrats" were in power, they could no longer blame Soviet communism for their economic woes. Faced with these crises, Yeltsin assumed that he had little time to think about the niceties of democratic governance. He and his new government focused instead on defining the borders of the state, slowing the economic free fall, and launching economic reform.

Helping Russia navigate the transition from a command economy to a market system was also the top priority for U.S. policymakers focused on Russia. The Bush administration shipped food to Russia and the other newly independent states, while Western advisors took up residence in Moscow to provide knowledge about market reforms. After seventy years of communism, building Russian capitalism on the ruins of a command economy presented a monumental challenge: the task had never been done before.

At the time, I agreed with the focus on economic assistance, but my NDI colleagues and I also tried to make the argument that successful economic transformation and deeper democratic reform were intertwined. We pushed our Russian colleagues to hold new parliamentary elections as soon as possible. The current members of Russia's legislative body, the Congress of People's Deputies, had been elected in the spring of 1990, when the Soviet Union still existed and Communist Party authorities had their fingers on the scales. We worried that the majority in this body would turn against market reforms once the going got tough. We also pressed for the immediate adoption of a new constitution and the formation of new political parties. Russia's democrats disagreed. They had just won: they'd stared down a coup attempt and prevailed. Why did they need another election? Amazingly, and perhaps naively, in the midst of the chaos of the Soviet Union's collapse, we at NDI organized a conference on December 14–15, 1991, called "Democratic Governance in a Time of Crisis." What a euphemism! Our Russian colleagues were dealing with the triple transition of empire to nation-state, command economy to market system, and autocracy to democracy — and we were there to discuss American, European, and Latin American experiences with democratic governance. Ten days after our seminar — December 25, 1991 — the Soviet Union collapsed.

Exchanges with our Russian partners revealed their frustration with our focus on democratic consolidation. In October 1991 I had a particularly sharp interaction with Yuri Afanasiev, one of the six cochairs of Democratic Russia at the time. The National Endowment for Democracy (NED) had agreed to provide Democratic Russia with a printing press, but wanted assurances that all democratic organizations would have access to it. NED was trying to be pro-democratic but nonpartisan. As we explained to Yuri our concerns about the governance structure for the printing press, he blew up. We ought to stop distracting him with such trivial matters, he said, predicting that he and his fellow democrats would soon control all media outlets. His reply suggested that he no longer considered our assistance necessary. They had won. They didn't need our ideas or support anymore.

We disagreed with his analysis. Believing that Russia's transition to democracy would be a protracted, tumultuous one, we at NDI made plans to be engaged for the long haul. In June 1992 I moved back to Moscow and opened the NDI office there. Several other American organizations in the business of democracy promotion followed in NDI's footsteps. If democracy could be consolidated there, then the Cold War would truly be history, Europe could unite, and the United States would gain a powerful ally in the pursuit of other international objectives. I was pursuing what academic theorists called the "democratic peace" theory. History shows that democracies rarely go to war with each other.[6] So a democratic Russia would no longer be an enemy of the United States. At the time, therefore, I firmly believed that rooting democracy in Russia was the most important challenge in the world.

In the summer of 1992 in Russia, I felt like a rock star. Everyone, it seemed, wanted to meet with my NDI colleague, Greg Minjack, and me simply because we were Americans and America was now an ally of democratic Russia. My job was to build relationships with Russian officials, party leaders, and civic activists; Greg's, to provide knowledge about parties and elections in the United States. The two of us attended all sorts of party congresses and political events as special guests. Because I spoke Russian, I was often asked to offer a few words of solidarity on behalf of the Democratic Party of the United States and the United States of America. I had no authority to speak for either entity, but I looked the part: a "typical" American — optimistic, passionate, idealistic. That was the energy and symbolism that many new political groups wanted represented at their meetings back then. We were ideological allies, committed to a common democratic cause.

Beyond cheerleading, our main job was to share information about democratic practices and institutions with party organizers, constitutional writers, government officials, or whomever our partners of the day happened to be. We met with politicians and did our best to help out — whether that meant obtaining and then translating the Michigan state charter, organizing training sessions on fund-raising, explaining the separation of federal and state powers, or connecting Russian civil society leaders with foundations that could give direct financial assistance. We were not "meddling" in Russia's internal affairs, but invited guests of the Russian government. We were not promoting regime change, but helping to consolidate the existing democratic system of government. Above all else, we were educators. And in the pre-Google era, we were also librarians, tracking down books, documents, data, and contacts in the United States requested by our Russian partners.

How to win elections became a major theme of many NDI workshops. In these seminars, we discussed door-to-door campaigning, grassroots fund-raising techniques, and the value of endorsements. We shared the American Democratic Party's experience of building alliances with trade unions, women's organizations, and other nongovernmental groups. Russians were always surprised to learn how few people actually worked directly for the Democratic or Republican Party and the extent to which nonparty allies bore the brunt of organizational work during campaigns.

Our workshops introduced participants to strategies for winning elections, but we believed that participants would have to see our elections in person to fully internalize our teachings. So in the fall of 1992 Minjack and I convinced our Washington bosses to sponsor a two-week trip to the United States for Russian party activists to witness our presidential election. The delegation I accompanied to Los Angeles met everyone from wealthy fund-raisers in Hollywood to get-out-the-vote organizers in a tough neighborhood in East L.A. All the American groups were eager to meet these Russian democrats who had destroyed communism. On election night, we celebrated Bill Clinton's victory at the Biltmore Hotel in the company of local Democratic Party dignitaries. That the Democratic Party was returning to the White House increased NDI's stock back in Russia.

United fronts *against* something are always easier to maintain than coalitions *for* something. Not surprisingly, with the shared enemy of communism and the USSR now gone, Democratic Russia splintered into several parties and groups. Those of us promoting democracy now faced the question of how and to what extent we should engage with communists and nationalists and how to encourage nondemocratic groups to engage with the political system — even if we did not share their ideals. In our view, wide participation in the democratic process was key to ensuring that this new political system took hold.

Since it was not our business to get involved in the substance of politics, only the process, I had no problem meeting with anyone, including communists, as long as they played by the rules of the democratic game. Guided by this logic, NDI organized a training program for a new political organization called Civic Union. This self-described "centrist" coalition brought one of our oldest partners, the Democratic Party of Russia (allegedly named after the U.S. Democratic Party), together with a group of industrialists — mostly directors of state-owned enterprises — as well as the Federation of Independent Trade Unions of Russia (FNPR). FNPR had been previously affiliated with the Communist Party, but was now independent. It seemed innocuous enough to me — who in Russia in 1992 didn't have some tie with former Communist Party structures? But for some in Washington — from Republicans accusing us of aiding communists to American trade unions suspicious of this communist-era trade union association — we were helping communists.

In response to this criticism, I wrote a detailed memo for the NDI home office, spelling out our rationale for developing a training program for Civic Union. I emphasized that we had an interest in the coalition's participation in the new political system: if members of the coalition opted out, they might undermine Russia's very fragile democracy. Still, we learned to tread more carefully with new communist and socialist parties, as well as with any groups affiliated with former Soviet institutions. Cold War mentalities didn't die overnight.

We faced similar issues with Russia's new nationalist parties. We quickly wrote off some, such as Vladimir Zhirinovsky's Liberal Democratic Party of Russia, because of its refusal back then to commit to democratic norms. The Christian Democratic Party of Russia posed more of a challenge for us. Headed by Viktor Aksiuchits, a handsome, charismatic, devoutly religious man, this party seemed committed to democracy but even more focused on restoring Russian conservative values after seventy years of soulless Soviet rule. Was restoring conservative values part of our mission? We continued to meet with the Christian Democrats and parties with similar ideological orientations, but also started to gravitate toward more liberal parties. Though we claimed only to support democratic practices, not particular parties, rigid commitment to the norm of neutrality was hard to maintain in practice in this first year of Russian democracy.

In addition to working with parties, we engaged dozens of members of the Russian Congress of People's Deputies. Specifically, we provided MPs information about different electoral systems, both by translating electoral laws from other countries and by hosting seminars with electoral-law experts flown in from the United States and Europe. It was a real thrill to meander the expansive hallways of Russia's White House — home to the Congress of People's Deputies at the time — with materials on democracy that had been translated into Russian tucked away in my briefcase. There was real demand among MPs for our ideas, our materials, and our help. It felt like we were making a difference — however small — in helping to nurture Russian democracy.

NDI also tried to assist the Constitutional Commission, which was tasked with amending the constitution in place from Soviet times. Formally, Yeltsin chaired this commission, but the secretary of the commission, Oleg Rumyantsev — a friend of mine at the time — really ran the show given Yeltsin's other responsibilities. There were dozens of issues that Rumyantsev and his colleagues sought to address, but one choice dominated the rest: should Russia adopt a presidential or a parliamentary system? Russia already had a president, but the powers of the executive branch remained ambiguous, since the presidency had been established by a referendum in March 1991, before lawmakers had ruled on the lines of authority between the parliament and the president. Spelling out the specific divisions between the two was now Rumyantsev's task.

Rumyantsev and his associates leaned toward a parliamentary system. As Yeltsin and his opponents became more polarized on issues of economic reform, the commission altered the draft constitution to give the parliament more authority to check presidential rule. They were gutting Yeltsin's power. The fight between the executive and legislative branches would ultimately end in tragedy in the fall of 1993, with friends of mine on both sides of the barricade. This constitutional crisis was exacerbated by economic dislocation. Had the Russians, with more help from the West, gotten the economy right — that is, turned Russia in a direction toward growth — political stalemate and ultimately a violent confrontation might have been avoided. Yeltsin's team did not succeed in achieving that goal. We did not do enough to help them.

As Russia's new leaders fought to preserve their state and stop an economic free fall, America turned inward. The United States had won the Cold War — so the argument went — so now was the time to focus on problems back home. Regarding Russia, candidate Clinton in 1992 championed contradictory campaign slogans, as often happens in presidential elections, arguing that President George H. W. Bush focused too much on foreign policy, but also chastising Bush for not doing enough in support of Russian reform.

Once in the White House, President Clinton realized the gravity of the moment in Russia. The reset in relations between Moscow and Washington started by Gorbachev and Reagan would only endure if Yeltsin succeeded in achieving three ambitious objectives: decolonization, marketization, and democratization. And if Russia's revolutionaries succeeded, Clinton argued, the nature of international politics would be changed forever.[7] Through bilateral assistance and multilateral programs from the international community, Clinton raised pledges of tens of billions of dollars in aid to Russia by the summer of 1993.[8] But by then, a year and half after the collapse of the Soviet Union, the assistance package seemed like too little too late. Had the recession been better handled, Russia's economic contraction may have been shorter, and animosity toward Russia's democrats in power at the time and us less severe.

Russia's market reformers advocated for a big bang approach: do everything at once and hope that radical reforms would be less likely to be reversed than gradual measures. After some hesitation, Yeltsin endorsed that approach and appointed a group of young economists — headed by Yegor Gaidar, his new deputy prime minister — to design and implement these reforms. Gaidar's team started with liberalization of prices and trade in January 1992. Overnight, the market — not the state — began setting prices for goods in stores. Gaidar and his government tried to avoid hyperinflation and massive government deficits by controlling the money supply and reducing government spending. Months after lifting controls on prices, the Gaidar team launched massive privatization. Gaidar's lieutenant, Anatoly Chubais, devised a program that allowed managers and workers to take majority ownership of their enterprises, while also giving every Russian a voucher to invest in Russia's newly privatized companies. On paper, it was the largest, most audacious, and fastest privatization program in history.

The day after Gaidar implemented price liberalization, I was working for *The MacNeil/Lehrer NewsHour* in Moscow, tasked by our producer with locating a place to film rioting in stores as the inevitable result of price reforms. That's what many expected. But nothing happened. Gaidar had predicted privately that his team would last only a few weeks after the shock of their economic changes, but the initial calm after their first set of reforms suggested otherwise. Maybe "shock therapy," as critics called Gaidar's plans, could succeed?

It did work — if only for several months. After an early inflationary spike in response to price liberalization, Gaidar managed to bring monthly inflation rates down from 38 percent in February to 9 percent in August 1992. His tenure, however, was short-lived: Gaidar's extreme austerity and tight monetary policy threatened directors and workers of large state enterprises, who used their political influence in parliament to press for his ouster. Yeltsin eventually retreated, dismissing Gaidar as acting prime minister in December 1992 and replacing him with the more conservative Viktor Chernomyrdin, the former head of the gas company Gazprom. That was the end of shock therapy.

Chernomyrdin muddled through, with poor results. The federal government failed to collect taxes, allowing deficits to grow. Subsidies to industry and agriculture increased, fueling inflation despite high interest rates. And as a handful of oligarchs acquired ownership of some of Russia's most valuable companies, privatization became synonymous with corruption.

Kto vinovat?—Who is to blame?—is a classic Russian question. Some blamed the shock therapists for going too fast and putting too much pressure on society; others blamed Gaidar's successor, Chernomyrdin, for going too slowly, allowing partial reform to fuel corruption, inflation, and inequality. Intellectually, I tended to side with the shock therapists and blame Chernomyrdin for Russia's economic mess. But Russians blamed the guy at the top, Boris Yeltsin. Over the summer of 1993 a growing majority in Russia's parliament decided that Yeltsin had to be stopped, and planned to use the Tenth Congress of People's Deputies, scheduled for October 1993, to adopt Rumyantsev's constitution and transform Russia into a parliamentary system.[9]

Yeltsin did not wait for his constitutional lobotomy. On September 21, 1993, he issued Presidential Decree 1400, which dissolved the Russian Congress of People's Deputies and called for elections to a new bicameral parliament. The Supreme Soviet reacted immediately and defiantly, denouncing the decree as unconstitutional and adopting a resolution terminating Yeltsin's presidential powers. The full Congress reconvened a few days later and approved Vice President Aleksandr Rutskoi as Russia's new president. In a tragic replay of August 1991, Russia again had two heads of state and two governments claiming sovereign authority over the same country.

Police and army forces loyal to Yeltsin's government cordoned off the White House, where Rutskoi and his supporters in parliament had barricaded themselves. Negotiations proceeded, but without a breakthrough. Moderates inside the parliament, including Communist Party leaders, eventually vacated the White House, and more radical factions showed up, including armed paramilitary groups and some neo-fascist organizations. On October 3 Rutskoi and his allies in the White House decided to end the stalemate, calling on their supporters, many of them armed, to seize control of the mayor's office adjacent to the White House and take over the Ostankino television studio, the country's main broadcast-news center. In response, Yeltsin ordered Russian soldiers to attack parliament. Later that evening, Yeltsin announced a state emergency and relieved Rutskoi of his duties as vice president. For the second time in two years,

Russian soldiers surrounded the White House, but this time they attacked, with tanks shooting at the parliamentary building, and later stormed the White House and arrested Rutskoi, Ruslan Khasbulatov, chairman of the Supreme Soviet, and several other MPs and activists, including my friend and former fellow democracy builder Oleg Rumyantsev. One hundred and fifty-eight people — 130 civilians and 28 from the police and army — died in the fighting, with several hundred more wounded.[10]

I was surprised and disappointed by Yeltsin's undemocratic decree. I sympathized with those who argued that economic reform couldn't get back on track with this Congress in the way, but the moment to change the balance of forces in parliament had been the fall of 1991, when Yeltsin was popular and economic reforms had not yet begun. By October 1993 it was too late. President Clinton defended Yeltsin. He explained to the American people, "There is no question that President Yeltsin acted in response to a constitutional crisis that had reached a critical impasse and had paralyzed the political process." He added that Yeltsin was "on the right side of history."[11]

After Rutskoi took armed action against the Russian government, Yeltsin had no choice but to respond. With that I agreed. But I lamented that the sides could not engage with each other peacefully to end the constitutional crisis. I watched on television as several people I knew well were hauled away in handcuffs. In this event, there were no white hats and black hats, no winners and losers. Everyone made mistakes, and everyone lost.

Back in Moscow a few weeks after this mini civil war, I was saddened to see the top of the Moscow White House blackened from fires sparked by tank shells. Over the last three years, I had spent countless hours in that building, watching and occasionally helping as the experiment of Russian democracy unfolded. The charred walls and broken windows silently testified to the failure to consolidate democracy — which meant that we foreign democracy promoters had failed as well. This outcome was not one any of us had predicted back in the euphoric days of the fall of 1991.

I still hoped that this tragic moment would lead to a new start — a genuine break with the communist past. But obstacles to democratic consolidation were now codified in the new constitution, including, first and foremost, strong presidential rule. The new document, which Yeltsin was now free to redraft as he and his aides saw fit, spelled out the basic rights guaranteed to all Russian citizens and established a new system of government, which included the office of the president, a prime minister and government, and a bicameral parliament consisting of a lower house, the State Duma, and an upper house, the Federation Council. Despite those positive developments, the enhanced powers of the president loomed largest in the new draft. The president was to be elected directly by the people for a fixed term of four years, and impeachment procedures were described — that was the good news. The bad news was that the new constitution gave the president more authority to issue decrees, greater control in selecting the prime minister and the government, and the ability to directly appoint the foreign minister, defense minister, minister of the interior, and head of the Federal Security Service (FSB, the successor to the KGB) — together, the so-called power ministries.[12]

Many analysts, including myself, described the new system as "superpresidential." But we differed in our assessments of just how authoritarian Russia had become. Some of my closest academic friends, Russians and Americans, argued that Russian democracy died during Yeltsin's military attack on the parliament. The new constitution, in their view, made Yeltsin an autocrat. I wasn't sure. I reasoned that the new constitution contained important checks on presidential power, and that Yeltsin had a basic commitment to democracy and would therefore not abuse that power. At the time, I accepted the argument that Russia needed a strong executive to implement economic reforms. In the hands of Yeltsin, this constitution provided just enough of that executive power without creating a full-blown dictatorship. But I also worried that this same constitution in the hands of another leader could move the country toward autocracy. And what could be worse than a thriving Russian economy in the hands of a dictator? As I wrote in 1995, "America's greatest national security nightmare would be the emergence of an authoritarian, imperialist Russian regime supported by a thriving market economy."[13] A decade later, that's exactly what happened.

In October 1993 my worries about a future new president abusing these new constitutional powers were eclipsed by more immediate events: elections. The upcoming contests would be carried out according to the new electoral system put in place by Yeltsin. Half of the 450 seats in Russia's new parliament, the State Duma, would be decided by a majoritarian system in newly drawn electoral districts: whichever candidate got the most votes in the district would win the seat.

The other half of the seats would be allocated through the national-party-list system of proportional representation: voters would cast a ballot for a party, not a person, and seats would be allocated to parties according to the percentage of the votes they received throughout the country. At NDI, we were confident that this system of proportional representation would help stimulate party development and thereby strengthen democracy in Russia. Initially, the new electoral system had the effect we hoped for, as politicians had an incentive to form electoral blocs — prototypes for political parties — to participate. The liberals who supported Yeltsin joined together to create Russia's Choice, headed by Yegor Gaidar and including former leaders of Democratic Russia. Self-labeled "democrats" who did not support Yeltsin — including those critical of his use of force to dissolve the Congress of People's Deputies — formed a new electoral bloc headed by Grigory Yavlinsky. That bloc was creatively named Yabloko, an acronym of the three founder's last names, and also the Russian word for "apple." A third group of government officials and older political parties combined to create what they called "centrist" blocs. Perhaps most importantly, leftist parties critical of Yeltsin, the Communist Party of the Russian Federation (KPRF) and its partner, the Agrarian Party of Russia (APR), decided to compete in these elections and not boycott them. I saw their participation as a good sign for democracy: they were playing by the rules of the game — even if the rules were dictated by their opponents — rather than challenging the regime through extraconstitutional means. Several national parties also agreed to take part.

The December 1993 parliamentary election, accompanied by a referendum on the new constitution, was the first post-Soviet vote in independent Russia. Political scientists call such an event a "founding election," one that locks into place a democratic political system.[14] Hopes were high concerning the possibility of the rebirth of Russian democracy.

The liberal election bloc, Russia's Choice, seemed to have the momentum during the fall campaign, with Yeltsin and his government giving it indirect support. Its leaders held national stature, and to us Americans observing, in public appearances and on television they said all the right things about the need for market reforms, democratic governance, and strategic partnership with the United States. The Clinton administration did not endorse a party in this election, but there was little question in my mind that Washington wanted Russia's Choice to win.

I always get excited on election days in democracies, no matter where they take place. There is something magical about citizens choosing their own leaders. December 12, 1993, in Russia felt especially important, part of the healing process after the tragic events of October. In Moscow at the time, as an international election observer, I was eager to watch democracy in action at two dozen precincts that day. Everywhere I went, I saw a free and fair process, babushkas running the procedures, buffets offering tea and sweets, and a generally orderly process. I decided against reporting the only election violations I saw: the occasional parent bringing a child into the booth to witness their vote.

That evening, I headed over to the campaign headquarters of Russia's Choice at the Liberal Club on Hertsen Street in the old part of Moscow.[15] Many of Russia's highest-ranking government officials, businesspeople, and television celebrities were beginning to congregate for a big celebration party planned in the downstairs restaurant. This was expected to be a historic night to commemorate: the final defeat of communism.

But then the first numbers trickled in from the Far East—Vladivostok, Khabarovsk, Magadan—and they were shocking. The Liberal Democratic Party of Russia (LDPR)—the far-right group headed by the militant nationalist Vladimir Zhirinovsky—was trouncing the other parties. Mikhail Shneider, the Russia's Choice official responsible for tracking vote tallies, kept yelling into his phone, "Are you sure?" He wrote down the totals from different regions: 25 percent LDPR, 28 percent LDPR, 40 percent LDPR. Everyone in the room stared blankly, stunned. We waited, hoping that these early results might be flawed, or that later results from other regions might be better. By that late hour in the evening, I had retired my credential as an international election observer and was now a concerned, small-*d* democrat, worried that I might be witnessing the end of democracy in Russia.

With time, the results improved. Russian voters closer to Moscow, especially in the big cities, supported the liberal parties, while traditional communist strongholds in rural areas in central Russia stayed true to form and backed the Communist and Agrarian Parties. By the end, the actual balance of power within the State Duma was divided relatively evenly between supporters and opponents of Yeltsin. But with nearly a quarter of the popular vote, Zhirinovsky was without question the symbolic winner. The same electoral system of proportional representation I had pushed for to stimulate party development had propelled a militant nationalist onto the national stage, and only a few years before the next presidential election. At the Liberal Club that evening, I overheard murmurs from worried Russian liberals comparing this election outcome with the rise of fascism in Germany and Italy.

Gradually, all the big shots left the Liberal Club. No one was in the mood to party. But I stayed late, in shock. Where had Russia's democrats gone wrong? Where had we democracy promoters gone wrong? Two years earlier, Russian and American believers in democracy had celebrated the fall of the Soviet Union and the collapse of communism together. Our ideas had won. But on that December evening in 1993, our triumphal pronouncements about the superiority and inevitability of worldwide democracy seemed premature.

In my first meeting with Zhirinovsky, in the summer of 1991, he did not seem like a serious politician. His views seemed too outlandish to attract popular support. Speaking with me at his modest apartment in Moscow, he was full of fire and hate, lambasting Jews, Balts, and people from the Caucasus and Central Asia as the cause of all of Russia's woes, and trying to impress on me the need for "Northerners" — that is, white people — to unite to prevent "the South" from dominating the world. I was not signing up; I had a hard time believing any rational person would. I was wrong.

If 1991 had not been Zhirinovsky's time, 1993 surely was. By then, the optimism about democracy many held in the spring of 1991 had been wiped out by economic depression and a two-day civil war. The widespread disillusionment had worked to Zhirinovsky's advantage. After his election victory, I believed the nationalist leader could — though not necessarily would — become Russia's Hitler.[16] The parallels went beyond racism and anti-Semitism. Zhirinovsky blamed the West for an unfair settlement after the end of the Cold War. He promised to reunite the millions of Russians stranded outside Russia, from Estonia to Kazakhstan, after the Soviet Union's collapse, just as Hitler had pledged to unite all Germans living beyond Germany's border. Zhirinovsky even threatened to take Alaska back, rebuild the Russian empire, and conquer enough territory in the south to open a Russian warm-water port.

My worries turned out to be grossly misplaced. Over the next three decades in Russian politics, Zhirinovsky rarely challenged the political rules of the game, but learned to play within them quite successfully. Some of his views were and remain undeniably vile, but he turned out not to be a fascist seeking to overthrow the existing political system. Two decades later, in my capacity as U.S. ambassador to Russia with a focus on outreach, I even invited him to our July Fourth celebration at my residence. But in 1993, there was a lot more uncertainty about his future trajectory.

Soon after Zhirinovsky's electoral victory, I witnessed actual Nazi salutes, young men goose-stepping, and red flags bearing swastika-like emblems at a rally for Russian National Unity.[17] These weren't middle-aged bureaucrats content with the way things were. They were young, mean men, full of hate and itching to overthrow the status quo. They truly frightened me, and made me and others ponder historical analogies from the interwar period even more closely.

In my academic work in the wake of these elections, I warned that economic reform without political reform would fail, and that an unstable, anti-Western, fascist regime armed with nuclear weapons would present a serious threat to American national security. It was in America's national interest to prevent such an outcome.

The election results had jolted me. Of course, in retrospect, I was completely wrong in my assessment. Fascism did not overthrow or even weaken Russian democracy. The real threat to Russian democracy would come from within. But that's not what I was thinking on December 12, 1993. Walking back to my apartment in the slush on Moscow streets as the sun began to rise the morning after this pivotal vote, I decided that I had to try to do something to prevent these worst-case scenarios from unfolding.

3

YELTSIN'S PARTIAL REVOLUTION

After Zhirinovsky's unexpected victory, I decided to delay my academic plans. I persuaded my new wife, Donna, to move with me to Russia and help fight for democracy. Today I feel foolish even reporting this arrogant and naive sentiment — how in the world was I going to contribute to the fight against fascism in Russia? Was there even a fascist threat? But that's how I felt at the time. I agreed to continue consulting on democracy promotion for NDI, but instead of rejoining its Moscow team, I took a job as the first senior associate in residence at the newly opened Carnegie Moscow Center, a branch of the Carnegie Endowment for International Peace. This fight, I believed at the time, was going to be a war of ideas, and Carnegie — a nonpartisan, analytic think tank — seemed a better fit for waging such a struggle.

In the summer of 1994, Russia was still midstream in a major social revolution. Donna and I rented from a former Bolshoi Opera singer an upscale apartment that stood in sharp contrast to its crumbling surroundings. Inside, our spacious Stalin-era digs came complete with pink walls, silk curtains, and a grand piano, all befitting a Soviet-era cultural icon. Outside, signs of state collapse were everywhere: piled garbage, shabby entryways, occasional evening gunfire, even packs of wild dogs. I understood why Zhirinovsky's message had appeal: he promised law and order at a time when there was little of it.

At the Carnegie Moscow Center, in keeping with the mission of the Carnegie Endowment, we convened seminars for politicians and analysts of all political persuasions — communists, liberals, nationalists, and everyone in between. In retrospect, these events must have been strange for our Russian participants. Why was a young American academic organizing discussions about Russian politics for a room full of Russian politicians, journalists, and professors? Who was paying for the good food and wine at these "tasty seminars," as one of my Russian friends called them?

Many people in Moscow at the time were convinced that Carnegie was a CIA front. For them, it was clear who was bankrolling the seminars and paying my salary. Nationalist newspapers railed against our activities; Zhirinovsky once even denounced me on the floor of parliament as a CIA agent and called for my expulsion. One weekend, someone fired bullets through the windows of my office; thankfully no one was inside at the time. Nonetheless, we forged on, guided by the belief that nonpartisan discussions of political issues could contribute to the development of Russian civil society. Back then, Russia had few independent think tanks. We were filling a niche.

At the time, I worried that too few Russians appreciated or valued democracy as a system of government. Maybe the transition from communist rule had occurred too quickly? Some democratic movements struggle for decades to gain power, giving constituents time to develop deep convictions about their ideas, often at considerable cost. In South Africa, for instance, the fight for democracy expanded over decades, with the movement's leader, Nelson Mandela, spending twenty-seven years in captivity—a huge price to pay for an idea. But these longer struggles did allow for democratic ideas to be discussed and debated, to develop and grow. With the exception of a handful of dissidents, most of Russia's new democrats had not significantly struggled or suffered in defense of democracy. In addition, most Russians were introduced to the concept and (alleged) practices of democracy at a time of enduring economic depression. As a result, democracy quickly became associated with economic hardship. Who wants a political system that makes you poorer? With no textbooks in schools on the subject of democracy, how could 150 million Russians learn how their new political system was supposed to work? We had to nurture a more philosophical discussion, or so I thought at the time.

My friend Andrey Bystritsky had a solution. He worked at RTR, Russia's second-largest television channel fully owned by the government, and shared my concern about the weak embrace of democratic values in Russian society. To promote more thoughtful public discourse on democracy, we teamed up to produce a prime-time show for RTR called *Lyudi,* meaning "people," in the fall of 1994. Each episode, Andrey would discuss a problem in Russian society — something as simple as potholes, for instance. I then explained how the American democratic system would tackle the problem, hinting at how individuals could use the democratic process to achieve concrete, desired outcomes. The message was simple: democracy can work for you; democracy can deliver. Andrey and I felt like we were making a real contribution to changing Russian mind-sets, and for a while the show did well: we had a prime-time slot and millions of viewers. Some wrote letters to newspaper editors to complain about my accent, but Andrey, ever the optimist, assured me that the show was striking a chord.

In December 1994 we crossed a line. That month, Yeltsin ordered Russian troops into Chechnya — a republic that was part of Russia but had been drifting toward greater independence since the collapse of the Soviet Union. He did so without consulting Russia's parliament. Following the invasion, we decided to do a show on how decisions to go to war are made in the United States. We devoted particular attention to the 1973 War Powers Act, a response to the Vietnam War, which requires the president to consult with Congress before taking military action, and regularly while such action is ongoing. The Kremlin did not like that episode, and removed our show from the air. It was an important lesson for me about what outsiders were allowed to promote and what was off-limits.

The parliamentary vote in December 1995 was another shocker. Just six months before the 1996 presidential election, the liberals, or "democrats," performed well below expectations for the second time in two years. The good news, from my point of view, was that the Communist Party of the Russian Federation (KPRF), led by Gennady Zyuganov, won the largest percentage of the popular vote, replacing Zhirinovsky's LDPR as the leading opposition party. Unlike Zhirinovsky, Zyuganov and his comrades did not give me cause for concern. They criticized Yeltsin, but also played within the rules. They did not seem as radical or revolutionary as Zhirinovsky, though over time Zhirinovsky would play a role similar to that of the Communists as a loyal opposition to the Kremlin after Putin became president. The specter of Russian fascism seemed to be waning. Some even wondered whether Russian democracy might benefit from a Communist victory in the presidential election the following year. The peaceful handover of power from one party to another by means of an election is the ultimate sign of democracy consolidating.

But I expected a Yeltsin win. Zyuganov had surged remarkably in the 1995 vote, but at the expense of Zhirinovsky. By my assessment, shared by other election analysts affiliated with the Carnegie Moscow Center, the Communists still didn't have the majority of the country on their side. The distribution of votes for the opposition and for pro-liberal, pro-Yeltsin parties remained roughly the same as in 1993.[1] Support had shifted between opposition parties, but their aggregate numbers were only slightly higher than they were two years before. Based on this analysis, I predicted that Yeltsin would win, as long as reformists united behind his candidacy.

If my Russian colleagues at the Carnegie Moscow Center agreed on Yeltsin's likely victory, few Americans shared our assessment, be they bankers or U.S. government officials. Most predicted a Zyuganov victory and worried that such an outcome would produce disastrous consequences for market reforms and democracy. Despite Yeltsin's flaws, most notably his decision to use force against the parliament in October 1993, many in the West still considered Yeltsin Russia's best bet for consolidating democracy and maintaining close ties with the United States. In Washington, the Clinton administration made clear its preference for a Yeltsin victory, believing that the election outcome would prove pivotal both for Russia's internal trajectory and the future of U.S.-Russia relations. As President Clinton remarked, "I know the Russian people have to pick a president, and I know that means we've got to stop short of giving a nominating speech for the guy. But we've got to go all the way in helping in every other respect."[2]

Clinton even excused Yeltsin's invasion of Chechnya in 1994, at one time comparing Yeltsin to Lincoln in his defense of federal unity.[3] The U.S. administration pressed for new International Monetary Fund aid to the Russian Federation in the spring of 1996, and delayed NATO expansion until after the Russian vote. U.S. officials also refrained from criticizing the massive insider sell-off of valuable state properties to friends of the Kremlin. Yeltsin justified this "loans for shares" program as necessary to fund his campaign, and liberal reformers defended the fire sale of Russia's most valuable assets to oligarchs as a hedge against a communist return to power. If Zyuganov did become president, he could not be allowed to control these companies, they asserted. Maybe the argument was logical, but loans for shares did irreversible damage to the Yeltsin administration's reputation among Russians.

Yeltsin took additional steps in the pursuit of victory. His initial takeaway from the results of the 1995 parliamentary vote was that in order to win he had to distance himself from liberals and their losing ideas and instead sound more like Zhirinovsky and Zyuganov. To help make that pivot, Yeltsin handed over his campaign to a group of intelligence officers who had never run a campaign before: his bodyguard, former KGB general Alexander Korzhakov, and Korzhakov's fellow hardliners.[4] Some even wondered if Yeltsin might be planning a coup in case of a failed election campaign.

By that time, I was back in the States, living in Palo Alto as an assistant professor at Stanford, but still working part-time for the Carnegie Moscow Center and traveling frequently to Russia. Our focus at Carnegie at the time was on election analysis, and Yeltsin's new campaign staff took an interest in our work. On a trip back to Moscow early in 1996, I got a call from Viktor (he never told me his surname), one of Korzhakov's deputies in charge of the campaign's analytic center, who explained that he had read one of my publications on the 1995 parliamentary vote and sought my advice for developing a winning strategy for Yeltsin's reelection. He asked if we could meet.

The Yeltsin campaign headquarters was at the Presidential Hotel, a late-Soviet-era facility that previously housed senior Communist Party dignitaries. When I met Viktor on the tenth floor of the hotel, metal detectors and soldiers armed with machine guns greeted guests at the elevators. The campaign workers were dressed in suits and gave off a bureaucrat vibe, and the halls were quiet. It was a very different ambience from the chaotic buzz of other Russian campaign headquarters that I had visited before.

Viktor acknowledged to me that they were new to this "campaign business," and asked me to discuss my theories about Russian electoral dynamics. In particular, he wanted me to explain factor analysis, a mathematical method we had used in our study of the 1995 parliamentary elections, the results of which suggested that Yeltsin could still win in 1996.[5] I tried, but the statistics were complicated for me to explain even in English, so I'm sure they were fairly incomprehensible as translated into my Russian. Still, he seemed intrigued and asked me to stay in touch.

Back in Moscow a few months later, in April, I received another call from Viktor (he seemed very well informed about my travel itineraries!). This time he proposed we meet at his boss's dacha, where we could speak without the distractions of the campaign office. It seemed like an odd proposal. I called my friend Valery Khomyakov, who had been active in liberal party politics and had initiated my first contact with Viktor. Valery's relationship with Viktor made me wonder if he had been working as an informant tracking me for the Russian intelligence services all along, going back to my days of helping Democratic Russia in 1990–91. Still, I needed his advice: I was nervous about a one-on-one meeting with a Russian intelligence officer and wondered if I should decline the invitation. Valery advised me that the consequences of not going would be worse than the alternative. Warily, I agreed to meet Viktor, with the condition that Valery accompany us.

As instructed, we met Viktor inside the Kremlin walls near the Tsar Bell, a 202-ton work of Russian craftsmanship that had been rendered inoperable by an accident during its construction. Looking at it, I hoped that our current ambitious project — Russian democracy — would not meet a similar fate. Viktor picked us up in a black Volga, the car of choice for senior Russian officials at the time, and we sped out of the Kremlin, blue siren flashing, the traffic stopped for us by police. Viktor clearly wanted to impress me.

Thirty minutes into our trip that morning, the Volga broke down. Embarrassed and angry, Viktor yelled at his driver, but then opened a compartment and started shouting instructions into a beige phone receiver connected by wire to the car. I was struck by how crude the technology was — both the car and the phone — but also by how quickly a new vehicle arrived at the scene. This was the KGB, now called the FSB. They had real resources, even if somewhat outdated.

We eventually arrived at the "near dacha" — a government-owned house in Kuntsevo, a suburb of Moscow. One of its previous occupants, Stalin, had died there. Now it was the country residence of one of the most important leaders in the Kremlin, Korzhakov, who was not just Yeltsin's bodyguard but one of his closest advisors. That he had Stalin's former dacha at his disposal underscored his status in the Kremlin court. Armed soldiers surrounded the compound and patrolled the premises, a reminder that Russia was fighting a war in Chechnya at the time.

We were given a brief tour of the compound, accompanied by an elderly caretaker who had worked all his life at this residence but had never entertained an American guest before. My presence seemed to make him nervous. After the tour Viktor got down to business. He grilled me on electoral support for various candidates, campaign strategy, the role of television ads, the utility of negative attacks, and so on. It was clear that he was indeed new to political campaigning. He was particularly interested in how to engage with Zhirinovsky and his nationalist supporters: should Yeltsin try to act more like him, or should they run a surrogate nationalist candidate to soak up some of Zhirinovsky's electoral support, who could then endorse Yeltsin in the second round?

Sometime later that morning, vodka was served. It was way too early in the day for me to be drinking, but I took part in the rounds of shots out of respect for my host. I began to wonder, however, about possible ulterior motives for the meeting.

Eventually we took a break from our discussions to watch the television coverage of a historic event: the signing of the Treaty on the Formation of the Community of Belarus and Russia, the first step toward an economic union of the former Soviet republic and the Russian Federation. As we watched Presidents Yeltsin and Lukashenka signing the accord, officiated by the head of the Russian Orthodox Church, my host hinted that this agreement might ultimately result in a new state uniting Russia, Kazakhstan, and Belarus. And if there were a new state, he speculated, the upcoming election for the Russian presidency might have to be postponed in order to properly form this new political entity. This scenario sounded kooky to me, but these were volatile times. As the day wore on, Viktor hinted at other scenarios that might compel Yeltsin to postpone the presidential vote: if it became clear that Zyuganov was going to win the presidency, for example, they might have to cancel the election to save Russia from communist restoration. If Yeltsin felt compelled to make this move, he of course would need American support. Surely, he prodded, we Americans agreed that a return to communism would be a worse outcome than a postponed election — right? Viktor looked to me for some affirmation of his logic. I clearly had no authority to offer any such validation. But we agreed that I would communicate this message to Clinton administration officials.

As we watched the signing agreement between Russia and Belarus, Viktor pointed to the TV screen, where his boss, Korzhakov, stood right behind Yeltsin, and another nondescript, middle-aged "comrade" stood behind Lukashenka. It was these two men, not the drunkard Yeltsin or the buffoonish Belarusian president, who had made this deal, Viktor explained. Just a few years earlier, they had both worked for the same agency: the KGB. The deep institutional, fraternal ties forged by being in the security service together allowed them to get things done, things like the treaty we were witnessing being signed at that very moment. Those kinds of ties were missing in U.S.-Russia relations, Viktor lamented.

It took me a while to grasp what he was suggesting. The vodka, his indirect language, and my imperfect understanding of Russian worked to garble an already bizarre message. I finally figured out that my KGB host wanted to establish a back channel of communication between his organization and "mine." Such a bond would help us manage our presidents and adhere to pragmatic policies — including, if need be, canceling Russia's presidential elections.

I pushed back. I did not work for the CIA! I insisted. I was an untenured junior professor at a university in faraway California. I was in no position to be of any help to him. Viktor didn't like my answer. Tersely, he told me that he knew *exactly* who I was — why else would he have invited me to this meeting? He warned that I needed to get serious about his analysis and proposition, and so did my government. Dangerous times demanded decisive leaders.

Now I was scared. Did Viktor want me to send signals to my friends in the Clinton administration? I did, after all, know Strobe Talbott, the deputy secretary of state; Tony Lake, the national security advisor; and Chip Blacker, Clinton's senior advisor on Russia at the NSC and my Stanford colleague. I could easily deliver his message to them. Or did he really think that I worked for the CIA? Many others in Russia thought as much at the time. To this day, I don't know the answer.

But I did reach two conclusions after my daylong briefing session with this senior official from the Yeltsin campaign's analytic center. First, those in charge of Yeltsin's reelection campaign were novices in the election business. Second, it was possible that they had been chosen for the job not to win an election, but to postpone one. As I reported to my colleagues in the U.S. government once I returned home, the administration needed to make some contingency plans. In my view, it also needed to get Clinton to lean hard on Yeltsin to proceed with the election. This was no time for ambiguous signals.

To his credit, Clinton did push back on the idea of postponing the vote. He sent Yeltsin a private message that registered his "strongest disapproval of any violation of the constitution." Talbott later recalled the dilemma faced by the administration: "[Clinton] felt we had no choice. He'd backed Yeltsin through thick and thin, always on the grounds that the U.S. was supporting not just the man but the principle, that Russia would work its way out of its crisis through elections, referendums, and constitutional rule."[6] Whether Clinton's interventions influenced Yeltsin's thinking is hard to judge. In his final memoir, Yeltsin recalls his flirtation with postponement, but credits his daughter, Tatyana Dyachenko, and campaign director and future chief of staff Anatoly Chubais—not President Clinton—for dissuading him from pursuing the antidemocratic plan.[7] But Yeltsin strongly valued his relationship with Clinton. Although it's impossible to know for sure, I believe that Clinton's prodding ultimately helped convince Yeltsin to do the right thing.

In the run-up to Election Day, Yeltsin took some dramatic actions, but not the ones I'd feared.[8] He reshuffled his campaign team, bringing in a brand-new group headed by Chubais. The personnel changes effectively marginalized Korzhakov and the other hardliners, which I saw as a positive sign, both for Yeltsin's electoral prospects and for the course of policy after the vote. The new team knew that Yeltsin couldn't win by trumpeting his record in office, so they focused instead on casting him as the lesser of two evils. Acknowledging the struggles of building democracy in recent years, they reframed the election as a vote about the future versus a return to the past.

In the election's first round, Yeltsin finished with a solid 35.8 percent of the vote, while Zyuganov came in second with 32.5 percent. The charismatic populist, General Alexander Lebed—the surrogate candidate chosen to soak up votes from Zhirinovsky—came in third with 14.7 percent of the vote, and then promptly endorsed Yeltsin. Most election experts estimated that most of Lebed's supporters would vote for Yeltsin in the second round. (If no candidate wins more than 50 percent in the first round of a presidential election, there is a runoff between the top two candidates in a second round.) The election seemed over.

The second-round vote proceeded as scheduled, with Yeltsin winning comfortably with 53.8 percent of the popular vote, compared to Zyuganov's 40.3 percent.[9] Some complained of falsification,[10] but the margin of victory was too large for the election to have been stolen.[11] Zyuganov himself readily accepted the result.

In the moment, the 1996 presidential election seemed to mark the end of the long transition from communism. That evening, the threat of Soviet restoration died. Both the nationalist Zhirinovsky and the communist Zyuganov went on to serve in the Duma (where they remain today), but they never seriously challenged the regime. Rather, both became fully integrated into the Russian political system, acting the part of opposition parties but never challenging the regime itself.

As an election, the 1996 presidential vote in Russia met several minimal definitions of a democratic process.[12] Russian voters — not a communist party, a royal family, or a military junta — decided who was going to lead Russia for the next four years. The election was competitive. The outcome before Election Day was uncertain. And the election results were not reversed after the vote. Yeltsin did not postpone the election; rather, he allowed the Russian people to choose their next leader. For a country with a thousand-year history of autocratic rule, that was an achievement.

At the same time, the way Yeltsin achieved victory did not feel virtuous. He had run a negative campaign, stoking voters' fears about the perils of communist restoration. He had used state resources to win — not exactly in keeping with "free and fair." And allegations of electoral fraud dampened the celebration about Russian democracy's resilience. More generally, the Russian political system still lacked many attributes of a consolidated, liberal democracy. This vote had been, more than anything, a referendum on communism, not an affirmation of democracy or Yeltsin and his policies. My mixed feelings about the vote were captured by two different headlines for the same exact op-ed. The title of my piece in the *Moscow Times* was "A Victory for Optimists." The title of the same piece in the *Washington Post* was "Russia: Still a Long Way to Go."[13] The task moving forward, then, was to strengthen the institutions that make democracy endure: political parties, civil society, and an independent judiciary.

After Yeltsin's reelection, many Americans believed that the project of building Russian democracy was over, and that the United States was now free to pursue other foreign policy interests. First up was NATO expansion, which President Clinton had delayed until after the Russian presidential election but now resumed with vigor. In March 1999 Poland, Hungary, and the Czech Republic formally joined NATO.

Yeltsin and his nationalist and communist opponents alike criticized the decision, as did many American analysts and former government officials. Ambassadors George Kennan and Jack Matlock were especially critical, with Kennan calling it "the beginning of a new cold war"[14] and Matlock lambasting President Clinton for reneging on past commitments, stating that "Gorbachev did get an informal, but clear, commitment that if Germany united and stayed in NATO, the borders of NATO would not move eastward."[15]

NATO expansion fueled uncertainty within the Russian elite about American long-term intentions regarding Russia. For the first several years of Russian independence, those in power in Russia had considered the United States a strategic partner. Yeltsin considered "Bill" a friend; NATO expansion seemed inconsistent with the idea of a strategic partnership, and offended Yeltsin personally. The Clinton administration, to its credit, took several parallel measures to take the sting out of NATO expansion. Most importantly, the administration convinced the alliance to create the NATO-Russia Council (NRC), whose mandate was "to serve as the principal structure and venue for advancing the relationship between NATO and Russia."[16] The Clinton administration also continued economic assistance to Russia throughout this period and encouraged the IMF to keep supporting Russia during its tumultuous economic transition. While publicly critical of the policy decision, Yeltsin did not overreact. Nonetheless, tensions over the decision to expand NATO did strain U.S.-Russia relations. The reset begun by Reagan and Gorbachev was not over, but optimism about deepening ties started to wane.

At the time, my view was not to stop NATO expansion — countries had the right to choose their alliance partners and Russia should not get to veto those choices — but to keep the door open for membership for *all* countries, including, prospectively, a democratic Russia. If Russia met the criteria for membership and both NATO and Russian leaders saw the wisdom of Russia joining the alliance, why not? Of course, I realized that such an eventuality was a long way off, but I never understood the logic of why countries like Ukraine or Georgia could be considered for membership but not Russia. Back in the 1990s, after all, Russian leaders were seeking to build democracy and markets at home and to join Europe. Needless to say, few officials in Washington or Moscow at the time listened too closely to my ideas about Russian participation in NATO.

Russia's financial crash in August 1998 provided another setback to U.S.-Russia relations, and to Russian economic and political development in general. The blow hit all the harder since Yeltsin had seemed poised to get serious about market reforms just a few months earlier, in March, when he replaced his Soviet-era prime minister, Viktor Chernomyrdin, with a young, new reformist government. The new team was headed by thirty-five-year-old Sergey Kiriyenko as acting prime minister and included, among others, two young, liberal stars, Boris Nemtsov and Boris Fyodorov, as deputies.[17] This new government pledged to break from years of budget deficits, insider privatization, and partial price and trade liberalization, and thereby end Russia's economic depression, reduce corruption, and implement the rule of law—or at least that was the promise.

Whether because of bad timing or years of inept governance before them, Kiriyenko's government failed quickly. In the early 1990s the Central Bank had printed money to cover the deficit, fueling rampant inflation. After the adoption of the new constitution in 1993, the Central Bank obtained more autonomy and began to pursue a more stringent monetary policy, but with negative consequences. The lack of liquidity in the economy compelled bartering and exacerbated the accumulation of debt between enterprises. By 1998, according to one analyst, "industry collected as much as 70 percent of its receipts in nonmonetary form, leaving many firms with too little cash to pay salaries and taxes."[18] The Russian government also ran up deficits, in part because the parliament approved unrealistic budgets with expenditures well beyond planned revenues, and in part because tax collection proceeded inefficiently.[19] At the time, Russia's oligarchs in particular were notorious for not paying taxes, which created real problems for raising revenue in the government. By 1998 the deficit had ballooned to more than 5 percent of GDP.[20]

So the Russian government had no choice but to borrow money, both domestically and internationally. The Russian Finance Ministry increased borrowing by issuing short-term bonds, known by their acronym in Russian as GKOs. But just as the GKO market was expanding dramatically in the summer of 1998, the Asian financial crisis and falling oil prices began to reverberate in Russia.

Some of the same people losing money in South Korea needed to cash out of their bond purchases in Russia, but the Kremlin could not repay its obligations on these short-term bonds. In a drastic, desperate move on August 19, 1998, the Russian government essentially defaulted.[21] The government also announced a new trading price for the ruble, 30 percent lower than the day before.[22] Two of Yeltsin's most important economic achievements—low inflation and a stable currency—were wiped away overnight. In turn, the Russian stock market collapsed, the ruble continued to fall, and banks began to close.

Russia's transition to capitalism seemed to be failing. Some, including me, worried that Russia's nascent democracy would collapse as well, as was the pattern in other new democracies around the world that had endured economic hardship. To rebuild confidence, Yeltsin fired Kiriyenko and his government and tried to bring back his former prime minister, Chernomyrdin. Yeltsin's opponents in the Duma successfully pushed back on reappointing Chernomyrdin, forcing Yeltsin eventually to appoint as prime minister Yevgeny Primakov, Russia's former foreign minister, who had close ties to the Communist Party.

Russia's financial collapse dampened optimism in the United States and Europe concerning the prospects for Russian internal reform and external integration. My friend Adam Elstein, at the time the head of the Bankers Trust Moscow office, proclaimed on the pages of the *Financial Times* that investors "would rather eat nuclear waste" than invest again in Russia.[23] Republicans in the U.S. Congress blamed the Clinton administration for wasting American tax dollars on the naive pursuit of promoting the free market and democracy in Russia. The title of the U.S. congressional study said it all: *Russia's Road to Corruption: How the Clinton Administration Exported Government Instead of Free Enterprise and Failed the Russian People.*[24] (Chapter 6 of this report is entitled "Bull****: Gore and Other Administration Policy Makers Systematically Ignore Evidence of Corruption of Their 'Partners.'") The report blamed Vice President Al Gore personally for supporting corruption in Russia. The actions of the IMF and the Treasury Department also came under tight scrutiny. "Who Lost Russia?" became a Washington parlor game.[25]

Throughout this economic meltdown, Yeltsin seemed checked out. Clinton administration officials were unsure about his state of health; some worried that he might resign. But paradoxically, the communist-leaning Primakov succeeded in drastically reducing government spending, slowing inflation and ending the economic free fall. Rhetorically, he promised to reverse market reforms, raise pensions and wages, curtail the activities of Western imperial agents such as the IMF and the World Bank, arrest corrupt bankers, and restore state control over prices and property.[26] In practice, Primakov and his Communist Party allies proved to be even more fiscally conservative than Gaidar or Kiriyenko.[27] Primakov had to be fiscally conservative, since no one was willing to lend to the Russian government unless the Kremlin committed to these austerity measures. For the first time since the collapse of the Soviet Union, the Russian economy began to grow in 1999.

Optimism about Russia's long-term prospects as a market economy, however, did not change quickly. Likewise, hope about closer diplomatic ties waned in Washington. Both within and outside the government, American critics began to challenge Clinton's bear hug of Yeltsin. The chorus doubting Russia's ability to change internally also grew, with many now openly arguing that Russians were not capable of practicing real democracy or real capitalism. As academic analyst and policy advocate, I stood my ground: Russia could change internally and become closer to us externally. All was not yet lost, I wrote at the time.[28] But my doubts were growing.

A year after the August 1998 financial crisis, U.S.-Russia relations suffered two more blows: major disagreements about the NATO bombing campaign against Slobodan Milošević in Serbia and Moscow's invasion of Chechnya.

In Serbia, NATO's military mission was clear: stop the ethnic cleansing of Albanians in Kosovo. As NATO's Secretary-General Javier Solana explained, "Our objective is to prevent more human suffering and more repression and violence against the civilian population of Kosovo."[29] Yeltsin, the Russian government, and most of Russian society, however, rejected NATO's explanation for intervention, seeing instead another instance of the United States using military force to expand its influence in Russia's neighborhood. To make matters worse, the United States was condemning fellow Slavs, fellow Orthodox believers. News about the genocide in Kosovo never reached most Russians.

I had few reservations: the United States and our allies had to end the killing in Kosovo. Milošević was a dictator who had to be stopped. Some Russian liberals quietly agreed with me, hoping that a quick, decisive American victory over a Serbian autocrat would help affirm the primacy of the West and the need for Russia to bandwagon with European democracies and integrate. NATO did win. A year later, Serbian society mobilized to protest a falsified election, eventually ousting Milošević from power. Momentum seemed to be on the side of the West and democracy.

Unfortunately, the negative effects on U.S.-Russia relations from the NATO military operation against Serbia lingered for decades. Yeltsin eventually made peace with Clinton; he felt betrayed by the NATO war against Serbia, but also understood that Russia's long-term interests were best served by maintaining close relations with the United States and the West. But not everyone in the Russian government agreed with the Russian president. For the *siloviki* — Russian military and intelligence officials — the NATO campaign in Serbia confirmed their theory about American imperial intentions. In their view, little had changed since the Cold War era, except that Russia was much weaker in 1999 and therefore lacked the means to counter American military aggression. The proper response, therefore, was not to kiss and make up, as the naive, aging Yeltsin opted to do, but to rebuild Russian military forces. One of the intelligence officers who held this view was Vladimir Putin. The following year, he became president.

Only a few months after the NATO campaign in Serbia ended, Yeltsin invaded Chechnya for the second time that decade. The attack was labeled an antiterrorist operation, and the formal reason for military intervention was to repel an incursion from Dagestan by Chechen fighters under the command of Shamil Basayev in August 1999.[30] The following month, terrorist attacks took place against residential apartments in Moscow, Volgodonsk, and Buinaksk, adding urgency to the military operation against Chechnya. The identity of the perpetrators of these attacks is still debated today: some say they were Chechen terrorists; others contend they were agents of Russia's Federal Security Service (FSB) seeking to help Putin — the law-and-order candidate — gain stature before the presidential election the following year.[31]

For Russian citizens at the time, there was no uncertainty: Chechen terrorists were attacking Russia. Yeltsin had to push back. Moscow deployed a massive ground force and a brutally destructive aerial campaign to reassert control in Chechnya.[32] Tens of thousands were killed and more than two hundred thousand civilians were displaced.[33]

U.S. officials were less forgiving than they had been during the first Chechen war. Clinton no longer compared Yeltsin to Lincoln, but instead took punitive steps — albeit minor ones, such as postponing a few loans from the Export-Import Bank. Business in most arenas continued, but the second Chechen war further affirmed the drift in U.S.-Russia relations, and raised new doubts about Yeltsin's commitment to human rights. The reset in bilateral relations started by Gorbachev and Reagan seemed a lot more tenuous a decade later.

On August 9, 1999, Yeltsin surprised everyone by naming Vladimir Putin as his new prime minister — his third head of government in only twelve months. With this appointment, the Russian president was making it clear that he considered the former KGB agent his heir apparent. I was puzzled by Yeltsin's choice. At the time, Putin was an obscure Kremlin bureaucrat. After St. Petersburg mayor Anatoly Sobchak lost his bid for reelection in 1996, Putin, Sobchak's deputy, had become unemployed (just like it's supposed to work in democracies). A former colleague had helped him secure a job in the Kremlin, where he bounced around until being promoted to director of the FSB — the successor organization to the KGB — and then to secretary of the Security Council.[34] But from that job to prime minister was a huge leap. Putin displayed little charisma, championed no clear set of ideas, had no political party behind him, and had never run for office. What did Yeltsin see in him that I didn't?

A few months later Yeltsin made another dramatic move. On January 1, 2000, he resigned his position as head of state and named Putin acting president — making it clear that he wanted Putin to win the March 2000 presidential election.

At 5:00 a.m. on that first day of the new millennium, Peter Jennings called from ABC News and asked me to go on live right then and there to explain Yeltsin's decision and Yeltsin's legacy, all in a few minutes (and before I'd had my coffee). Still half-asleep, I struggled then to explain Yeltsin's thinking, and I still struggle now. I did not respect Yeltsin's decision to step down. Although constitutional, the decision violated the spirit of democracy. I understood that Yeltsin or his entourage wanted to elevate Putin to improve his chances for election in March 2000. But it seemed like a cheap stunt unbefitting a world leader, let alone the first democratically elected president in Russia's thousand-year history.

On the other hand, Yeltsin willingly relinquished the power to rule Russia. That also had never happened — not in czarist Russia, not in the Soviet Union, and not up to this day, in post-Soviet Russia. He had to be applauded for establishing the precedent. I can't prove it, but I believe that Yeltsin's decision to relinquish power in 2000 played some role in convincing Putin not to run for a third term in 2008: Putin did not want to defy the Russian constitution after Yeltsin had established the precedent of retiring after two terms.

On the large legacy question, Yeltsin did deserve credit for his historic role in transforming the Soviet Union and then Russia. Despite many shortcomings, Yeltsin made major strides toward decolonization and marketization. No major empire in the world had dissolved so peacefully, and Russian capitalism in 2000 seemed battered but ensconced. There was too much corruption, the oligarchs had accumulated wealth through unjust means, and the state owned too many assets. But two fundamental concepts of the market economy that were illegal in the Soviet era — private property and free prices — seemed to be uncontroversial attributes of Russia's economic system. At the time, I celebrated these achievements.

Yeltsin's legacy regarding democracy was much more ambiguous. The day Yeltsin stepped down as president, the formal institutions of democracy were still in place. The constitution adopted in 1993 had not been violated. Political parties, civil society organizations, and some independent media all existed, and the state was not actively seeking to repress their activities. And most Russians, according to public opinion polls, supported the ideas of democracy — a revolutionary achievement, given Russia's long history of autocratic rule.[35] The Russian political system met the minimal definition of a democracy: an "electoral" democracy, to use the jargon of political science. But Russia was far from reaching a more stable and permanent system of "liberal" democracy, or "consolidated" democracy.[36]

The most troubling flaw in the political system was what I called at the time "Russia's privatized state." The oligarchs wielded too much power over Kremlin decisions and controlled too much of Russia's media, especially television. Meanwhile, Russian political parties remained weak, and nongovernmental organizations — civil society — could wield only limited power to influence election outcomes, legislation, or policy implementation. Russia's court system also lacked genuine independence. Finally, many institutions of the Soviet autocratic system remained in place a decade after independence, including, first and foremost, the KGB. Renaming it the Federal Security Service (FSB) did not fundamentally alter the norms or practices of this powerful institution. In retrospect, I underestimated the deleterious consequences for democracy of this lingering Soviet legacy, especially when one of its own — Putin — became head of state. Yeltsin's legacy would be judged by what came next. Would democracy gradually deepen and consolidate, or would the thin inheritance of democratic governance from the Yeltsin era not be enough to take root?

Democracy promoters — myself included — had to share in the blame of Russian democracy's uncertain future. By 2000 I was focused on my academic career, yet remained an active participant in the policy debate in the United States about democracy promotion in Russia, and remained friends with Russians who had helped end communism and dissolve the Soviet Union a decade earlier. I still wanted the democrats to succeed. NDI and other democracy-promoting organizations continued to work in Russia, and the U.S. government helped fund many of these groups — though less than was necessary to make a difference. It was nearly impossible to measure in any scientific way the effectiveness of these American activities. One thing was clear: Russian democracy seemed much more fragile in 2000 than it had been a decade earlier. American efforts to support reform were being overwhelmed by bigger, more powerful factors undermining democracy inside Russia.

I was frustrated by Russian efforts to consolidate democracy and our efforts to help that process. The United States emerged from the Cold War as the world's only superpower. Yet this super-superpower proved unable, inept, or unwilling to influence domestic change in Russia. As Yeltsin stepped down at the beginning of the new millennium, I worried that we had not done enough to encourage Russian reform internally or to assist Russia's integration into the West. We — Russians and Americans who believed in democracy and closer ties between our two countries — had missed crucial opportunities.

And yet I also saw no alternative to continuing to try working toward our common mission. In a formal assessment of NDI's activities in Russia conducted in 2001, I pushed for greater engagement of the communists, more attention to strengthening Russia's leading election-monitoring organization, Golos (the Russian word for "voice" and "vote"), and less work with Russian government officials. But my main message was to stay engaged with democracy promotion, as the process of democratic consolidation in Russia was going to take much longer than in other postcommunist countries to Russia's west. In those days, I also stressed that one person — Vladimir Putin — would play the key role in determining the path of Russia's political system. That the fate of Russian democracy in 2000 was so tied to the ideas and actions of one individual underscores the failure of Yeltsin and the democrats to institutionalize democracy.

4

PUTIN'S THERMIDOR

All revolutionary change generates backlash. If Gorbachev tried to reform the ancien régime, and Yeltsin led the revolution to destroy Soviet communism and replace it with democracy, markets, and Russian independence, Putin pressed for a partial counterrevolution — restoration of some aspects of Soviet rule in response to Yeltsin's radical changes. Putin did not seek to roll back all changes. Instead, he aimed to fuse aspects of the old and new. In the French Revolution, Thermidor (referring to the eleventh month of the French Republican calendar) marked the end of the Reign of Terror. The term has come to be associated by historians of revolutionary theory with a period of pushback in which the new order accommodates itself to features of the old regime. Putin's reintroduction of the Soviet national anthem embodied his approach: he replaced the new hymn, which nobody liked or knew, with the communist-era song, then added post-Soviet lyrics. More consequentially, he altered Russia's governing structures through a similar strategy that melded old and new, restoring autocratic rule and strengthening coercive Soviet-era organizations such as the FSB — the successor organization to the KGB — while maintaining new democratic-looking institutions such as the State Duma and the new constitution.

In undermining liberal, democratic institutions at home, Putin eventually reduced the prospects for cooperation with the United States and integration into the West more generally. Yeltsin wanted to join the West, and understood that the price of admission was building democracy and markets at home. After some early flirtations with cooperation and integration, Putin developed a more suspicious interpretation of Western intentions. At times, American actions helped to reinforce those suspicions. By the end of his second term in 2008, Putin had redefined Russia's role in the world, largely, but not completely, in opposition to the United States and the West.

Putin and Putinism were not predetermined. Innate, structural forces did not produce Putin; Yeltsin selected Putin as his successor. The Russian people merely ratified Yeltsin's choice. Putin did not rise to power through a groundswell of popular support for his leadership style or political program. He did not plot a path to the Kremlin over the course of decades. He had never participated as a candidate in an election until he ran for president in March 2000. He was simply in the right place at the right time.

Years later, Putin and his Kremlin team would propagate an idea about Russian societal demand for his leadership, claiming that Putin was not Yeltsin's handpicked successor, that he hadn't supported Yeltsin's policies, and that Putin was indeed the antithesis of Russia's first president. But the facts of Putin's ascendance run counter to these assertions. Especially mythical is the revisionist claim about mass demand for Putin and his ideas. How could there have been such demand? No one knew anything about Putin in the spring of 2000.

Moreover, Yeltsin had other choices for a successor. Putin's selection was not inevitable. Most intriguingly, Yeltsin flirted with choosing Boris Nemtsov as his heir.[1] Yeltsin had admired Nemtsov for years, seeing traits of his younger self in the smart, charismatic, plainspoken, and ambitious democrat from the Urals. In March 1997 Yeltsin appointed Nemtsov — then governor of Nizhny Novgorod — first deputy prime minister in a move that many interpreted as the next phase of Yeltsin's cultivation of him as his successor. Nemtsov took on the oligarchs, and championed popular policies such as income declarations for all government officials. Nemtsov stayed on after Yeltsin fired Prime Minister Chernomyrdin and replaced him with a new, young head of government, Sergey Kiriyenko. Prime Minister Kiriyenko and his government lasted only a few months, ousted from power by the August 1998 financial meltdown. Although Nemtsov had no personal responsibility for this economic crisis, his reputation suffered like everyone else's in the Russian government at the time. Without that economic collapse, Yeltsin might have very well selected Nemtsov as his successor, and the world might never have heard of Vladimir Putin.

Most Russians only remember Nemtsov as an opposition leader who organized popular demonstrators against Putin's autocratic regime. But I knew him for nearly two decades in a different role, when he was a popular, successful, democratically elected governor in Nizhny Novgorod. Nemtsov boldly pursued radical economic reforms in that role, yet managed to maintain popular support and protect his reputation.[2] I also remember First Deputy Prime Minister Nemtsov, who challenged the oligarchs and made tackling corruption one of his highest priorities.[3] He had the skills and charisma to have become a successful president—a successful *democratic* president.

The 1998 financial crisis interrupted Nemtsov's rise to power. But another factor played a role: Yeltsin's fear of retribution after he left office. Yeltsin opted for a KGB colonel instead of Nemtsov or other candidates because he wanted a successor who could protect his family and their assets. As a member of the *siloviki,* the power ministers, Putin was more likely to put up a strong defense than was Nemtsov or someone else with a more democratic orientation.

In Yeltsin's defense, he also could have had some reason to believe in 1999 that Putin might adhere to market and democratic principles. Putin, after all, had worked for five years for St. Petersburg mayor Anatoly Sobchak, one of the most committed democrats of his time. He had multiple career opportunities after he left the mayor's office in 1996, but took a job working for Yeltsin, first in the Kremlin's Presidential Property Management Directorate and eventually as the head of the FSB. His backer for his first Kremlin job was Alexei Kudrin, a liberal reformer.[4] If, back then, Putin aimed to roll back democratic reforms and renationalize private property, he was keeping his plans quiet.

In his first years in the Kremlin as president, Putin took steps to deepen market reforms. He restructured Russia's debt, introduced a 13 percent flat tax on individual income, and reduced the corporate tax rate, winning him praise from business circles inside and outside Russia.[5] Many of Putin's staff working on economic policy at the time, including his chief economic advisor in the Kremlin, Andrei Illarionov, were pro-market, pro-Western liberals.[6]

Putin's initial statements and actions also suggested continuity with Yeltsin's pro-Western foreign policy. During a trip to the United Kingdom in March 2000, Acting President Putin seemed eager to demonstrate his Western orientation, hinting in an interview at the possibility of closer cooperation between Europe and Russia and even NATO and Russia. "Russia is part of the European culture," he explained. "And I cannot imagine my country in isolation from Europe and what we often call the civilized world. So it is hard for me to visualize NATO as the enemy." When asked point-blank if Russia might join NATO, Putin replied, "I don't see why not. I would not rule out such a possibility."[7] At the time, I was shocked. Years earlier, I had made the case for keeping the door open for Russia — that is, a democratic Russia — an idea that was roundly ridiculed as naive and idealistic. Now, Russia's new president hinted that he might agree.

The idea of Russia joining NATO, however, was never seriously tested, in part because of Putin's clearer preferences and policies regarding democracy. In 2000 all members of NATO were democracies, but Putin was moving Russia in an autocratic direction. To strengthen the state, Putin believed that he had to weaken checks on his power. He moved first against independent media. Soon after his inauguration, his Federal Security Service raided the company that owned Russia's most popular independent television station, NTV. The station's owner, Vladimir Gusinsky, left the country after being incarcerated briefly on fraud charges, and was later forced to cede control of his media empire to Gazprom, Russia's state-controlled natural gas monopoly.[8] A similar drama unfolded over ORT, Russia's largest television network, controlled by another Russian billionaire, Boris Berezovsky. Although Berezovsky was close to the Yeltsin family and was one of the biggest advocates for Putin's appointment as prime minister and then president, Putin turned on him almost immediately, seizing full control of ORT and pushing him into exile. Since the state already owned RTR, Russia's second-largest television network, Putin had acquired de facto control of Russia's three largest and most important television channels before the end of his first year in office.

I was concerned about Putin's antidemocratic proclivities even before his election. In an op-ed in the *Washington Post,* the first week in March 2000, I wrote, "Not since the August 1991 coup attempt has the future of Russian democracy been more uncertain than it is today."[9] I chided the Clinton administration for saying and doing nothing about Putin's rollback of democracy, and argued that greater autocracy in Russia would produce foreign policy challenges down the line. "Arms control [that is, the Cold War treaties regulating U.S. and Russian nuclear arsenals] did not end the Cold War. Rather, it was the collapse of communism and the emergence of democracy within the Soviet Union and then Russia that suspended the international rivalry between the United States and the Soviet Union. If a new nationalist dictatorship eventually consolidates in Russia, we will go back to spending trillions on defense to deter a rogue state with thousands of nuclear weapons." The morning the piece was published, I received a call from Stephen Sestanovich, at the time Clinton's ambassador-at-large and special advisor to the secretary of state for the newly independent states of the former Soviet Union. Steve is a close friend, and our chat was predictably warm and cordial. But he pushed back on my claim that his administration was doing nothing to shore up Russian democracy. For me, that conversation marked the beginning of a decade of combat with U.S. government officials about America's Russia policy. Throughout this period, I maintained that we were not doing enough to push back on democratic erosion under President Putin. I'm not sure my arguments influenced policymakers in Washington, but I am certain that my criticisms made me no friends inside the Kremlin.

At the beginning of the Putin era, few analysts in Washington devoted much attention to his autocratic tendencies. In fact, few devoted much attention at all to anything Russian. Back then, most American national security specialists considered Russia a declining power and therefore a peripheral foreign policy problem. Was Putin for or against Milošević? Would Putin tolerate more NATO expansion? How would Putin react to President George W. Bush's decision to pull out of the Anti-Ballistic Missile Treaty? The answer to each of these questions was, Who cares? Putin could do little to influence any of these issues. Russia was weak, so the argument went; Russia didn't matter anymore. Just as managing the Soviet collapse was a central challenge for George H. W. Bush in 1991, some analysts in the early 2000s believed Russia's weakness represented a central U.S. foreign policy challenge, as Russia's decline could cause instability in surrounding regions.

I disagreed. Russia was no longer a superpower, but it was not weak, and certainly not destined to be sidelined. I believed that Russia, just like other post-communist countries, would recover economically. Russia also had one of the two largest nuclear arsenals in the world, diplomatic and economic relations with many countries, and a centuries-old tradition of playing a central role in European and, later, world affairs. What mattered most to us were not calculations of Russia's power capabilities but Russia's internal political development. If democratic institutions in Russia took hold, then Russian foreign policy would become more pro-Western and pro-American, irrespective of the size of Russia's economy or the strength of Russia's military. If Russia's faltering democracy gave way to new authoritarian tendencies in the Kremlin, Russia would once again become our competitor and eventually even enemy. Of course, the United States and Russia would always have some competing interests around the world. But a democratic Russia — strong or weak — was a more likely partner of the United States than an autocratic Russia — whether strong or weak.

Our abilities to predict long-term power trajectories are limited.[10] In 2000, Russia was still recovering from a major economic depression triggered by the transition from communism to capitalism, but over the long haul, Russia's natural resources, land, educated population, and industrialized economy would most certainly propel the country to a position of power in the international system, or so I argued at the time.[11] And a rising power ruled by a strong autocrat would eventually create problems for the United States and our democratic allies. Our problem was not Russian weakness, but Putin's autocratic ways.

In May 2001, I had the chance to rehearse these arguments with President George W. Bush. On the eve of his first trip to Europe as president, the new U.S. leader hosted five outside experts on Europe and Russia at his residence. It was an impressive group: Oxford professor Timothy Garton Ash, *Financial Times* editor Lionel Barber, former ambassador to France Felix Rohatyn, and Tom Graham, a former U.S. diplomat (and future NSC staffer in the Bush administration) and one of the United States' top experts on Russia.

Before the two-hour discussion session kicked off precisely on time — Bush was known for his punctuality — the president gave us a quick tour of his new home: the Truman Balcony, the Lincoln Bedroom, and the Queen's Bedroom, where, the president joked, his mom liked to stay. Bush was friendly and laid-back, engaging in conversation on topics ranging from presidential history to Stanford baseball. If one of the objectives of the tour was to generate in us a feeling of fondness for the president, for me, it worked.

After the tour, we convened on the second floor of the residence in the Yellow Oval Room, joined by Vice President Dick Cheney, National Security Advisor Condoleezza Rice, and many members of Bush's new foreign policy team. Bush asked all the questions. With subtle references to some of our publications, he signaled that he had done some homework, and seemed eager to listen and learn. He was new to foreign policy and understood the importance of preparing for his first trip to Europe, and his first meeting with Putin. During our discussion, he flubbed a few historical points and didn't know an acronym or two, but projected confidence.

After a first hour on Europe, the discussion shifted to Russia. Tom Graham argued forcefully that the new president needed to focus on managing Russia's decline. Later, in print, Graham quoted Primakov, who "at the time of his confirmation as prime minister in September 1998 warned that there was a growing danger of Russia splitting up and vowed to take tough steps to avert it."[12] When Bush turned to me, I did not challenge Tom's analysis. But I urged the president to consider integrating Russia into our liberal international order as the central task ahead. We had to support democracy inside Russia and encourage Russia to join the international institutions set up by the United States after World War II. We wanted Russia inside our tent, not challenging us from the outside. I asked President Bush to imagine having visited Moscow in 1927. A decade after the Bolshevik Revolution, the Soviet Union would also have looked like a declining power, and after a revolution followed by brutal civil war, the Red Army did not pose a threat to the world's great powers. Yet two decades later the USSR had emerged as the second-most-powerful country in the world. Russia might well recover from the 1990s as a major power, I asserted, and if it did, we wanted it on our side.

Bush agreed with me, or that was my impression. But then he added that we needed Russia on our side because someday we would all be dealing with China's rise. His words perplexed me. He may have meant that all the democracies of the world would have to confront a more powerful Chinese autocracy. Or perhaps he meant that all the world's great powers, autocracies or not, would have to band together to counter China's rise — a classic realist way of thinking about the balance of power in the international system. I left the White House that day believing he'd meant the latter; I had lost the argument.

After September 11, 2001, Bush rethought many of his assumptions about the world, and became a passionate advocate for democracy promotion. That passion for pushing freedom and liberty around the globe, however, applied only episodically to Russia.

A few weeks after our meeting with Bush on Europe and Russia, the new American president met President Putin for the first time, in Ljubljana, Slovenia. Back at the White House, Bush had told us that he intended to establish a close relationship with Putin. He had found personal relationships key to making deals as a businessman, and that was going to be his approach to foreign policy. And so Bush tried hard to connect with the Russian president during their first meeting. A devout Christian, Bush was impressed with a story Putin shared with him about the cross he wears, which he had saved from a fire at a dacha.[13] At the press conference, perhaps trying to tighten their bond through praise, Bush remarked: "I looked the man in the eye. I found him to be very straightforward and trustworthy. We had a very good dialogue. I was able to get a sense of his soul."[14] That was too much. My guess at the time was that even Condi thought the new president had gone too far in trying to read Putin's soul after one meeting. Vice President Cheney and his team most certainly thought so;[15] they remained skeptical about rapprochement with Putin for the entire eight years of the Bush administration. Quoted in the *New York Times* two days later, I called Bush's remarks a "rookie mistake." I went on: "I can understand the strategy on rapport, but it went too far . . . I think there is plenty of good reason not to trust President Putin. This is a man who was trained to lie."[16] I was never invited back to brief President Bush again.

Years later, during Obama's first encounter with Putin in July 2009, I listened to the Russian leader fondly describe his relationship with Bush. I think Putin genuinely grew to like the American president after several years in office; their closeness, however, did little to reduce the policy differences between our two countries.

Tragedy — September 11, 2001 — abruptly altered the dynamics of U.S.-Russia relations. In the wake of this horrible terrorist attack, American and Russian leaders forgot about their differences seemingly overnight, and pledged to join together to fight a common enemy. Prophetically, Putin had authored a *New York Times* op-ed two years earlier, in which he warned that Americans might someday experience the horror of terrorism as Russians had that year: "Imagine ordinary New Yorkers or Washingtonians asleep in their homes. Then, in a flash, hundreds perish at the Watergate or at an apartment on Manhattan's West Side."[17]

Putin was one of the first world leaders to reach out to President Bush after the attacks, offering sympathies and promising support for the U.S. fight against terrorism.[18] Putin then provided greater assistance for the Northern

Alliance fighting the Taliban in Afghanistan, offered to share intelligence with the United States about terrorist organizations, and supported the opening of American military bases in the former Soviet republics of Uzbekistan and Kyrgyzstan.[19] After September 11, Bush divided the world into those against us and those with us, and Russia fell clearly in the latter camp. For a while, it appeared that the United States and Russia had entered a new, more cooperative period in bilateral relations.

Bush was quick to test the strength of this new relationship. He genuinely wanted to work closely with Putin on fighting terrorism, but not at the expense of other national security priorities. Topping this list was Bush's intention to withdraw from the Anti–Ballistic Missile Treaty, a decision he announced in December 2001.[20] The ABM Treaty put stringent limits on the deployment of missile-defense systems, thereby ensuring mutual assured destruction in the event of a nuclear war. In pulling out of this treaty and announcing his intention to deploy new missile defenses, Bush threatened to deploy weapons that would allow the United States to win a nuclear war, or so Russian officials and analysts argued at the time. Russia lagged behind in the development of its missile-defense programs, so Bush's move was likely to be seen as highly destabilizing.[21] But Putin's statement in response to the U.S. withdrawal wasn't as strong as I had anticipated: "The Treaty does indeed allow each of the parties to withdraw from it under exceptional circumstances. The leadership of the United States has spoken about it repeatedly and this step has not come as a surprise to us. But we believe this decision to be mistaken."[22] That was a muted response. By the end of the Bush administration, missile defense would emerge as a major issue of contention between the United States and Russia, but in 2001 Putin was trying hard to nurture closer ties with America. The following month, Putin's foreign minister, Igor Ivanov, went so far as to compare the new level of cooperation to the alliance between the United States and the Soviet Union during World War II.[23]

Bush was equally effusive about Putin. At a G-8 summit in Canada in June 2002, Bush praised Putin as "a stalwart in the fight against terror . . . He understands what I understand, that there won't be peace if terrorists are allowed to kill and take innocent life. And, therefore, I view President Putin as an ally, [a] strong ally in the war against terrorism."[24] I don't recall any president since Roosevelt using the word "ally" in reference to a Kremlin leader.

During this new honeymoon in U.S.-Russia relations, Bush and Putin got some things done. They signed the Strategic Offensive Reductions Treaty, also known as the Moscow Treaty, which committed both countries to reducing strategic nuclear weapons to between seventeen hundred and twenty-two hundred warheads by December 2012.[25] At the 2002 NATO summit in Rome, Putin and the NATO heads of state created the NATO-Russia Council, a new forum for consultation and cooperation between NATO and Russia.[26] Bush and Putin also pledged to work together to stop nuclear proliferation around the world.

I shared the optimism of the moment and applauded the new cooperation between Russia and the United States.[27] But, worried that Putin did not share our democratic values, I also saw the need for caution. If Putin continued to push Russia in a more autocratic direction, I believed that our two countries would clash again.[28] Throughout American history, our deepest and most reliable allies have been democracies. Conversely, our adversaries have been dictatorships. The United States has occasionally worked closely with autocracies on mutual security interests, but those relationships never endured, either because autocrats changed their minds, or because they fell from power. In Russia's case, I was worried about the former.

In the Bush-Putin era, it was not only growing Russian autocracy that worked to reverse the rapprochement between Russia and the United States. Some of Bush's foreign policy decisions dismayed the Kremlin. Putin's policies in turn accelerated the pace of the negative spiral.

In November 2002, NATO invited Bulgaria, Estonia, Latvia, Lithuania, Romania, Slovakia, and Slovenia to join the alliance; two years later, these states became NATO members. At the time, Putin called NATO obsolete and ridiculed the security gains of expansion.[29] Putin and his inner circle were bothered in particular by the admission of three Baltic states — former Soviet republics — to the alliance. But U.S.-Russian relations survived that second round of NATO enlargement. Putin did not criticize Bush personally, nor did he say the move would trigger a new era of confrontation with the West. As with Putin's temperate response to the American withdrawal from the ABM Treaty, I was surprised by his measured tone. Years later, Putin would resuscitate NATO expansion as a major issue of contention in Russian-American relations. In 2002, however, just a year after September 11, Putin remained committed to working with the United States. U.S.-Russia relations were strained but not broken by this latest wave of NATO expansion.

President Bush's decision to invade Iraq delivered a more devastating blow to bilateral relations. Putin adamantly denounced Bush's justification for invasion, asserting in chorus with many other leaders around the world, including several American allies, that there was no evidence of an Iraqi nuclear weapons program. Even if Iraq did have nuclear technology, however, Putin would have protested. In 2003 — before Russian interventions in Georgia in 2008 and Ukraine in 2014 — the Russian president championed the inviolability of state sovereignty as a central principle of the international order. He especially detested the use of force for regime change and America's brazen disrespect for the United Nations Security Council. Russia had a veto within the Security Council, a key instrument of Russian foreign policy at a time of lesser Russian economic and military power; by invading Iraq without the council's approval, Bush was also diminishing an instrument of Russian power.

Putin resented Bush for dismissing his warnings about the risks of an invasion. Saddam Hussein, in Putin's view, held Iraq together with an iron fist; his removal would destabilize the country and the surrounding region.[30] Being repeatedly ignored on this point fueled Putin's frustration with the United States. Had Bush engaged with Putin, as he explained to President Obama when they met in Moscow in July 2009, the United States and Russia might have been able to cooperate even on Iraq. Above all else, Putin wanted to be a player, a key decision maker, in global affairs, but Bush denied him that role, or so Putin perceived, especially regarding Iraq. The American invasion also validated Putin's earlier assumptions about the United States. Bush had deceived him with friendly talk about the United States and Russia being allies in the fight against terrorism. Instead, Putin now returned to his original theory about the United States as an imperial hegemon, which uses force to achieve its objectives. As leader of one of the world's few powers not controlled by the United States, Putin believed that Russia had to resist and counterbalance American hegemony and defend state sovereignty. The flirtation with partnership after September 11 had been a mistake.

• • •

In addition to concern over U.S. action in Iraq, Putin was feeling increasingly threatened by U.S. support for regime change closer to home. More than any other issue — more than the cancellation of the ABM Treaty, NATO expansion, or the invasion of Iraq — the so-called color revolutions in Georgia and Ukraine renewed tensions in U.S.-Russia relations and erased the cooperative spirit sparked by September 11.

On November 2, 2003, supporters of Georgian president Eduard Shevardnadze won a disputed parliamentary election. Election observers uncovered electoral fraud, and thousands of demonstrators took to the streets in Tbilisi and other major cities in protest. After three weeks of standoff between the demonstrators and the government, President Shevardnadze tried to open a new session of parliament on November 22. A group of protesters led by Mikheil Saakashvili barged into the hall with roses in their hands and seized control, claiming to be the actual winners of the election. Unable to mobilize government troops in his support, Shevardnadze resigned the next day. Six weeks later, Saakashvili ran unopposed in a hastily organized presidential election and won with 96.2 percent of the vote.[31] Some Western governments called these events a democratic revolution. Russian officials called it a coup.

In the years before the Rose Revolution, the U.S. government had funded NGOs in Georgia. The U.S. Agency for International Development (USAID) oversaw a major program on democracy and governance, and the usual American democracy-promoting organizations, including my old employer, the National Democratic Institute, were active there. These organizations could have a greater impact in a small country like Georgia than in Russia. Putin most certainly recognized that, and didn't like it. In his view, the United States had actively promoted regime change in a former Soviet republic. Putin had no warm relationship with Shevardnadze, but despised Saakashvili, whom he considered an American puppet brought to power by American actions.

The following year, the Orange Revolution in Ukraine represented an even bigger setback for Putin. This change in government unfolded much like the Rose Revolution had. Election monitors, exit polls, and parallel vote tabulations pointed to irregularities, this time in a presidential vote. The U.S. government funded many of those election-related activities. The exposure of electoral falsification sparked massive demonstrations in downtown Kyiv and set off a two-week standoff between protesters and the government, which nearly ended in violence. The government eventually capitulated and allowed for a new election. In the new vote, Viktor Yushchenko, the opposition candidate, defeated Viktor Yanukovych, the victor in the original election.

Putin did not remain on the sidelines in the Ukrainian struggle for power. Endorsing Yanukovych, Putin even appeared at a campaign event with the Ukrainian prime minister in Kyiv to show his support.[32] Russian television channels, with big audiences in Ukraine, also made their support for Yanukovych clear. Russian sources provided tens of millions of dollars in cash to the Yanukovych campaign.[33] American and Russian actors battled on two different sides in the Ukrainian election and in the ensuing standoff. The American side won; Putin's allies lost.

President Bush and most leaders of the democratic world celebrated these events as a breakthrough for democracy. Putin and his associates denounced the Orange Revolution as another American-backed coup, which had installed an anti-Russian leader in a country of vital strategic interest to Russia. Putin asserted later that Ukraine was "a made-up country"[34] that should not exist, let alone pose a threat to his rule by trying to build democracy on Russia's border.

Putin responded to the Orange Revolution by becoming more anti-American in his rhetoric and policies, and by asserting greater control over independent political actors inside Russia to prevent similar processes from unfolding at home. Before 2004, he already had been weakening checks on his power: the 2003 parliamentary election was much less competitive than the 1999 contest, and he had abolished gubernatorial elections in 2004. Constraints on independent media also tightened, with companies sympathetic to the Kremlin assuming greater control over most large Russian newspapers.[35] In October 2006, investigative journalist and Putin critic Anna Politkovskaya was assassinated in the lobby of her Moscow apartment. Civil society also faced new pressures. The government enacted a law that placed demanding new reporting requirements on NGOs, severely limiting the activities of Western actors in Russia.[36] Putin also expelled the Organization for Security and Cooperation in Europe from Chechnya, suspended the Peace Corps program in Russia, and prevented Irene

Stevenson, the longtime director of the AFL-CIO's Solidarity Center in Moscow, from reentering the country.[37] Protesters found it increasingly difficult to organize public events.

During this same period, Putin also limited the political power of independent economic actors, while gradually increasing the economic clout of his personal friends. Most dramatically, on October 25, 2003, he arrested Mikhail Khodorkovsky, chief of Russia's largest oil company and Russia's wealthiest person at the time. Formally, Khodorkovsky was convicted on charges of fraud, tax evasion, and embezzlement; informally, his crime was getting too involved in politics. The foundation he had launched, Open Russia, was supporting dozens of NGOs in Russia at the time, and gearing up to back independent candidates in upcoming parliamentary elections in December 2003. For Putin, those challenges were unacceptable. Khodorkovsky was also just too rich at a time when many of Putin's KGB buddies wanted wealth themselves. They had acquired next to nothing during the Yeltsin era, and with one of their own in the Kremlin, they now wanted to cash in. Eventually, most of Khodorkovsky's oil company, Yukos, ended up under the control of Rosneft, a state-owned energy company run by Igor Sechin, one of Putin's closest personal associates (the same Putin aide with whom I had organized seminars on democracy in St. Petersburg in the spring of 1991). Putin put another KGB officer, Vladimir Yakunin, in charge of Russian railways, and gave Gazprom to his associate Alexey Miller.[38] Another Putin buddy, Gennady Timchenko, who was virtually unknown when Putin first came to power in 2000, was exporting Russian oil and oil products worth $30 billion by 2006.[39]

Throughout this period of growing autocracy and property redistribution in Russia, I became increasingly pessimistic about the nation's course. Putin was rolling back everything I had fought for in the previous decade, and my friends were being pushed to the periphery of Russian politics. And it all happened so quickly. I was in Moscow on the day that masked men stormed Khodorkovsky's jet at dawn at an airport in Siberia to arrest him. Discussing the day's events that evening with Leonid Gozman, a liberal political party leader, we agreed there seemed to be no one who could stop Putin. Nor could we predict how far he might go. If Putin could arrest the richest man in Russia, he could arrest anyone.

If I was concerned about events inside Russia, I was also frustrated by how little the Bush administration seemed to care. Bush had embraced a "Freedom Agenda" for the world after September 11, but it didn't seem to apply to Russia. At times, senior U.S. officials appeared reluctant to acknowledge what was happening inside Russia. Five years into Putin's presidency, Secretary of State Colin Powell was still saying that after the chaos of the 1990s, President Putin had "restored a sense of order in the country and moved in a democratic way."[40] I hoped Powell was simply uninformed; I feared he was saying such things to appease Putin.

Would a different U.S. policy have mattered? Without question, Putin was much weaker in the early years of his presidency than he is now; pressure would have had a greater likelihood of success back then. Those political and economic actors seeking to check Putin's power were also much stronger in the early Bush years than in the late Obama years. Empowering them, even simply through rhetorical support, would have had more impact then than later. But how much more of an impact? The United States always was going to be a marginal player in the fight to shape Russia's political system. Especially after the Iraq invasion in 2003, President Bush enjoyed little respect or authority within Russian society. Had Bush criticized Putin more stridently, who would have cared? In addition, Putin had high approval ratings, sustained by a strong economy for which he claimed credit. Putin was also a young, energetic leader, a stark contrast to the ill and old Yeltsin. Also, he was not grossly violating the constitution: no martial law, no cancellation of national elections (though he did suspend gubernatorial elections), no mass arrests. Putin's system of "managed democracy"—as his aide Vladislav Surkov labeled it—was just democratic enough, especially as the Kremlin appeared to be providing other public goods in greater demand.

In my writings and discussions with policymakers at the time, I focused on debunking the myth that Putin's autocratic ways helped the Russian economy grow. In the 2000s that was the alibi for Putin's rollback of democracy: Russia needed a strong hand to end economic depression and prevent state collapse. Russia was emulating the Chinese model, not Western democracies, so the argument went. It was a compelling claim, especially inside Russia. After a decade of brutal depression, Russians yearned for better economic conditions. For many, sacrificing access to some independent media sources in exchange for economic growth was well worth it. But the actual relationship between Putin's strengthening autocracy and robust Russian economic growth was spurious. The rise of autocracy did not cause economic growth; higher oil and gas prices did, as well as a few good economic reforms implemented during Primakov's months as prime minister and Putin's early years in office.[41]

Kathryn Stoner and I made this argument in an article in *Foreign Affairs* called "The Myth of the Authoritarian Model: How Putin's Crackdown Holds Russia Back." After its publication, the editor got a call from the Russian Embassy complaining about our "propaganda." Allegedly, the diplomat on the phone was conveying a personal message from Putin. When I met Putin two years later with President Obama, I wondered if he remembered what I had written about him. My guess is yes. Putin carries grudges.

In February 2007, Putin further clarified his thinking about the United States and the West with a shocking address to the Munich Security Conference. It seemed as if he finally decided to say out loud what he had been thinking for years, maybe decades. He lambasted America for attempting to be the world's "one master, one sovereign." He criticized the U.S. government's use of military force, contending, "Unilateral and frequently illegitimate actions have not resolved any problems. Moreover, they have caused new human tragedies and created new centers of tension . . . plunging the world into an abyss of permanent conflicts . . . One state and, of course, first and foremost the United States, has overstepped its national borders in every way. This is visible in the economic, political, cultural, and educational policies it imposes on other nations. Well, who likes this? Who is happy about this?"[42] Putin perceived American hegemonic, unilateral military power as a threat to every country in the world, including Russia. He also expressed anger about NATO expansion and American missile-defense plans in Europe — abandoning the restraint he had shown several years earlier when these proposals were first announced. Putin went on to recommend ways to check American power. Most importantly: "The use of force can only be considered legitimate if the decision is sanctioned by the UN." He called upon the entire world to help him constrain the American threat.

For me, Putin's Munich speech presented a moment of clarity about the Russian president's worldview: the United States was Russia's rival and his enemy, and what was good for America was bad for Russia. Of course, not everyone in

Russia, and not even everyone in Putin's government, shared this view. But in terms of the future of U.S.-Russian relations, only Putin's view mattered.

Not long after Putin's infamous Munich speech, he announced his decision not to seek a third term as president. The Russian constitution limits the president to serving two consecutive terms, but Putin had the votes in parliament and support throughout the regions to amend the constitution if he wanted to stay on the job at the Kremlin. Instead, he decided to abide by the existing constitution, and endorsed as his candidate for president Dmitry Medvedev, who was serving at the time as prime minister. After being elected easily as president in the spring of 2008, Medvedev and Putin switched jobs.

In the run-up to Putin's decision about succession, many in the chattering classes in Moscow and Washington had assumed that Putin would either run for a third term or endorse his defense minister, Sergei Ivanov, to become Russia's next president. Ivanov was also from the KGB, but a general, outranking Putin. Ivanov was a charismatic, strong, and decisive leader. He spoke fluent English, having served for years in the KGB in London (not Dresden, like Putin), which lulled some into believing he had pro-Western proclivities. But compared to Medvedev, Ivanov was the hardliner choice. That Putin did not select Ivanov as his successor signaled the possibility of a new, more liberal period in Russian politics, or so some hoped.

In my view, Putin judged Ivanov as too strong a successor, too much a threat to Putin's plans to stay active in governing Russia from the prime minister's office. Medvedev was a safer choice. Medvedev had no KGB rank, and he came to the job enjoying no support from loyalists within the Russian elite. He was much younger than Putin and owed his entire political career to his mentor. He would never challenge Putin, even after obtaining an electoral mandate directly from the Russian people.

But Medvedev was different from Putin. He was a lawyer, from an educated family. He spoke a more sophisticated style of Russian than Putin, and seemed more oriented to, or at least familiar with, Western ideas. He said obvious, banal things like "Freedom is better than the lack of freedom,"[43] but after eight years of Putin rule even this statement seemed fresh. As a lawyer, Medvedev emphasized the importance of the rule of law. His young age made him truly post-Soviet in a way that Putin could never be. His election allowed some democrats inside

Russia and their supporters abroad to consider the possibility of political liberalization in the Medvedev era, albeit slow and within constrained parameters.

Most of Putin's opponents inside Russia, however, viewed Medvedev's appointment/election skeptically: he was just window dressing; the real power in Russia remained Putin. Their hypothesis seemed to be confirmed just a few months after Medvedev's inauguration, when Russia invaded Georgia. Medvedev was president, but Putin took charge of this operation, which clearly signaled who still ruled Russia.

A year after Putin condemned the United States for disrespecting sovereignty in his Munich speech, he violated his own recommendation by sending Russian soldiers into Georgia, a former Soviet republic, in August 2008. Who holds responsibility for starting this six-day war remains a point of contention to this day. Georgian president Saakashvili made a mistake when he ordered his troops to attack Russian peacekeepers in the separatist region of South Ossetia on August 7, 2008.[44] But Russia had set a trap: Putin wanted Saakashvili to attack to give Russia an excuse to invade. Russia's invading army defeated Georgia's armed forces, although not as quickly as expected. Thankfully, Putin decided against marching into Tbilisi, the Georgian capital, a move that would have produced massive casualties. But he did leave soldiers in Abkhazia and South Ossetia, regions that declared their independence from Georgia soon thereafter.

In Washington and in European capitals, a blame game ensued. Some criticized Saakashvili for provoking the Russian bear. As Condoleezza Rice, secretary of state at the time, wrote in her memoir, there "was a lot of anger at Saakashvili, whom many Europeans accused of provoking the crisis."[45] A smaller group of critics chastised President George W. Bush for pushing NATO to recognize Georgia's membership aspirations earlier that year.[46] Some members of the Bush administration close to Saakashvili also came under fire for possibly having sent mixed signals and encouraged him to fight.

For me, Russia was responsible, a belief I emphasized in testimony before Congress on September 9, 2008: "Georgian military action within its borders can in no way be equated with or cited as an excuse for Russia's invasion and then dismemberment of a sovereign country. Russia's actions were disproportionate and illegal. The tragic loss of life — soldiers and civilians alike — on all sides was regrettable, unnecessary and avoidable."[47] But assigning blame did not clarify proper American foreign policy responses. Surprisingly, the Bush administration did little to punish Russia for invading Georgia. It suspended the NATO-Russia Council and tabled potential cooperation between American and Russian companies on nuclear energy. That was it. Not a single Russian official or Russian company was sanctioned. Georgia requested, but never received, lethal military assistance. NATO offered no signals of an accelerated timetable for Georgian membership.

The feeble Bush response, however, did not reflect a desire by the White House to try to get relations with Russia back on track. On the contrary, by the time the Bush administration ended, American and Russian officials were barely talking to each other. The reset in relations between Washington and Moscow that had been launched by Reagan and Gorbachev at the end of the Cold War was over. During the previous two decades, the relationship had endured two rounds of NATO expansion, the Russian economic crash in 1998, the NATO war against Serbia in 1999, the U.S. invasion of Iraq in 2003, and the Orange Revolution in 2004. Yet after each of these challenges, leaders in the two countries still returned to the belief that engagement and cooperation served both Russian and U.S. interests. Over the Bush-Putin years, commitment to that theory zigged and zagged, but then finally died in the Caucasus in August 2008.

A year later, the Obama administration would attempt to revive it.

5

CHANGE WE CAN BELIEVE IN

Early in 2007, Tony Lake called me out of the blue and asked me to join the Barack Obama presidential campaign. Lake, who had served as President Clinton's national security advisor, explained that he and another former Clinton administration official, Susan Rice, were assembling a team of unpaid advisors for the campaign. Lake said he had worked for many talented politicians, but that Senator Obama was the most brilliant, inspiring, and transformational leader he had ever met. He urged me to join them in making history. I told him I would think about it. An hour later Susan Rice called me with the same request, stating emphatically that Obama would be the next president of the United States. In her typically blunt fashion (Susan and I had known each other since our undergrad days at Stanford, so formalities were largely dispensed with), she told me to "get my shit together" and join this historic ride. So I did. I knew next to nothing about Obama at the time. While driving on a back road in Wisconsin, I had listened on the radio to his soaring oratory at the 2004 Democratic National Convention, and thought at the time, *This guy knows how to give a speech.* That was it. I didn't have a clue about his foreign policy views. But Susan was on board and that was enough for me. I had no idea that saying yes out of loyalty to a friend that day would alter my professional life for the next seven years.

In the early days of the campaign, my advice was not really needed. Our working group on Russia and Eurasia prepared a DPG — a daily press guidance — wrote a few policy papers, and held the occasional conference call with the candidate. It was a talented group of people. Celeste Wallander, my deputy, would go on to perform major jobs in the Obama administration. But our work did not seem in high demand. American voters were not that interested in Russia.

Russia's invasion of Georgia changed everything. On the morning of August 8, 2008, the day the world learned of the Russian military intervention, I was slow to submit the DPG, in part because I was on the West Coast and still sleeping when the news broke. The McCain campaign responded decisively, denouncing Russia's actions as a "clear violation of international law [that has] no place in 21st century Europe."[1] Our campaign's first reaction was more tepid, urging a cessation of hostilities without assigning blame. When I read that first press statement, I urged our bosses in Chicago to put out a new release — a highly unusual move. The second statement sounded more like McCain's: "Russia has escalated the crisis in Georgia through its clear and continued violation of Georgia's sovereignty and territorial integrity," it read, stating explicitly that "Russian military forces invaded Georgia." We called for the "deployment of genuine international peacekeeping forces in South Ossetia and Abkhazia."[2] We were doubtful that any country would send soldiers into the conflict zone at that moment, but we had to propose something concrete — that was the style of our campaign. In contrast to McCain, we fancied ourselves not as ideologues but as policy wonks offering solutions to foreign policy problems.

We released new statements daily, but we were playing defense. The Russian-Georgian war highlighted perfectly McCain's campaign narrative: he was the candidate with deep foreign policy experience, compared to the lightweight newcomer Barack Obama. McCain also knew Georgia's President Mikheil Saakashvili well, had traveled to Georgia several times, and had issued strong statements against Russian aggression well before this crisis. Our candidate had little to no track record in this part of the world. Obama had the additional disadvantage of being on his way to a vacation in Hawaii when the story broke, so while McCain delivered his messages in blue suit and red tie, standing in front of the American flag, Obama made his comments in khakis and a windbreaker, against a backdrop of tropical trees. Polls at the time underscored that most Americans thought McCain would be more able than Obama to deal with Russia.[3]

With the Obama campaign facing the prospect of losing voters as a result of this international crisis, I began receiving numerous phone calls from campaign headquarters. Every day, I was talking to Denis McDonough, Mark Lippert, and Ben Rhodes — three close advisors to Senator Obama — working with them to develop talking points for our candidate. We aimed to convey the message that we were going to be tough but pragmatic in dealing with the Russians.

On August 15 Obama returned from his poorly timed Hawaii trip. His first appearance back on the campaign trail was at Saddleback Church, a massive evangelical ministry in Southern California founded by Pastor Rick Warren. I linked up with the campaign there, planning to brief Obama during the flight to his next stop in Reno, Nevada. It was there that I first met Obama in person, as well as David Axelrod, his campaign strategist; Robert Gibbs, his press secretary; Lippert, his foreign policy advisor; and a handful of other young, dynamic, idealistic campaign aides who would later become my colleagues and friends at the White House. The energy of the Obama team was electric and tense, especially after Obama appeared to blow his Saddleback interview. He was behind in the campaign; it was a tough moment in the race.

On "O-Force One," I settled into a big leather seat across from Obama and next to Lippert and Axelrod, and wondered what kind of session this was going to be. I had briefed presidents, senators, and candidates before, and they generally fell into one of two categories. One type wants to know what to say in front of the cameras. A second wants to tell you everything they know about a topic and then seek your approval. I was guessing that Obama would be in the first category.

I was wrong. As I launched into my talking points about what the campaign should say, he interrupted to explain that Axe would take care of the politics; my job was to provide knowledge, context, and explanation. He asked me to explain how Russia and Georgia got into this whole mess. When I began to recount the events of early August, he interrupted again, asking me to go back years and decades, not weeks. He really wanted to understand the historical context; nothing else was on the schedule until we landed in Reno. Our conversation felt more like a Stanford seminar than a candidate briefing. We discussed Soviet history, the personalities of Putin and Saakashvili, and even some international relations theory, including the difference between two big paradigms in IR: realism and liberalism. We drilled down on whether the domestic organization of political systems—democracies versus autocracies—influences their international behavior. Realists argue no; liberals (in the academic use of this word, not the political one) say yes. We also chewed on whether states can successfully pursue win-win outcomes, as liberals suggest, or whether the interaction between states — especially great powers — is fundamentally a zero-sum contest, as most realists contend. Obama understood these two theories easily, but rejected both as overly simplistic. In the real world, he argued, you have to be both part realist and part liberal, depending on the circumstances. We would return to this discussion several times during my years in government.

Sometime during the conversation, knowing that Obama was a basketball player and fan, I tried my hand at humor, playfully bragging that I had played for a national championship basketball team at the collegiate level. He looked at me skeptically until I added a key fact: I played as a graduate student for Oxford in the UK league. That brought a hearty laugh and big smile from the candidate. I had heard it said that Obama was aloof, hard to talk to, professorial, and difficult to connect with. I came away from this trip with a very different impression. In fact, as the plane began to descend, I was hooked. He was smart, engaging, funny, and a contemporary, just a few years older than me. I decided that day that if he won, I wanted to work for him.

The Russian-Georgian war compelled our campaign to develop a comprehensive policy toward Russia. Through conversations with the campaign staff —including former Clinton administration officials who had jumped on our bandwagon as Obama's electoral prospects improved—the essence of what would become known as the Reset started to crystallize. The major working paper we wrote stated bluntly: "Improved relations with Russia should not be the goal of U.S. policy, but a possible strategy for achieving American security and economic objectives in dealing with Russia." This framing of means versus ends was adopted as policy the following year. Over the next five years in the government, I repeated hundreds of times, both in internal deliberations and in public interviews, that difference between means and ends, to underscore that good relations with Russia was not the goal of Reset but a strategy for pursuing American interests regarding Russia. It was a tough sell.

Our Reset policy paper specified five major objectives:

(1) Support democratic partners and uphold principles of sovereignty throughout Europe and Eurasia, and address tensions between countries before they escalate into military confrontations.
(2) Strengthen the NATO alliance in order to deal with Russia with a unified voice.
(3) Help decrease the dependence on Russian energy of our allies and partners in the region.
(4) Engage directly with the Russian government on issues of mutual interest, and also reach out directly to the Russian people to promote our common values.
(5) Keep the door open to fuller integration into the global system for all states in the region, including Russia, that demonstrate a commitment to act as responsible, law-abiding members of the international community.[4]

Regarding specific ways to achieve these objectives, we recommended upgrading the NATO-Ukraine and NATO-Georgia Commissions as ways to demonstrate our commitment to deepening security and economic ties with these two countries; ratifying the Comprehensive Test Ban Treaty (CTBT); and working with Russia to delay Iran's nuclear program. In congressional testimony that fall after the Georgia crisis, I tried to summarize our proposed approach to Russia in one paragraph:

> Instead of business as usual or isolation, the United States must navigate a third, more nuanced, more complicated, and more comprehensive strategy that seeks to bolster our allies and partners, check Russian aggression, and at the same time deal directly with the Russian government on issues of mutual interest. The long-term goal of fostering democratic change and keeping the door of Western integration open for countries in the region, including Russia, must not be abandoned. American foreign policy leaders have to move beyond tough talk and catchy phrases and instead articulate a smart, sustained strategy for dealing with this new Russia, a strategy that advances both our interests and values.[5]

Part of our strategy was for Obama to connect directly with the Georgian government, both to help him learn firsthand what was happening inside the country and to offset public perceptions of McCain's ties to Georgian officials. It was also important that our Georgian colleagues not appear to be too pro-McCain, since they might be working with us after January. I worked with Georgia's ambassador to the United States at the time, Vasil Sikharulidze, to organize a call between Obama and President Saakashvili, which seemed to go well. As another signal of support, Obama met with the Georgian delegation attending the Democratic National Convention in Denver as observers — on the day he accepted the nomination, no less. I was amazed the meeting happened and thrilled to be there with Obama on such a historic day. Obama even added a shout-out to Georgia in his acceptance speech. That reference, in one of the most important speeches of his life, as well as the meeting with the Georgian delegation, felt like victories, however small, for my policy ideas. I was mindful of the fact that my opportunity to become more involved in the Obama campaign was a by-product of the Russian invasion, but it was still exhilarating to contribute to this historic candidacy. Like others in the crowd waving CHANGE WE CAN BELIEVE IN signs in Mile High Stadium that day, I was brimming with optimism.

As the weeks went on, we were closing the foreign-policy-experience gap between our guy and McCain. Our progress became clear at the first presidential debate, held on September 26 at the University of Missouri, which focused on foreign policy. Alongside Iraq, Iran, and Afghanistan, Russia was a major topic.[6] Our working group had labored hard on the talking points for this debate, and Obama delivered them eloquently. To my biased ear, he sounded like a seasoned world leader. As I watched the debate from my living room in Palo Alto with a campaign colleague, Derek Chollet, we fired out rebuttals to McCain's answers on our BlackBerrys, our missives then circulating from Chicago to the national press. It was exciting to be not just observing a presidential debate, but indirectly participating in it.

Days later, the campaign was over. In September, the U.S. economy spun into free fall: the beginning of the worst recession since the Great Depression. McCain decided to suspend temporarily his campaign and instead participate in negotiations on the proposed financial-industry bailout in Washington. He was trying to demonstrate leadership, but the gambit backfired. The press jumped on him for having no ideas.[7] Voters blamed President Bush and the Republican Party for the nation's economic woes, locking in momentum for the candidate of change. The financial crisis had transformed a tight race into a blowout, partly due to McCain's erratic response. By the end of September, the email chatter among our campaign team grew increasingly confident. We were going to win. It was time to think seriously about moving to Washington.

Election night — November 4, 2008 — was a tremendous moment of celebration for me, my family, and millions of other Americans. As the polls closed in California, the networks immediately projected Obama the winner. Obama's message of hope and change, ratified by the American people, seemed to signal the beginning of a new era of renewal. I was filled with the promise of possibility, for my country, of course, and also for myself. Obama's election meant that my family and I would be leaving the paradise that is Stanford University for a few years and moving to Washington, D.C.

Members of our foreign policy team in Chicago — McDonough, Lippert, Rhodes, Tony Blinken (Biden's foreign policy advisor), and a few others — changed their emails from @barackobama.com to @ptt.com (presidential transition team), moved into temporary offices in downtown Washington, and began to assemble a new government. Some outside advisors, including the person most responsible for bringing me on board the Obama team, Susan Rice, also joined the transition headquarters. As instructed, I began drafting transition papers, assuming that I would soon be in the government helping to implement them. I hoped to work at the National Security Council or the State Department. But then Obama surprised everyone from our campaign team when he named James Jones, a retired four-star Marine general, as his pick for national security advisor, and Hillary Clinton as secretary of state. These two individuals had had nothing to do with our foreign policy team during the campaign. Both appointees, I feared, would staff their offices with their own people, not with members of Obama's campaign team, as indeed President-elect Obama had promised them. As Susan said to me glumly one evening in the transition headquarters in December, only two people were getting the jobs they wanted — Obama and Eric Holder, slated to be attorney general. The rest of us — including even the vice president (who was briefly a presidential candidate) — had to settle for second-best options.

McDonough and Lippert had two backup options for me: special assistant to the president for democracy at the National Security Council or ambassador to Ukraine. The special assistant job would allow me to engage on issues and regions beyond Russia, an aspiration I had been trying to realize for several years. But when I talked to people well versed in Washington bureaucratic politics, most warned that "functional" jobs had little juice; it was geographically defined jobs that conferred power. Kyiv intrigued my family and me. Marilyn Pifer, the spouse of former ambassador Steve Pifer, convinced my wife that the school for American diplomats there was a good fit for our sons. Our younger son, Luke, only six at the time, saw the photos of the giant ambassador's residence and started lobbying for Ukraine.

But then I interviewed with Jones — Obama had persuaded him to take on board some of us campaign hacks — and it seemed to go well. For a four-star Marine general, he had an easy conversational style and seemed genuinely interested in trying to do something big and new on Russia. Several days later, I was in the audience at a concert of Zimbabwean music at Stanford when my phone lit up with a conference call. Jones's first deputy Tom Donilon, McDonough, and Lippert were on the other end of the line to offer me the position of special assistant to the president and senior director for Russian affairs at the White House. In a bureaucratic turf fight, I lost part of my portfolio from the campaign — the Caucasus, Ukraine, Belarus, and Moldova — to the European directorate. But I was elated to be going to the White House to work for this president. I stepped out of the auditorium to accept, returned to the concert hall, and tried unsuccessfully to settle back into the African rhythms. My mind raced, imagining what work at the White House with Barack Obama would be like on one of the most difficult assignments in foreign policy — the Russia account. I was both excited and terrified.

Soon after Obama was elected, leaders from all over the world bombarded us with requests for congratulatory phone calls. From my campaign portfolio of countries, I recommended that President-elect Obama take calls from Russian president Medvedev, Ukrainian president Yushchenko, and Georgian president Saakashvili, and that Vice President–elect Biden call Kazakh president Nazarbayev — though Lippert later upgraded "Naz" to a POTUS-elect call. The significance of a call with the president-elect — not offered to any other leader in

Central Asia — was not lost on Nazarbayev. I was beginning to understand how much these little pieces of prestige and status mattered, and how careful the people who doled them out had to be.

In his first phone call with Medvedev, Obama advanced his opening argument for the Reset. I was at my son's soccer game on a Saturday in Palo Alto when I was patched in to the call to listen to two of the most powerful people on the planet talk on the phone. Obama kept it short. He had a long call list and several hundred decisions to make during this transition period. But from that first conversation, it was obvious to me that the two presidents shared some common ground. Medvedev sounded eager to get off to a good start with the new American president and move beyond the U.S.-Russia standoff sparked by Russia's intervention in Georgia just a few months earlier. Obama was careful not to ignore Russia's aggression against Georgia, but also signaled his intention to engage Medvedev on a wider range of issues.

The Obama-Saakashvili call was also positive. The Georgian president had a completely different style from Medvedev; he spoke emotionally, quickly, and fluently in English, jumping from one big idea to the next. Styles aside, I left the call thinking there was a way to pursue Saakashvili's agenda, Medvedev's agenda, and our agenda all at the same time. Reset with Russia would not have to mean disengagement with Georgia.

Around this same time in the transition, in a *Meet the Press* interview, Obama first publicly used the term "reset," which would come to be associated with his administration the way "détente" evokes Nixon and Kissinger:

> It's going to be important for us to *reset* U.S.-Russian relations. Russia is a country that has made great progress economically over the last several years. Obviously, high oil prices have helped them. They are increasingly assertive. And when it comes to Georgia and their threats against their neighboring countries, I think they've been acting in a way that's contrary to international norms. We want to cooperate with them where we can, and there are a whole host of areas, particularly around nonproliferation of weapons and terrorism, where we can cooperate. But we also have to send a clear message that they have to act in ways that are not bullying their neighbors.[8]

I had worked on the talking points for that interview with McDonough and Rhodes. After it aired, I chatted with Denis about the Russia portion of the interview, and we both agreed that the word "reset" described well what we aimed to do with Russia. Words matter. This one felt new and modern. It evoked change. As a way to describe our future policy, "reset" was worth deploying again.

6

LAUNCHING THE RESET

On January 20, 2009, my family bundled up in multiple layers, met our friend Larry Diamond, and made our way to the Washington Mall to join two million other witnesses to history: the inauguration of America's first African American president. As a reward for my work on the campaign, we got to stand fairly close to the podium to hear the new president give his address. In terms of my future work for the administration, the line that most resonated with me was this: "To those who cling to power through corruption and deceit and the silencing of dissent, know that you are on the wrong side of history, but that we will extend a hand if you are willing to unclench your fist." That message was intended for countries like Iran, Cuba, and Burma, but also for Russia, at least in my mind. It was going to be my job, starting January 21, to translate those words into policy.

The next day I reported for work at the Eisenhower Executive Office Building (EEOB), the majestic nineteenth-century building in the French Second Empire style, on the White House compound next to the West Wing. The hallways were wide and long — real "corridors of power." I acquired my temporary blue badge, a pile of paperwork, a list of instructions for how to access my three computer accounts — unclassified, secret, and top secret — and the multiple codes necessary to gain access to my office. That morning, I joined other new White House employees squeezed into a conference room, where President Obama read us the oath of office. And just like that, dozens of the most important jobs in the

U.S. government changed hands from the Bush team to the new administration. Change You Can Believe In — it seemed to be happening that day.

My staff, however, had not changed. The day before, these career foreign service and civil service officers had worked for President Bush. That was weird. Things were worse at other agencies, where no one at my level — the assistant secretary level — had been confirmed. For the first several weeks serving in the Obama administration, my counterparts at State, Defense, and elsewhere were Bush appointees. How were we going to pursue policy changes in the absence of personnel changes?

From day one, my strategy for changing policy was to write. As requested by my superiors, I initiated a policy review, quickly drafting the discussion paper on Russia. Before getting specific, we had to settle on fundamentals. Were we going to seek to isolate, contain, ignore, or engage Russia? Critics of the Reset have argued that we would have achieved better results with the Russian government in 2009 had we sought to punish it for invading Georgia. But the arguments in favor of isolation, containment, or punishment in January 2009 were less compelling than those for principled engagement, for several reasons.

First, we had several major foreign policy objectives that required Moscow's cooperation: replacing the expiring Strategic Arms Reduction Treaty, addressing the Iran nuclear threat, and reducing our dependence on Pakistan as a supply route to our soldiers in Afghanistan as we geared up to extend the counterterrorism fight into Pakistani territory. Second, we had no consensus in the U.S. government about the causes of the Russian-Georgian war. When Russia annexed Crimea several years later, there would be no disagreement about who was at fault and how severe the international crime, making the case for sanctions and isolation much easier. In 2009, however, many of my new colleagues in the U.S. government, including those who had served in the Bush administration when Russia invaded Georgia cited Saakashvili's recklessness as a contributor to the war.[1] Giving up on collaborating with Russia on Iran because of the Russian-Georgian conflict the year before was not a trade most of our senior leaders were prepared to make. Third, the Bush administration had responded tepidly to the Russian invasion of Georgia. If it had not implemented sanctions or isolated Russia further in the immediate aftermath of the conflict, how were we supposed to do so now, several months later, especially as most of our European allies would not have followed us?

At another extreme, some in the administration recommended unqualified rapprochement with Moscow: let's forget about past disagreements, even Georgia, and try to establish rapport with Medvedev quickly. It was time for better relations with Moscow, some proposed. I adamantly rejected this approach. It sounded like a replay of Bush's first year in office when he went out of his way to embrace Putin, looking into his soul and all that. Rather, the question for me was whether a warmer rapport with Russia would help the United States achieve its objectives. We needed to have concrete deals to change the mood music in our bilateral relationship, not the other way around.

I advocated a third approach: engage with the Russian government to pursue our objectives (not theirs), while also criticizing it for bad behavior, doubling down on support for frontline states like Georgia and Ukraine, and keeping true to our values. I took inspiration from George Shultz and his approach to the Soviet Union—specifically, the communist hardliners in power when Shultz joined the government, not reformers like Gorbachev, who came to power in 1985. (Engaging Gorbachev was easy.) Isolation and confrontation of the Soviet regime had been the Reagan administration's approach before Shultz became secretary of state in the summer of 1982. After joining the administration, however, Shultz pushed for a strategy of "realistic reengagement"—to encourage the Soviet government to pursue common strategic interests without compromising U.S. values or other interests.[2] That seemed like the right approach for our era, too. We had to deal with the leaders in the Kremlin, but guided by a series of clear assumptions and principles. Collectively, these became the Reset.

The Reset's first assumption was that the United States and Russia had some common interests, especially regarding our biggest security challenges. We reasoned that Russia might be willing to cooperate to prevent Iran from obtaining nuclear weapons, that we might be able to harness Russian support for our war effort against the Taliban and al-Qaeda in Afghanistan, and that, more generally, the U.S. and Russia could unite over the common goal of combating terrorism worldwide. We also assessed that neither the United States nor Russia would gain from a lapse of the Strategic Arms Reduction Treaty (START), a bilateral nuclear-arms-reduction agreement signed in 1991 and scheduled to expire in December 2009. I also believed that increasing trade and investment between our two countries was a mutual interest, though not everyone in our new government assigned this goal a high priority. Of course, there were some issues on which the United States and Russia would never agree — the demarcation of Georgia's borders, for instance. But we did, I believed, have an interest in avoiding a second Russian-Georgian war, first and foremost because such a conflict would further weaken Georgia at a time when I thought strengthening Georgian democracy would help to contain Russian imperial ambitions.

A second core assumption of the Reset, closely related to the first, was that we could achieve win-win outcomes on many of these shared interests. We were not destined to compete with Russia; we could cooperate in ways that would make both countries better off.

Third, we aspired to develop what we called a "multidimensional" relationship with Russia, beyond a singular focus on security issues. Without question, security issues, and specifically nuclear security, topped our list of priorities; President Obama had a special interest in reducing nuclear weapons, and even wrote his undergraduate honors thesis on arms control agreements between the U.S. and the Soviet Union.[3] But the president seemed equally passionate about developing new dimensions of our relationship with Russia: deepening economic ties, building contacts between our societies, managing the global economy, sharing our climate-change agenda. If all we did with the Russians was talk about nuclear weapons, the president once observed, then our time in government would be no different from the Brezhnev-Nixon era.

A fourth assumption of the Reset was that engagement was the primary means for realizing these outcomes. Our new team reasoned that our options for changing Russian behavior either through isolation or coercion were limited. Instead, we had to create incentives to induce cooperation, which could only be realized through direct engagement with Russian government officials. Therefore, we aimed to create more connections both vertically and horizontally between our two governments. At one point during the Russian-Georgian war, I was told, Admiral Mike Mullen, chairman of the Joint Chiefs of Staff, maintained the only open line of U.S.-Russia communication, with his Russian counterpart General Nikolay Makarov. That sounded imprudent to me, and proof of the importance of a new strategy to develop more lines of communication between our two governments. Shultz had pushed for open channels with his Soviet counterparts even during the most confrontational moments in the early Reagan years. Initially Shultz was dismissed as a *détentenik* by hardliners in the Reagan administration, but he helped implement that change well before Gorbachev came to power, and over time Reagan and Shultz demonstrated that talking was not a sign of weakness, but a strategy for advancing American national interests.[4] That lesson from Reagan informed our Reset policy.

Again, improved relations with Russia was not the *goal* of Reset. Rather, deeper engagement was defined as the *means* to achieve concrete American security and economic objectives. Our first Reset strategy paper articulated a tangible list of objectives: a new arms control treaty; reducing nuclear proliferation; facilitating nuclear energy cooperation; expanding a northern supply route through Russia to Afghanistan; denying Iran a nuclear weapon; exploring modalities for missile-defense cooperation; enhancing European security; fostering democratic development; helping Russia join the World Trade Organization; deepening economic cooperation; promoting energy security and diversification among our European allies; restarting military-to-military cooperation; combating anti-Americanism inside Russia; and deepening ties between our two governments through the creation of a new bilateral commission. In the spring of 2009, that was the list.

Fifth, in parallel to greater engagement with Russian government officials, our Reset strategy advocated increased engagement with Russian society. We sought to stimulate more interaction between U.S. government officials and Russian NGOs, as well as between Russian and American civil society leaders through a process we called dual-track engagement. The more connections between our two societies, the better.

Respect was a sixth component of the Reset. Given all the acrimony about the lack of respect between Putin and Obama that came later, it's important to remember that we didn't start out that way. Throughout his time in office, President Obama repeatedly emphasized our aim of engaging with other countries based on mutual interests and mutual respect. When I first heard this phrase, it sounded like a platitude: don't all countries seek to engage with other countries based on mutual respect? But the longer I worked with the president, the more I appreciated the salience of this concept in his dealings with foreign leaders and other societies. In part because his father was from Kenya, his stepfather was from Indonesia, and he had lived years of his childhood abroad, Obama had a deep understanding of how others saw Americans. More than many on our national security team, he had empathy for his foreign interlocutors, including the Russians.

Along these lines, Obama was not afraid to acknowledge aspects of Russian greatness. For instance, in his speech in Moscow in July 2009 he went out of his way to praise the essential role that Soviet soldiers played in defeating Hitler: "Just as men from Boston and Birmingham risked all that they had to storm those beaches and scale those cliffs, Soviet soldiers from places like Kazan and Kyiv endured unimaginable hardships to repel an invasion, and turn the tide in the east."[5] More boldly, Obama acknowledged Russian hardships after the collapse of the Soviet Union.[6] His voice — our voice — would not be one of triumphalism but of respectful partnership.

Respect for Russia and the Russians, however, did not mean avoiding criticism of Russian government actions. A seventh principle of the Reset was that we would engage with our Russian counterparts without checking our values at the door. Obama was not the lecturing type; he was not going to tell Medvedev what to do. At the same time, our Reset strategy explicitly called for defending what we called "universal values," even when it was inconvenient to do so.

An eighth tenet of the Reset was that the U.S. would not forfeit positive relations with Russia's neighbors in the pursuit of improved relations with Russia. Moreover, we believed that renewed conflict between Russia and Georgia or state breakdown in Ukraine would adversely affect our interests. Avoiding these outcomes through invigorated diplomacy with all involved — in Kyiv, Moscow, and Tbilisi — was our approach. This concept was easy to enumerate, but much harder to put in practice, especially in a part of the world that tended to frame bilateral relations in zero-sum terms.

A ninth component of the Reset, closely related to the last two, was the rejection of "linkage," a negotiating strategy in which unrelated issues are purposely tied together in an attempt to create leverage from one domain to another. The idea of linkage had developed in the 1970s, when President Nixon and his secretary of state, Henry Kissinger, sought to persuade the Soviet Union to stop supporting communist revolutions abroad in exchange for greater cooperation over issues like arms control and the economy.[7] In the post-Soviet era, some foreign policy strategists, especially human rights advocates, pushed for a different kind of linkage: making cooperation with Russia on economic or security issues conditional on democratic progress. I sympathized with these arguments, but eventually rejected them as unhelpful for our purposes. I believed that linkage would slow down if not altogether halt progress toward the outcomes we sought to achieve with the Russians. If we linked Russian progress on democratic reform to our pursuit of a new arms control treaty, for instance, we would never get a new agreement in place — and the current treaty was soon to expire. Another reason to keep conversations on these issues on separate tracks was that we had to refuse Russian attempts at linkage. We could not let the Russians pressure us to reduce our support for Georgia in exchange for their support on Iran, or make turning a blind eye to human rights violations the price for allowing us access to Russian supply routes to support our soldiers in Afghanistan. The Russians loved this kind of deal making; we had to reject it.

These nine principles eventually were adopted as our new strategy for dealing with Russia. In the back of my mind, however, I always believed three additional hypotheses about the Reset that weren't spelled out in policy papers. First, I believed that deeper integration of Russia into the global economy, including through increased trade and investment with the U.S. and the West, would have a positive effect on Russian foreign policy in noneconomic areas. Chinese-American relations offered a useful model: we had to manage several difficult issues, including the defense of Taiwan, territorial disputes in the South China Sea, clashes over Tibetan sovereignty, and competing views on climate change. Yet our differences had not developed into confrontations, in part because of our deep economic ties: in 2009 bilateral trade between China and the United States exceeded $365 billion,[8] while U.S. investment in China topped $50 billion.[9] This degree of economic interdependence, I believed, created ballast in our bilateral relations with China. By contrast, U.S.-Russian trade in 2009 totaled only $23.5 billion,[10] and U.S. investment in Russia $20.8 billion.[11] We had work to do.

Economic ties did not provide the same check on bad policy in U.S.-Russia relations as they did in U.S.-China relations. For instance, when Russian leaders considered the potential economic costs of strained relations with the United States after invading Georgia, they must have quickly concluded that there was not a lot to lose. I wanted to increase the cost of doing stupid things. If Russians owned pipe-making companies, steel plants, and basketball teams in the United States, they would be motivated to lobby for cooperative political ties with Washington. Likewise, if Americans owned parts of Russian oil companies, opened car plants in St. Petersburg, or hired armies of Russian programmers to provide technical support for U.S.-developed products, then these American business leaders would have an interest in pushing the White House to maintain constructive relations with the Kremlin. Beyond economics, I believed that closer ties between the United States and Russia more generally could help reduce tensions between Russia and its neighbors.

I also believed that a Russian economy more integrated into international markets would create pressure for market reforms inside Russia. If Russian companies floated their shares on the New York Stock Exchange, they would have to become more transparent and focus on creating value for their international shareholders, rather than simply serving the interests of Russian state officials. To join the World Trade Organization, the Russian government would have to liberalize the Russian economy—lower tariffs, eliminate restrictions on foreign participation in several service industries, remove import-licensing requirements inconsistent with WTO rules, and reduce industrial subsidies.[12] A more liberal economy would lead to growth in sectors beyond oil, gas, and metals, which in turn would help expand Russia's middle class, which in turn would demand greater democratic changes. Far-fetched? Maybe. But my aspirations for how Russia might change were supported by social science research about the positive relationship between trade liberalization and development, and development and democracy. Development did not necessarily trigger democratization, but it most certainly helped sustain democracy once a transition occurred.[13]

A final, even more private assumption that I held about the Reset was that a more benign international environment for the Russian government would create better conditions for democratic change internally. The most powerful historical analogy on my mind was political change inside the Soviet Union, when Reagan's embrace of Gorbachev helped make democratization within the USSR safe for the Soviet leader to pursue. While always assessing the odds as long—at the time, few academics had written more about growing autocracy in Russia than I—I reasoned that the same logic could pertain to U.S.-Russia relations in 2009. Confrontation with the Kremlin would impede democratization, while engagement with the Russian government would make it easier for societal actors to push for political change. President Medvedev was no Gorbachev. He was more cautious, and of course less powerful, since Prime Minister Putin remained the key decision maker, especially on domestic matters. Nonetheless, Medvedev had expressed some desire for political modernization, however modest, and I reasoned that engagement would make it safer for him to pursue these reforms.

Not everyone in the U.S. government — maybe not even Obama — agreed with this last aspirational hypothesis. For many on our team, promoting democracy inside Russia was not a priority. Still, the naysayers supported the same strategy of engagement that I did, albeit for different ends. As long as encouraging democratic change in Russia did not impede negotiations over the new arms control treaty or new sanctions on Iran, they would not object. In the early days of the Reset, it seemed as if all good things could go together; we could engage with the Kremlin to produce outcomes on arms control or Iran, all while continuing to talk about human rights and maintaining democracy-promoting programs inside Russia.

Reset was consistent with the strategic aims for all U.S. administrations — Democratic and Republican — since the late years of the Cold War. To varying degrees, Ronald Reagan, George H. W. Bush, Bill Clinton, and George W. Bush all aspired to integrate Russia into international institutions and the global economy as a way to promote democratic and market reforms within Russia and to reduce tensions between Russia and the West. To varying extents, those administrations also believed, as I did, that democratic change within Russia served American national interests.

As an academic, I was comfortable thinking and writing about abstract strategy. I was less adept, at least initially, at translating conceptual ideas into concrete policies. Simple tasks such as "flipping" a document from one classification system to another seemed baffling. Why was one white paper marked SENSITIVE BUT UNCLASSIFIED but another SECRET? I also was confused about who should be included in the policy deliberations at my level. I chaired the Interagency Policy Committee (IPC) on Russia, which consisted of assistant secretaries from various agencies throughout the government. But each meeting, someone new appeared, especially people in uniform from the Pentagon. I was told that my commission as a special assistant to the president was the formal equivalent of a three-star general, but I wasn't about to kick a real general out of my meeting! I took the opposite approach — the more people looped in, the better.

And I was in a hurry. I wanted to send our new policy up the chain for President Obama's approval before his first meeting with Medvedev, planned for the first week in April on the sidelines of the G-20 summit meeting in London. Once the president signed off on our strategy paper, it would guide policy development and implementation throughout the government. As codified in our Russian strategy papers, I also worried that the honeymoon period created by the election of Obama and Medvedev in the Kremlin might be short-lived. We had to act as quickly as possible to get things done before conditions changed.

I was surprised by how easily the IPC reached consensus. For those professionals in the government who had focused on Russia for most of their careers, the Reset offered an opportunity to think anew about the United States' relationship with Russia, following a long period of decline that almost ended with a complete cutoff of contacts after the Russian invasion of Georgia. Even those who had advocated a more confrontational approach after Russia's invasion of Georgia saw the value of using the moment to try to engage with Moscow again, as a means to avoid renewed conflict in that region. Some cautioned against too much enthusiasm. I most appreciated advice from Dan Fried, who had served as Condoleezza Rice's deputy for Europe, first at the NSC and then at the State Department, and who was still assistant secretary of state for Europe and Eurasian affairs while Obama's appointee, Phil Gordon, awaited confirmation. Fried encouraged me to pursue Reset, but predicted we would fail. He warned that our Russia policy would follow a similar arc as Bush's policy — cooperation in the beginning, confrontation in the end. Dan is a wise man.

Dan's immediate boss at the State Department, Under Secretary of State for Political Affairs Bill Burns, was a critical ally in developing and getting approval for our new policy. I had first met Bill on the basketball court at the U.S. Embassy in the mid-1990s when I was at the Carnegie Moscow Center and he was heading the political section at the embassy. By the time I joined the government in 2009, Bill's reputation in the State Department was already legendary. One of the greatest diplomats of his generation, he had a keen interest in being part of the Reset from the outset. He had served as the U.S. ambassador to Russia from 2005 to 2008, which gave him stature among the Russia hands in the State Department and across the government. He became Clinton's point person for all things Russian (and eventually for all things!). Over my five years in government, Bill would prove to be an invaluable partner, sharing both the victories and failures of our Russia policy.

The Deputies Committee (DC) — one rung up from the IPC — approved our Reset policy paper on March 18, 2009, teeing up a Principals Committee (PC) meeting two days later on Russia made up of cabinet secretaries, the vice president, and the national security advisor. At the latter meeting, there were no critics of our approach, just some words of caution from one important voice: Secretary of Defense Robert Gates. As the one holdover from the Bush administration at the cabinet level, Gates had participated in the deterioration of our bilateral relationship with Russia over the last two years. He did not disagree with the strategy, but expressed little enthusiasm to engage personally; he was busy fighting two wars at the time, after all. Although Gates was not a fan of the Reset, he also cautioned against overreaction to Russian bad behavior. He viewed Russia as a declining military power, capable only of symbolic gestures, not actual force projection. For instance, when we discussed how to react to a joint naval exercise between Russia and Venezuela, his response was no response. "Let them knock themselves out," I remember him saying, "and hope that their ships don't sink along the journey." Secretary of State Hillary Clinton embraced the strategy with more enthusiasm, expecting to play a role in its execution through engagement with Foreign Minister Sergey Lavrov. In fact, it was decided that she should meet with Lavrov and float some of the general ideas of the Reset before President Obama met with Medvedev in London. However, she also cautioned to keep expectations low and not give away anything for free. As Clinton wrote in her memoir about Russia policy, "Straight up transactional diplomacy isn't always pretty, but often it's necessary."[14] I agreed. After PC consent, President Obama then signed off on our new policy. The ideas of the Reset had been approved at the highest levels well before Obama's first meeting with Medvedev.

I also wanted to get our Reset policy out to the public as quickly as possible, knowing that stating policy on the record would help lock it in within the administration. Vice President Biden's national security advisor, Tony Blinken — both a friend and an ideological ally — was charged with writing the first major foreign policy statement of the Obama administration: the vice president's speech to the Munich Security Conference set for February 7, 2009. I had no formal role in writing that speech, until Tony emailed me to ask for some language on Russia, which I eagerly provided. With that invitation, the essence of the Reset made it into the speech. Biden explicitly stated, "It's time to press the reset button and to revisit the many areas where we can and should be working together with Russia." Biden then spelled out two key assumptions of our strategy: the potential for achieving win-win outcomes with Russia "where our interests coincide . . . and they coincide in many places," and the importance of not abandoning our relationships with other countries in the region in pursuit of closer ties to Moscow: "We will not agree with Russia on everything. For example, the United States will not — will not — recognize Abkhazia and South Ossetia as independent states. We will not recognize any nation having a sphere of influence. It will remain our view that sovereign states have the right to make their own decisions and choose their own alliances."[15] With Biden's speech, only a few weeks into the Obama administration, Reset went from bumper sticker to policy.

To further advance our new strategy, Bill Burns proposed an old-fashioned idea: have Obama send a letter to Medvedev, spelling out the logic behind the Reset. General Jones and Tom Donilon endorsed the tactic, so I drafted the missive. This letter spelled out the concrete goals that we aimed to accomplish, and ended with a hope that two new young presidents with a different mind-set not shaped by the Cold War might be able to chart a new course in U.S.-Russia relations. The letter was comprehensive and strategic, not a get-to-know-you communication.

Burns then proposed that he and I travel to Moscow to deliver it. At first, the idea of two senior officials flying eleven hours to deliver a letter seemed silly and inefficient. We were policymakers, not postmen, and there is this thing called email. But Burns's instincts were spot-on. The letter had to be transmitted with proper fanfare to have its intended effect, and by traveling to Moscow in February 2009, Bill and I would have the chance to expand verbally on what Obama had outlined in written form. We did not see Medvedev on that trip, but met with most of the key players in U.S.-Russia relations, including Foreign Minister Lavrov, Deputy Foreign Minister Sergey Ryabkov, Medvedev's foreign policy advisor, Sergei Prikhodko, and Prime Minister Putin's foreign policy advisor, Yuri Ushakov.

All of these meetings were full of optimism. Lavrov, I would later learn, rarely shows positive expressions of emotion, yet he was downright ebullient during our meeting. Prikhodko was a man of few words, but praised Obama's letter and reported that Medvedev had too. Maybe this was a moment for genuine change in the dynamics of U.S.-Russian relations? I returned to Washington convinced that Reset could work.

On that trip, we also started the practice of dual-track engagement. I requested that a handful of political opposition leaders, including Garry Kasparov, the former World Chess Champion turned political activist, be invited to a reception at the ambassador's residence at Spaso House (my future home). Some at the embassy were skeptical: We were the first Obama delegation to travel to Moscow, and Garry was one of the government's biggest critics. He had a reputation for being "prickly," as one State Department official put it, and might offend guests from the Russian government with his combative ways. But I insisted. I wanted to show a credible commitment to the principle of engaging both government and society. At one point during the evening, I noticed Garry in the corner, sparring with Medvedev's chief economic advisor, Arkady Dvorkovich. Arkady's father was a big shot in the chess world, so I hoped they were talking chess, not foreign policy. If not, maybe I'd been mistaken in insisting on bringing Russian officialdom and the Russian opposition together at one reception. But both Arkady and Garry said farewell to me with smiles on their faces. I remained hopeful that dual-track engagement was worth pursuing. Maybe it was also useful for Kremlin officials to meet with their critics from time to time.

Biden's speech introduced the world to Reset as a label for our Russia policy; our February visit to Moscow marked our first attempt to explain Reset in detail to our Russian counterparts. But the Reset brand became official and famous, if not infamous, during the first meeting between Secretary Clinton and Foreign Minister Lavrov on March 6, 2009, in Geneva.

I joined Secretary Clinton in Brussels en route to the meeting. The briefing session on her plane went well. Clinton seemed eager to demonstrate her ability to execute the president's policy and showcase her skills as a diplomat. She was just as new to foreign policy making and implementation as I was, but she had prepared well, and asked many hard, important questions of me and the other members of the Russia team on board, which included Under Secretary Burns and Assistant Secretary Fried. When someone reported that Lavrov had a history of difficult relations with women counterparts, Secretary Clinton shrugged off the warning. If new to diplomacy, she was not new to dealing with difficult male counterparts.

As we finished the briefing shortly before the plane was scheduled to land, a staffer I had never met before asked me how to spell "reset" in Russian. I wasn't sure. I hadn't studied Russian formally since college and I hadn't lived in Russia in over fifteen years, long before the word "reset" had even made its way into the Russian language. But I gave it my best shot: *peregruzka?* None of the other Russian speakers on board disagreed, and if it was for official purposes I assumed the staffer would verify the spelling with a translator once we were on the ground. I didn't know at the time his reason for making the query.

A few hours later, I found out. At the opening press spray, a common ritual before a private diplomatic meeting, Clinton handed Lavrov a button with RESET written out in both English and Russian. But the Russian word was missing two letters: *perezagruzka* means "reset," but *peregruzka* means "overload." Lavrov pointed out the spelling error in front of dozens of camera crews. I was mortified, as were the other Russia hands in our delegation.

I had no idea that Clinton's team had been planning this PR stunt. It seemed out of keeping with the seriousness of this first meeting with Lavrov. Why hadn't our more experienced State Department hands quashed the idea? But once the decision had been made to go ahead with it, we should have at least executed it correctly. Having a professional translator on the plane would have helped. Clinton laughed off the mistake in front of the cameras, and was equally forgiving with our team when we assembled later that evening. Other senior officials would have raged at their staffs for such a mistake, but instead she chuckled and sipped a glass of wine — and I knew right then that I was going to like working with her. The meeting with Lavrov had ended on a positive note. And on Russia, Clinton remained my personal supporter and intellectual partner for the rest of her time in Obama's administration.

Despite making our team the butt of many jokes, the "reset button" fiasco generated tremendous attention for our new Russia policy, and locked in "Reset" as its name. Not since Kissinger's "détente" had one word come to represent an entire mood and moment in U.S.-Russia relations.

"Bilats" — one-on-one meetings between the president and another head of state — are often just an hour or two long, but they present rare opportunities for generating policy and making decisions. They happen during state visits or on the sidelines of multilateral summits, such as the G-8, G-20, APEC, and NATO. Still, even a ten-minute conversation between heads of state on a policy issue requires months of preparation at lower levels of the government. Especially early on in our new administration, I believed that these meetings were too important to waste on casual chitchat.

The first Obama-Medvedev bilat was set for April 1, 2009, at the American ambassador's residence during the G-20 summit in London. I wanted deliverables for this first meeting that would reflect the Reset: specifically, a detailed joint statement about our new agenda in U.S.-Russian relations. Along with Gary Samore, Obama's coordinator for arms control and weapons of mass destruction at the National Security Council, I wanted to secure a second joint statement about starting negotiations for a new arms control agreement to replace the one expiring later that year.

We drafted separate statements on both topics and sent them to Ambassador Sergey Kislyak, the Russian ambassador to the United States, while U.S. ambassador John Beyrle delivered the same documents to his closest counterpart at the Russian Ministry of Foreign Affairs, Deputy Foreign Minister Ryabkov. I thought the drafts were straightforward statements of our mutual aspirations to work more closely together. I predicted we would receive edits from our Russian colleagues in one or two days, we'd respond in a day or so, and we'd be done. Did I ever miscalculate! It was my first experience negotiating with the Russians. Even if we were entering a new era of optimism, we were going to fight over every word.

I first met with Kislyak in Washington to discuss our joint statements. Sergey was old-school, more Soviet than post-Soviet. He had joined the Soviet Foreign Ministry in 1977. He had deep experience sparring with Americans, first during the Cold War and then over NATO expansion when he served as the Russian ambassador to NATO in the 1990s. He knew how to say *nyet.* In my early briefings with U.S. government officials, people often rolled their eyes when I reported earnestly that I wanted to develop a direct, working relationship with Kislyak. For our first meeting, Sergey had his guard up. Years later, he confessed to me that he had read all my critical articles about Putin and his foreign policies prior to that meeting in Washington, and he had a large briefing book on my conservative views about foreign policy and activities in Russia, particularly my work with the "democrats." He viewed me as a hawk — a neocon — someone with whom you had to battle, not cooperate. He was tough, but that was his job. Moscow didn't send him to Washington to become friends with U.S. government officials. His writ was to advance Russian national interests, and he remained true to that mission, at least in his interactions with me. With time, we did learn to work closely together, and I eventually used our bilateral channel to advance significant policy. I came to enjoy our lunches at his residence, just a few blocks from the White House.

Despite the sharp edges and tough talk Kislyak displayed in our first encounters, he and his bosses back home wanted to seize the opportunity afforded by the change in the U.S. administration to stop the negative spiral in our relations. Kislyak blamed the Bush administration entirely for the breakdown in U.S.-Russia ties: it was the U.S., therefore, that had to do the resetting. I pushed back. Yes, some Bush administration policies had contributed to tensions, including, most importantly, its decision to invade Iraq. But the Bush administration did not invade Georgia. If we were going to agree to a common agenda for the future, we were going to have to agree to disagree about some of our competing explanations of history.

My first draft of the joint statement intended for release at the London meeting was eight pages. For the Russians, it was too much, too fast. We were setting aspirations for cooperation too high. And of course, they aggressively trimmed all of the language about democracy and human rights. But we did reach agreement on several key paragraphs — it looked like we would have something concrete for the Obama-Medvedev meeting.

Samore's arms control one-pager, however, ran into greater trouble. The Russians wanted to insert language about the relationship between offensive and defensive weapons into the initial statement about the START negotiations. We worried that this seemingly innocuous material was really a Trojan horse for future language designed to limit U.S. missile defenses. As a general description of warfare, the Russians were right: if you can protect yourself against my offensive weapons with your defensive weapons, then you have an advantage. But we knew that such a defensive capability was decades away, if not impossible altogether. We also knew that any discussion of limits on missile defenses would guarantee the treaty's death upon arrival for ratification at the U.S. Senate. So we played a game of chicken, insisting we were better off with no statement at all than with one referencing missile defense. I was surprised at how hard we struggled to get this short, ambiguous document ratified. If we couldn't agree to this "chickenshit" (as Samore called it) one-page joint statement, how were we going to negotiate an entire treaty?

Days before the London meeting, we had several outstanding issues to address in both joint statements, so Deputy National Security Advisor Tom Donilon instructed me to join Secretary Clinton, who was scheduled to meet with Lavrov in The Hague, to achieve agreement on both documents. Howard Solomon, then a senior official on the Russia desk at the State Department, joined me for the trip. Howard and I spent several hours cleaning up the joint statement with a talented Russian diplomat, Igor Neverov, who clearly had instructions to be creative and helpful. We were doing real diplomacy, and I liked the feeling. We left a few tough sections for Clinton and Lavrov to tackle on their own, which they managed to do; they were still in a cooperative mood after their positive meeting a few weeks earlier in Geneva.

Once completed, the "Joint Statement by President Dmitry Medvedev of the Russian Federation and President Barack Obama of the United States of America" codified a commitment to do things differently, beginning with a bold first sentence: "Reaffirming that the era when our countries viewed each other as enemies is long over, and recognizing our many common interests, we today established a substantive agenda for Russia and the United States to be developed over the coming months and years." First and foremost on our common agenda was "nuclear arms control and reduction," with Russia agreeing — amazingly, I thought at the time — to commit "our two countries to achieving a nuclear free world." The joint statement also pledged mutual commitment to combating nuclear proliferation — especially in North Korea and Iran — fighting al-Qaeda and other terrorist groups, reducing drug trafficking in Central Asia, rebuilding Afghanistan, strengthening Euro-Atlantic and European security, deepening economic ties between the U.S. and Russia, and "mak[ing] efforts to finalize as soon as possible Russia's accession into the World Trade Organization." The joint statement underscored our disagreement about the Russian-Georgian war (aka the Russian invasion of Georgia) but noted, "We agreed that we must continue efforts toward a peaceful and lasting solution to the unstable situation today." Importantly for me, the statement also included the following lines: "We also discussed the desire for greater cooperation not only between our governments, but also between our societies — more scientific cooperation, more students studying in each other's country, more cultural exchanges, and more cooperation between our nongovernmental organizations. In our relations with each other, we also seek to be guided by the rule of law, respect for fundamental freedoms and human rights, and tolerance for different views."[16]

I was pleased with the substance of this first joint statement. The goals were comprehensive and multidimensional — exactly those that the Reset aspired to achieve. They were also ambitious, including such difficult issues as missile-defense cooperation. But I wanted to start big, with a clear map of the road we were prepared to travel with Russia's cooperation.

Within our government, tension built as the Obama and Medvedev meeting approached. We had been public about our desire to reset relations, so we were worried that Medvedev might embarrass President Obama by rejecting our new framework. David Axelrod, Obama's political advisor, traveled with us on that trip; he wanted to avoid a repeat of the first Kennedy-Khrushchev meeting in June 1961 in Vienna, at which Kennedy had looked inexperienced, weak, a pushover. I was worried that Obama would look too conciliatory. Less than a year before, Russia had invaded Georgia. The new American president had to express a desire to cooperate with Russia on issues of mutual interest without downplaying our differences.

After arriving in London, I spent the small hours of the morning tweaking the talking points over and over again. This was my first meeting of heads of state, so I wanted to perfect every part of the "package": the background briefing materials and the bullet points that the president of the United States would use in speaking to the Russian president for the first time in person. This was also my first encounter with blue tents inside hotel rooms — temporary SCIFs, or sensitive compartmentalized information facilities. Our security team had put them in place to prevent hidden cameras from viewing our classified email systems. We had to check our phones with the Marines in the lobby before entering the SCIFs, where we sat in front of temporarily installed "top secret" computer terminals while some eerie sound loop played in the background to interrupt listening devices. I learned that this was the way we always rolled around the world while traveling with POTUS. It was a giant, expensive operation, all so that the Russians would not know today what the president planned to say to them tomorrow.

By the time I finished late that evening, I liked the script. Obama would lay out his theory of the Reset in some detail — the "win-win" focus, the need for a multidimensional relationship, his commitment to developing new relations based on common interests and mutual respect, all while standing firm in our support for Georgia. In addition, Obama planned to raise two sensitive issues with Medvedev at the end of the meeting when the two of them met one-on-one with only translators in the room. First, Obama was going to tell Medvedev about our new policy of engagement with Iran, including that the president had reached out to Iran's supreme leader in the form of a letter. Not everyone in the administration agreed with the idea; few people in our own government knew about the letter. It had not been made public. What if Medvedev spilled the beans? But we wanted to win Russia's support for our new approach to Iran, and we also wanted to test Medvedev's commitment to developing a relationship with Obama based on trust. If the Russian president kept quiet about the letter, that would be a good sign.

The second issue to be raised was also controversial within our White House team: whether to discuss human rights with Medvedev in this first meeting. The day before, a Russian human rights activist, Lev Ponomarev, had been beaten up by thugs outside his Moscow home. Lev was an old friend of mine, a former leader of the Democratic Russia political movement back in the 1990s who had played an instrumental role in the end of communism and the fall of the Soviet Union. When I learned of the attack, I wanted to use my new government role to bring attention to his case, so I added a few talking points on Ponomarev into the president's brief. Initially, few in the White House inner circle expressed much enthusiasm for this addition. It didn't seem important enough — "presidential" enough — especially if it might dampen the mood of our first meeting with Medvedev. I called Lev and learned that he was in the hospital, but doing fine. It seemed highly unlikely that Medvedev had ordered this attack — so couldn't it wait? I understood my colleagues' argument, and was not going to fall on my sword over this issue just three months into my new job at the White House. We compromised: include it in the talking points about human rights in the memo, but leave it to Obama to decide if he wanted to raise Lev's case during his one-on-one with Medvedev.

We met Medvedev at the American ambassador's residence in London — an impressive estate, which prompted me to wonder why the American taxpayers footed the bill for such extravagant quarters. We were a democracy, after all, not a kingdom! I did not know then that I'd be living in one of those government-supplied mansions — Spaso House — three years later.

Soon after our two delegations assembled around a long table in the ambassador's dining room, it was obvious to me that Medvedev had come prepared to embrace Reset. He went out of his way to present himself and Obama as young, new leaders, not burdened by Cold War thinking or even constrained by actions taken by their predecessors, Putin and Bush. He also noted their shared academic training as lawyers, and claimed that he had read some *Harvard Law Review* articles that Obama had edited. Obama and Medvedev also shared similar communication styles: both were soft-spoken, analytical, and rational. I quickly concluded that this relationship was going to click.

Obama and Medvedev spent the first part of the meeting laying the groundwork to try to finish negotiations on a new strategic arms reduction treaty before the existing treaty (START I) expired on December 5, 2009. In that conversation, they committed both countries to reducing their arsenals to levels below those stipulated in the 2002 Strategic Offensive Reductions Treaty (the Moscow Treaty). Getting a new treaty done would be the focus of most of my energies for the rest of the year.

As the two presidents turned to other issues, Medvedev indicated to Obama that Russia now agreed with the American assessment of the Iranian threat, particularly regarding Iran's development of ballistic missiles. This was a startling statement. The Russian leader was admitting that the Iranian program was advancing faster than Russian analysts had initially reported. I could not believe what I was hearing. It was one thing to have issued an incorrect assessment; it was quite another, especially for a Russian leader, to admit it. This was a gift to the U.S., given by a leader wanting to make a good first impression.

And then Medvedev delivered an even bigger gift. To help our military operations in Afghanistan, Medvedev offered to allow the U.S. to fly military equipment and troops through Russian airspace. This bold, unilateral move was completely inconsistent with the briefings we had received about Russian negotiating behavior. They never gave away anything for free. They were always "transactional," or so we were told. Medvedev's offer further confirmed my suspicion that Obama and Medvedev might have a chance at developing a new kind of relationship between our two countries.

The conversation then turned to Kyrgyzstan. Just a few weeks earlier, Kyrgyzstan's President Kurmanbek Bakiyev had visited Moscow and unexpectedly announced plans to shut Manas Air Base, an American military installation near the Kyrgyz capital, after receiving a pledge for $2 billion in economic assistance from Russia.[17] As our only air base in Central Asia, Manas was a key logistical hub for our war efforts in Afghanistan, providing a transit point for American and coalition military personnel as well as supplying fuel for U.S. aircraft operating inside Afghanistan. At the time, the majority of our soldiers on their way to Afghanistan transited through Manas. We could live without the base, but not easily, especially on the eve of a major troop surge into Afghanistan.

In this part of the discussion, I remember Obama deploying the Reset playbook to challenge the wisdom of Russian pressure on the Kyrgyz government. He questioned whether the nineteenth-century concept of spheres of influence applied to twenty-first-century politics, when territory no longer played such a critical role in security, and asked why Russia felt threatened by American soldiers spending a few days at Manas before deploying to fight the Taliban and al-Qaeda. The United States was not seeking to counter Russia's "sphere" in Kyrgyzstan. On the contrary, our military mission served Russian national interests; if we were not fighting these terrorist groups, they might be trying to attack Russia. After all, al-Qaeda had been involved in financing, supporting, and even fighting against Russian soldiers alongside Chechen separatists in the Caucasus during the Chechen wars in the 1990s. The Manas base was helping us fight a common enemy.

As Obama spelled out his fresh logic about U.S.-Russia relations in the twenty-first century, Medvedev nodded in agreement. What Obama was saying seemed to resonate with the Russian president both because of the logic of the argument and Obama's manner of presenting it. Rather than lecturing Medvedev about Russian efforts to close Manas, he was trying to explain why it was not in Russia's interests to see the American air base closed. That our presence in Manas was good for the United States and good for Russia was a radical reframing. On the spot, Medvedev was not ready to commit to reversing Russian policy regarding the base; he was relatively new to diplomacy, and was at the table with several more-experienced Russian negotiators, some of whom, as we later learned, still supported closing Manas. And Putin, of course, was not in the room. Still, I could tell Medvedev agreed with Obama's win-win way of thinking about U.S.-Russia relations. Reset that day moved from an Obama idea to a joint venture.

During their one-on-one, Obama also decided to bring up the attack on Ponomarev. This was an important signal — to me and to the rest of our national security team — that Obama was not going to ignore human rights issues. When I asked him afterward how Medvedev reacted, Obama answered that Medvedev had learned about the incident on the internet (which was interesting in and of itself). He'd expressed concern, and did not react defensively to Obama's mention of the attack. That was intriguing, too. Maybe Medvedev himself wanted to see greater political change in Russia? Our idea of dual-track engagement with both the government and society might not be as contentious as many had assumed.

Obama and Medvedev also disagreed on many issues in this first meeting. They shared no common ground on Georgia, and Medvedev took a big swing at criticizing our missile-defense plans for Europe. Afterward, speaking anonymously as a "senior administration official," I described the meeting to the press, from the basement kitchen of the ambassador's residence, in the following terms: "In the meeting today, particularly when talking about Georgia, when talking about Abkhazia and South Ossetia, when talking about spheres of influence, when talking about missile defense, we were talking about disagreements, not agreements."[18] Nonetheless, there was more agreement than disagreement, and a genuine desire from both presidents to explore a new way of doing business together.

At the press conference after the meeting, the spirit of the Reset was also on display. Obama referred to the drift between the U.S. and Russia, but pivoted to our positive progress on a host of common interests, and ended his remarks on an optimistic note: "My hope is that given the constructive conversations that we've had today, the joint statements that we will be issuing both on reductions of nuclear arsenals, as well as a range of other areas of interest, that what we're seeing today is the beginning of new progress in the U.S.-Russian relations. And I think that President Medvedev's leadership is — has been critical in allowing that progress to take place."[19] Obama's decision to praise Medvedev personally was his call — I did not write that line into the talking points. But those words reflected Obama's belief that Medvedev represented a break with the past: someone with whom he could engage as a like-minded leader. I agreed with the president's analysis. But expressing too much love for Medvedev was going to create problems for us with Putin, and probably for Medvedev himself back home. Managing this delicate balance would remain a central challenge of the Reset for the next four years.

Medvedev also echoed the principles of the Reset, ending his remarks, "After this meeting, I am far more optimistic about the successful development of our relations, and would like to thank President Obama for this opportunity."[20] We were in business. The next step was to see if we could translate rhetorical aspirations into concrete outcomes.

7

UNIVERSAL VALUES

Before joining the government, I had spent decades of my academic career writing about democratization and democracy and the lack thereof in Russia and elsewhere. So when I walked into the EEOB on January 21, 2009, I was determined to maintain my focus on these issues. Of course, I knew that these concerns could not be my only focus — I had the new strategic arms treaty, supplying our troops in Afghanistan, and securing sanctions on Iran in my inbox as well. I also had to acknowledge that the Russian intelligence agencies with whom we would work to track down international terrorists were also suppressing civil society activists within Russia. Still, I started the job confident that it was possible to balance security and economic interests with the task of promoting democratic values. I left government five years later with those same convictions, but sobered by how hard it is to work with autocratic regimes and promote universal values within them at the same time. Autocrats, from Beijing to Moscow to Tehran, will never agree that democracy is a universal value, even as they sometimes pay lip service to it.

In retrospect, Obama is often labeled a realist in the tradition of George H. W. Bush. While working with him, however, I rejected that oversimplification. I knew from my very first discussion with then senator Obama on his campaign jet in August 2008 that he rejected the idea of realism and liberalism as being mutually exclusive. Leaders in the "real world," he told me, must employ a mix of both, both in theory and in action. Unlike Bush, Obama did not seek to end tyranny in the world, nor did he embrace all elements of the Bush strategy of promoting freedom.[1] But neither did he reflexively shy away from speaking out when other countries violated democratic values. His approach provided just enough cover for me to push my ideas about advancing democracy and democratic ideals abroad.

In my first year in the administration, one ally was most important: Obama's speechwriter and deputy national security advisor at the White House, Ben Rhodes. More than most staffers at the White House, Rhodes was interested in crafting a new message about advancing democracy and human rights — what we eventually called "universal values" — consistent with Obama's philosophy and voice. That's why Ben reached out to me. As one of the few professors at the NSC, I knew these arguments about democracy promotion well, and so served as an informal academic resource for Rhodes — a source for the arguments about why promoting democracy and defending human rights abroad served American national interests.[2] Ben wrote the first draft of every major foreign policy speech given by President Obama, which gave Ben incredible power to shape that message. Quietly, sometimes without others knowing, Rhodes allowed me to "chop" — that is, read, edit, add — many of Obama's foreign policy speeches.

Our first serious debate on these issues took place before Obama's Cairo speech in June 2009. This address was one of Obama's first major foreign policy speeches and his first shot at articulating his take on the relationship between interests and values. Regarding what we were promoting, some argued for a focus on "dignity," a noble *d* word less charged than "democracy," while others advocated for attention to economic development as a precursor to democratic change — classic modernization theory. I had a different view. I did not oppose advocating for dignity, but I made the point that voting or freely expressing oneself were dignified acts. I also did not oppose support for economic development, but I tried to inject into the debate more recent data about the relationship between democracy and development, which showed that democratic change could foster economic growth.[3]

After one heated exchange about the relationship between democracy and development, one Obama loyalist argued that it really didn't matter what the data showed if it's not what the president thinks. Reacting to this blatant display of opinion trumping facts, I shot back, "I don't care if the guy walks on water; if that's what he thinks, he's wrong." Everyone in the room turned to stare at me, startled into complete silence by my flippant remark about the leader of the free world. I went home that night wondering how long I'd last at the White House! I did not speak in that manner about the president ever again.

In the end, I was not fired, and Obama and Rhodes devoted three of the final seven sections of the Cairo speech to issues of democracy and human rights. In his address, Obama stated bluntly, "The fourth issue that I will address is democracy," and then argued, "I do have an unyielding belief that all people yearn for certain things: the ability to speak your mind and have a say in how you are governed; confidence in the rule of law and the equal administration of justice; government that is transparent and doesn't steal from the people; the freedom to live as you choose. Those are not just American ideas, they are human rights, and that is why we will support them everywhere."[4] That was not realism! Kissinger never made any such statement.

Obama made a similar statement in his graduation-ceremony speech at the New Economic School in Moscow during the July 2009 summit, which I helped to write along with Rhodes and his talented deputy, Terry Szuplat. In this speech, Obama spoke about the link between democracy and prosperity and security, saying, "Governments that promote the rule of law, subject their actions to oversight, and allow for independent institutions are more dependable trading partners. And in our own history, democracies have been America's most enduring allies, including those we once waged war with in Europe and Asia — nations that today live with great security and prosperity."[5] Nixon never uttered these words in Moscow.

In a speech in Accra, Ghana, just a week after Moscow, Obama spelled out the relationship between anticorruption and democracy, proclaiming, "No person wants to live in a society where the rule of law gives way to the rule of brutality and bribery. That is not democracy, that is tyranny, even if occasionally you sprinkle an election in there."[6] That was a tough-love message, one that some African leaders found condescending. It could not be confused with realpolitik.

Obama's first address to the United Nations General Assembly, on September 23, 2009, provided the next test of whether his statements about democracy in Moscow or Accra were just one-offs or a critical component of his vision for American foreign policy. As I talked to Rhodes in the weeks before this speech, it became clear to me that the president wanted to use his remarks to outline the main tenets of an Obama doctrine. Democracy promotion was not making the cut.

Obama's speech rejected realist approaches to foreign policy: "Power is no longer a zero-sum game," he stated. Instead, the president promoted strengthening international institutions like the United Nations to deepen cooperation between countries in the pursuit of common interests. Guided by this framework, he spelled out four major pillars of his policy: (1) "stop the spread of nuclear weapons, and seek the goal of a world without them"; (2) "the pursuit of peace"; (3) "take responsibility for the preservation of our planet"; and (4) "a global economy that advances opportunity for all people."

I admired and supported all four goals. Some of us at the White House, however, wanted to add a fifth pillar: support for democracy and human rights, or, if the concept needed to be softened, the promotion of universal values. Rhodes was sympathetic, and Samantha Power, our senior director for multilateral affairs and human rights at the National Security Council at the time, was also a passionate proponent of including this goal in the speech. As we got close to the UN General Assembly meeting, however, I sensed that some other senior White House officials opposed adding it.

The night before the speech, I got a call from Deputy National Security Advisor Denis McDonough, instructing me to meet him and Rhodes in a hotel room at the Waldorf Astoria, where our presidential delegation was staying. I was in New York to staff the Obama-Medvedev meeting the next day, and had no responsibilities for the president's events at the United Nations, so I assumed my late-night meeting with McDonough was going to involve either a review of my briefing memo for the Medvedev meeting or beers in the bar.

On our way to the bar, we detoured to another room to do final edits to Obama's speech the next morning. Ten minutes into our session, Obama walked in. I signaled to Denis that I perhaps should leave and meet him downstairs later, but he told me to sit tight, so that's what I did while the three of them worked through some of the trickier portions of the speech. They spent the most time hashing out the section on the Middle East peace process. Their review of that part of the speech occasioned the first time I'd ever heard the president swear — and not just once. When they'd finally made their way to the end of the speech, Rhodes told the president that "some people" wanted to add additional language about democracy and human rights. That's when I realized why I was there.

For thirty minutes that felt like hours, I got to do what every academic dreams of: walk the president of the United States through every argument of the book I had just published, titled *Advancing Democracy Abroad: Why We Should and How We Can*. I explained democratic peace theory, which holds that the United States has a security interest in seeing the proliferation of democracies around the world, since democracies do not go to war with each other. I reminded the president that all of our enemies — past and present — were autocracies, while all of our strongest and most enduring allies were democracies. I spelled out my reasoning about the false promise of autocratic stability. I summarized the most recent research on the relationship between democracy and economic growth. I even squeezed in a mention of why the spread of democracy can benefit the American economy. At times, Obama pushed back. We wrestled, for instance, with the differences between evolutionary democratic change and revolutionary change. McDonough and Rhodes jumped in as well. Like my discussion with Obama on "O-Force One" back in August 2008, I felt like I was back at a seminar at Stanford. Even with all the things swirling in his head, I sensed again that night how much Obama loved the intellectual, analytic debate.

At the end of our conversation, the president gave Rhodes some instructions for how to add some of these ideas to the end of the speech, and then, ready to decompress, he left to go watch ESPN. As we digested the president's instructions for revisions, Obama walked in again, this time wearing just a sleeveless white T-shirt, to give us one more edit about democratic values. Eventually, we did make it to the bar and Obama to his television set, but even while we sipped beers downstairs, Obama emailed McDonough one more idea about democracy and human rights. He may have been watching ESPN, but he was also chewing on the conversation we had just finished. Such uninterrupted time to discuss big ideas with the president was rare for me, but even McDonough and Rhodes, who spent significant time with Obama, reflected on what a unique moment we had shared.

Per the president's instructions, Rhodes made promotion of democracy and human rights the last theme of the speech, not as a fifth pillar, but a premise for pursuing the other four. The result was sweeping language about dignity, democracy, and human rights — principles, Obama said, "which ensure that governments reflect the will of the people." He went on: "These principles cannot be afterthoughts — democracy and human rights are essential to achieving each of the goals that I've discussed today, because governments of the people and by the people are more likely to act in the broader interests of their own people, rather than narrow interests of those in power." He warned those standing against democratic progress that attempts to "muzzle dissent" or "intimidate and harass political opponents" were anathema to true leadership and would not be tolerated. He ended by invoking the UN Charter to explain why all countries of the world had the responsibility to defend the human rights of all individuals, arguing that "no nation should be forced to accept the tyranny of another nation, no individual should be forced to accept the tyranny of their own people."[7]

The next day, standing beside David Axelrod in a room adjacent to the giant United Nations General Assembly stage, I watched the president deliver this speech. Obama always finished his addresses strong, but this ending seemed particularly powerful to me, probably because the conclusion focused on themes about which I cared deeply. Axe agreed. It was a great speech, he said, especially the part about all that "values stuff."

Speeches state lofty ambitions and worthy objectives; they rarely spell out concrete strategies for achieving them. This disconnect was most certainly true regarding Obama's speeches that referenced democracy promotion. Over time, however, our administration did articulate a strategy for advancing democracy abroad, even if at times we failed at executing it. As I saw it, our strategy had seven key features.

First, Obama emphasized the importance of getting our own house in order as a necessary condition for inspiring others to emulate our system of government. Second, we promoted international integration, believing that deeper incorporation of autocratic states into international institutions would have a positive influence on democratic development in those countries.[8] Strengthening multilateral organizations also could have a positive influence on promoting universal values. In addition, we promoted new multilateral initiatives to encourage accountable government, including the Open Government Partnership (OGP), launched by President Obama at the 2011 UN General Assembly meeting. Jeremy Weinstein, at the time NSC director for democracy and development, came up with this creative initiative, whose aim was to inspire governments to develop national plans for increasing openness and thereby foster new international norms about transparency.[9]

Third, our early thinking about democracy promotion placed an emphasis on democracy that delivers. New democratic governments had to benefit their citizens, especially economically, if they were going to succeed. Fourth, our administration believed—or more precisely, some of us in the Obama administration believed—that direct engagement with autocratic governments could sometimes but not always be useful in pushing them in a more democratic direction. If these regimes had a stake in engaging with us, they might be less likely to crack down on their own citizens. Even while engaging governments, we were going to call them out for human rights abuses; at least, that was our stated policy. Fifth, in parallel to greater engagement between governments, we pressed for deeper engagement with societies, both between our government and civil society leaders and between American NGOs and like-minded organizations in other countries. We called it peer-to-peer dialogue.[10] Sixth, we continued to provide direct assistance to civil society, independent media, legal organizations, and other kinds of nongovernmental actors around the world. Some in our administration wanted to cut budgets for democracy assistance. They succeeded later on, but in the first years of the Obama administration, a group of us within the government fought to preserve Bush-era budget levels for "governing justly and democratically."

Finally, Obama made clear that the United States was getting out of the business of imposing democracy on other societies. Obama stated emphatically, "Democracy cannot be imposed on any nation from the outside. Each society must search for its own path, and no path is perfect. Each country will pursue a path rooted in the culture of its people and in its past traditions."[11] Obama not only rejected the use of force to topple autocratic regimes, he did not support the use of coercive power to pressure dictatorships into democratizing.[12] Even when we imposed sanctions on Iran and Russia, it was not for the purpose of regime change.

I participated in developing this general strategy for promoting universal values, and then worked to apply the framework in developing a strategy for supporting democracy and human rights in Russia. It took time, many IPC meetings in the White House Situation Room, but our administration eventually adopted a new strategy for advancing democracy and human rights in Russia.

We started from the premise that the United States could do very little in the short run to effectively influence the trajectory of Russia's internal political development. Our strategy, therefore, had to be incremental and long-term. We identified two main objectives: promote more transparent and accountable government and strengthen Russian civil society actors pressing for democratic reforms and human rights.

The hardest question was whether we could engage with the Russian government to pursue common interests and at the same time press for reform. My answer was yes. Before joining the government, I had written several essays about the conditions under which engagement with an autocratic government can facilitate democratic change.[13] The story was not a simple one. In some cases, engagement helped democratic change; in others, engagement impeded change. For me, Reagan's engagement with Gorbachev in the late 1980s was a powerful example of the former, while Reagan's "constructive engagement" of the apartheid regime in South Africa during this same period illustrated the latter.

Regarding the specific circumstances in Russia in January 2009, I concluded that Medvedev wanted to push his political system in a democratic direction, albeit slowly and carefully. Medvedev said as much, arguing that "a weak democracy" was a significant problem "even for a state like Russia"[14] and that "in the previous years we didn't do enough to overcome the problems we inherited."[15] He even recognized the importance of dialogue between the government and civil society, pointing out, "The quality and effectiveness of democracy depends not only on political procedures and institutions, but also on whether the government and civil society are prepared to listen to each other."[16] Medvedev also argued in favor of more political freedom and pledged to fight a war against legal "nihilism."[17] In a major statement about his political vision published in 2009 called "Go Russia!," Medvedev acknowledged that "as a whole democratic institutions have been established and stabilized, but their quality remains far from ideal. Civil society is weak, the levels of self-organization and self-government are low."[18] Putin had never said that. Medvedev therefore argued that Russia needed to undergo a major modernization, based for the first time ever on democratic ideas and institutions.[19] To engage with him on these issues, therefore, was not to reinforce autocracy, but to subtly, incrementally, embolden his push for democracy.

Not everyone shared my assessment about Medvedev's desire for democratic change. In fact, many leaders of Russia's democratic opposition — people whom I respected — disagreed with me. They told me so in private and at times ridiculed our administration in public. I listened closely to these skeptical voices, and consequently never became overly optimistic about the probability of Medvedev's success in pushing Russia in a more democratic direction. Even if Medvedev was a closet liberal — a second Gorbachev — he was operating in a very constrained environment. He worked for Putin, and Putin clearly was not interested in political liberalization. In the margins, however, I believed that Obama's personal engagement with Medvedev might nudge him in the right direction. I also believed that we had little to lose in trying, and no other option. Through sanctions or other coercive instruments, we were not going to pressure Russia into democratizing. We had neither the means nor the desire to do so.

In parallel, however, we adopted as policy a commitment to criticize human rights abuses and democratic erosion inside Russia. In the early years of the Obama administration, many of these critical statements were released by the White House, not the State Department, since I could write them and get them approved more quickly through the NSC.

We also encouraged meetings between U.S. government officials and civil society leaders. In February 2009, Under Secretary of State Burns and I met with civil society leaders. Obama did so again in July 2009; Clinton in 2010, in both Moscow and Washington; Biden in 2011; and later, when I was ambassador, Clinton held meetings in 2012 in St. Petersburg, Kerry in May 2013 at Spaso House, and Obama in September 2013 in St. Petersburg.

Our new policy placed greater emphasis on supporting vigorous peer-to-peer dialogues between American and Russian civil society leaders. We wanted Americans and Russians to engage as equals — not Americans lecturing Russians — on everything from public health to community development, anticorruption to prison reform. At the same time, the Obama administration continued the U.S. government practice of providing technical and financial assistance to Russian actors, both inside and outside government, for the purpose of promoting a more accountable government. In the 2010 fiscal year, we increased our spending on democracy assistance for Russia by $3.5 million over what the Bush administration had been allocating. But we shifted our emphasis away from trying to change government institutions by supporting "reformers" from within, and instead devoted greater resources to strengthening NGOs, which in turn would pressure from without government institutions to change.

One of our biggest debates while reviewing our democracy-promotion strategy concerned election assistance. In 2009, the next Russian parliamentary election was still two years away. Most Russian experts both inside and outside government already were predicting that this vote had little chance of being free or fair, and therefore the ruling party, United Russia, would win by a landslide. So why should we waste our shrinking resources on activities that required competitive, free, and fair elections to produce results? Specifically, why should USAID provide financial assistance to election-monitoring organizations when exposing falsification would make no difference?

I entered this debate with some historical experience — or, as I'm sure some of my government colleagues would say, with a lot of bias. My old employer, the National Democratic Institute, in parallel with another democracy-promoting organization, the International Republican Institute, used U.S. government funds to train political party operatives in Russia. NDI had developed a reputation for providing quality technical assistance and financial support to nonpartisan electoral-monitoring organizations around the world. At the time of our policy review in 2009, however, NDI programming in Russia had not produced major results. Independent political parties remained weak, and Golos, Russia's largest electoral-monitoring organization, was considered by many democracy-assistance experts to be incapable of organizing a national election-monitoring effort in the largest country in the world. Some in our government thought it was a complete waste of taxpayers' money to continue to fund Golos. Instead, they argued, we should devote greater resources to nonelectoral activities. I pushed back. I shared some of the skepticism about past results and had few expectations about impact in the upcoming election, but still pressed for greater resources, not fewer, for Golos. I believed that Golos would not have the capacity to monitor the 2011 parliamentary elections effectively, but that its participation would allow the organization to gain valuable experience and increase capacity for future elections. I was focused on 2015 or 2019, not 2011. And importantly, Golos was nonpartisan. The Obama administration was not going to finance partisan political activities in Russia, but we should remain committed to nonpartisan efforts, or so I argued at the time. Free and fair elections are a goal that all Russian political actors should endorse, including Putin.

Eventually, my side of the argument won, and USAID increased the budget earmarked for Golos by $1 million, making our contribution to the organization $3 million in 2011. Our total expenditures for all election-related activity that year were nearly $9 million.[20]

To the surprise of many of us in Washington, and most certainly to many in the Kremlin as well, Golos did some amazing work in December 2011, exposing electoral fraud that in turn stimulated massive protests. In 2009 no one predicted that protests two years later triggered by voter-fraud exposure would be a factor in ending the Reset. In 2009, it was possible to engage the Russian government and Russian society at the same time. By 2012, that had become nearly impossible.

8

THE FIRST (AND LAST) MOSCOW SUMMIT

During my first week at the White House, I made two dozen color-coded files—the old-fashioned, paper kind—which I stored in a big steel file cabinet secured with a mechanical combination lock. Each file was labeled with a concrete policy outcome regarding Russia that I hoped to achieve while in government: NEW START TREATY, DENYING IRAN THE BOMB, MISSILE DEFENSE COOPERATION, REPEAL JACKSON-VANIK, et cetera. I made the DEMOCRATIZATION file red, as that one was going to be the most difficult to achieve. Our job in government was to get things done—"land some planes," as Chief of Staff Rahm Emanuel used to say, or "close out accounts," in the words of Denis McDonough, one of my NSC bosses at the time. We were in government *to do* things. I had a long, ambitious to-do list, and I was determined to cross off some items during Obama's time in the White House. Some of those planes touched down gently, others barely got off the ground, and a few crashed.

After Obama's first meeting with Medvedev in London in April, we shifted into a higher gear to produce some results for the Moscow summit. In London, we had achieved broad agreement on our Reset approach, both within our administration and, I thought at the time, with the Russian president. There was still one wild card: Prime Minister Putin. He never traveled with Medvedev, so we couldn't be sure how much of the positive atmosphere at the London meeting reflected Putin's thinking. That's why, in my view, we had to get to Moscow as soon as possible and meet him. Presidential time is the most precious commodity in the U.S. government, so I scored a huge bureaucratic victory in getting Moscow onto Obama's calendar in July 2009 (months before he made his first trip to China). To make the trip worthwhile, we needed to produce some tangible outcomes. We focused on three: a proposed U.S.-Russia Bilateral Presidential Commission, a "lethal transit" agreement, and a framework agreement for the new strategic arms treaty.

A first priority, almost a precondition for making progress on other issues, was to configure a better way for our governments to interact. During the Bush administration, direct lines of communication between senior American and Russian officials had atrophied. I wanted to rebuild these channels and open new ones: compel our governments to engage. I also wanted to widen the range of issues that we discussed with our Russian counterparts to include nonsecurity matters such as trade and investment, health, human rights, even sports.

Veterans of the Clinton administration recommended modeling this communications channel on the Gore-Chernomyrdin Commission, officially the U.S.-Russian Joint Commission on Economic and Technological Cooperation. Chaired by Vice President Al Gore and Prime Minister Viktor Chernomyrdin, the commission created working groups on topics such as defense conversion, agriculture, and space exploration. The group met once every six months, alternating between Washington and Moscow, and then reporting on progress to Presidents Clinton and Yeltsin. I liked the scope of Gore-Chernomyrdin, but worried about the logistics of forming senior-level working groups and the pressure to produce new results every six months. The "Gore-Cherno" model, however, had one big upside for us: it would allow us to engage directly with Putin without undermining Medvedev. Vice President Biden said yes to the idea. Putin said no. Putin did not consider the vice president his peer. And that was that: opportunity lost.

After reviewing a few other modalities, we eventually settled on something new: the Bilateral Presidential Commission, or BPC. In this configuration, Obama and Medvedev would serve as chairs, and other ministers and cabinet secretaries would cochair the working groups. I was excited when Secretary Clinton told our team at the NSC that she wanted to take a leading role, serving as "commission coordinator" along with Lavrov. To pull her deeper into diplomacy with Russia would be good for the Reset.

Working with our Russian counterparts to create the BPC working groups was no cakewalk. For instance, they did not want a separate group on "defense policy," in part because Russia has a weak tradition of civilian oversight of defense activities and in part because they did not want to have to engage Sandy Vershbow, then assistant secretary of defense for international security affairs. Vershbow had earned a reputation in Moscow as a hardliner while serving as ambassador there from 2001 to 2005. For that very reason, I wanted the group to be set up! We needed people like Vershbow to be directly involved in difficult security matters.

The most contentious debate with the Russians, however, involved what would later become the most famous — or infamous — working group, the Civil Society Working Group, given the unwieldy acronym CSWG, in true U.S. government fashion. We quickly agreed to create a working group on educational and cultural exchanges, and our Ministry of Foreign Affairs colleagues wanted to fold civil society issues into its agenda. I objected. Although not everyone at the State Department agreed with me, I insisted on a stand-alone working group as a signal of the importance we assigned to civil society. Moscow resisted the CSWG proposal until just before the July summit, and then proposed a provocative compromise: we could establish such a working group, but its cochair would be Medvedev's deputy chief of staff, Vladislav Surkov. The Russian counteroffer seemed sinister, since Surkov was known as the author of concepts such as "managed democracy," and later, "sovereign democracy," which civil society activists considered extremely hostile to their work. Surkov also had created Nashi, a pro-Kremlin youth group, which some Russian civil society leaders compared to Hitler Youth.[1] It would have been hard to find someone in the Russian government with a worse track record on civil society issues. Moreover, Surkov worked at a very senior level in the Kremlin, and at the time was considered one of the most important decision makers in the Russian government. We had no other cochairs in the BPC from either the Kremlin or the White House. We considered dropping our request for the working group, but ultimately decided that it was better to engage with people like Surkov, if only to learn how they thought, than to ignore them.

One of my directors at the NSC proposed a clever idea: have me serve as Surkov's cochair. In rank, I was not his equal. To the extent that Surkov was viewed as a foe of Russian civil society, however, I was known as its champion. I agreed to take on the role. I was intrigued by this shadowy personality, who was known as the mastermind behind Putin's 2000 election campaign, the suppressor of independent media and opposition activities, and, as he himself claimed, "the author . . . of the new Russian system"[2] — a system that others characterized as one of "democratic rhetoric and undemocratic intent."[3] He also wrote novels under a pseudonym, and, I had heard, had a portrait of Tupac Shakur hanging in his Kremlin office — a mysterious fact I confirmed on my first visit. If we wanted to engage with all segments of the Russian government, we could not avoid contact with people like Surkov.

By the time of the formal launch of the BPC at the July summit in Moscow, we had created thirteen working groups; we later added three more, on innovation, the rule of law, and cybersecurity.[4] This would mark the first time we were engaging our Russian counterparts on some of these topics; and some of them had previously been taboo. That felt like progress.

At his first meeting with Obama in April, Medvedev expressed his government's willingness to allow the United States to transport lethal military equipment across Russian territory to Afghanistan. The overture was not simply a symbolic gesture, but a first step toward an agreement that could diversify our supply lines as well as speed up and reduce the costs of supplying our soldiers. The Bush administration had started to explore new supply routes to Afghanistan in 2008, but only a trickle of materiel was moving through Russia when Obama entered the White House, and we wanted to substantially increase the volume and kind of supplies moving along what we called the Northern Distribution Network. The NDN included other countries — Kyrgyzstan, Uzbekistan, and Kazakhstan, as well as Georgia and Azerbaijan — but Russia was pivotal given its size and location. Transporting through Russia would diversify our supply lines and reduce the costs of moving heavy equipment.

We had another specific motivation for developing the NDN. We were conducting a major review of our Afghan policy, and planned to unveil the new "AfPak" strategy, which placed greater emphasis on the Pakistani dimension of the Afghan war. Because Pakistan offered a safe haven for the Taliban, al-Qaeda, and, as we would later learn, Osama bin Laden himself, we planned to disrupt operations in Pakistan and eventually pursue al-Qaeda and Taliban leaders inside Pakistan.[5] Violating Pakistani sovereignty in the hunt for terrorists, we predicted, would create tensions in our relationship with the Pakistani government, and threaten disruption of our supply routes through Pakistan, through which 80 percent of the supplies for American and coalition forces in Afghanistan flowed in 2008.[6] We had to reduce that vulnerability.

Secretary Clinton assigned one of our most skilled diplomats, Steve Mull, to translate Medvedev's offer into a real agreement. Of course, there were bumps along the negotiating path, especially concerning Russia's demand for inspections of American planes operating in Russian airspace, which we refused to allow; and transit fees, which we insisted had to be waived. But it was clear to me that Medvedev had given firm instructions to his team to complete the agreement by the time Obama arrived in Moscow in July. That signal from above made all the difference. Steve and his counterpart, Igor Neverov, got it done. In addition to allowing lethal transit by rail, the new agreement also permitted forty-five hundred flights a year, which we calculated would save $133 million in fuel and transportation costs. In signing the agreement, we had a concrete result from the Reset: not just a joint statement or cooperation on an obscure topic, but an achievement that enhanced our war effort in Afghanistan, one of our most important national security interests at the time.

We were also under tremendous pressure to get a new strategic arms treaty in place by December 6, 2009, when the existing START treaty expired. Our Republican critics in the U.S. Senate already were hammering us over the danger of allowing the inspections regime of the old treaty to lapse. Senator Jon Kyl warned, "On December 6 . . . for the first time in 15 years, an extensive set of verification, notification, elimination and other confidence building measures will expire. The U.S. will lose a significant source of information that has allowed it to have confidence in its ability to understand Russian strategic nuclear forces."[7] To avoid that outcome, we wanted to secure a preliminary commitment —what would eventually become known as a framework agreement—about the basic contents of the treaty before President Obama's trip to Moscow. The agreement negotiated in time for the Moscow summit was thin, but did specify some important target limits on strategic forces, including a range of 1,500 to 1,675 strategic warheads and 500 to 1,100 strategic delivery vehicles, the bombers and intercontinental missiles (launched from land and submarines) that deliver nuclear warheads. Importantly for us, the "joint understanding" contained no mention of limits on defensive weapons, the subject of so much discord before the London meeting. The START framework agreement was another achievement only months into the Reset.

As I boarded *Air Force One* for the first time, I was more nervous than excited about the trip to Moscow. I had never organized a summit before. I did not know if Obama would be satisfied with the agreements we had reached in advance, or if he wanted to dig as deeply into difficult issues as I had outlined in his talking points. We were just five months into our respective jobs, still learning what works and doesn't work for these meetings with heads of state. I also was worried about how Obama would react to the other events I had planned. He had signed off on the agenda for the trip, but did he really want to meet with political opposition and civil society leaders? My White House superiors had given me more control in organizing the trip than I had expected, but that was a double-edged sword. If anything went wrong, it would be my fault. Ronald Reagan's historic visit to Moscow in May 1988 was my model. But at a minimum, I hoped for a trip that would not be remembered for its mistakes. Sometimes nonmemorable summits are good enough.

A few hours after takeoff, we assembled in the conference room to do the pre-brief for the Medvedev meeting the next day. Secretary of State Clinton had fractured her elbow in a fall a few days before the trip, so she sent Under Secretary of State Bill Burns in her place. Bill's presence was comforting to me, even if it meant that I had to share some of the "Russia expert" limelight. We had an experienced diplomat on board, and one who fully embraced Reset. As I looked around the table that evening, I realized that everyone else besides Bill—Obama, Jones, McDonough, Lippert, Rhodes—was attending a Russian summit for the first time. In this briefing, however, Burns respectfully let me lead. This was my trip.

I had hoped for a few hours of sleep on the ten-hour journey, but got none—preoccupied with too many things that could go wrong during the president's visit. I checked and rechecked everything, every talking point, every event, every manifest for every meeting. As we descended, we changed from jeans to suits, and then were off, our motorcade racing to the Kremlin on empty streets cleared for Obama. I had spent thousands of hours of my life stuck in Moscow traffic. Not this day!

Our first stop was a controversial one: a wreath-laying ceremony at the Tomb of the Unknown Soldier. I had pushed hard for a visit to this solemn site. Nothing is more sacred to the Russians than their victory over fascism in the Great Patriotic War, known to us as World War II. Politicos at the White House worried that such a gesture might generate criticism from Republicans back home. They didn't want a new president with a left-leaning image to appear to be honoring communists or Stalin. Robert Gibbs, the White House press secretary, asked us to find a photo of a previous American president, preferably a Republican, laying a wreath at this same place, and we did — George W. Bush had paid his respects there in 2002 — thus making it safe for Obama to do so as well. This was my first lesson in a different kind of dual-track diplomacy: everything you did for an audience abroad also had an audience back home.

After the wreath laying, we zoomed around the Kremlin walls to attend our meeting with Medvedev. As would become tradition for the next three years, Obama and Medvedev met first with just a small group of aides present. Obama, National Security Advisor General Jones, Under Secretary Burns, and I represented the United States; Medvedev, Foreign Minister Lavrov, and Sergei Prikhodko, aide to the president of the Russian Federation on foreign policy (making him the closest Russian equivalent to the U.S. national security advisor), were present on the Russian side. The Russians generously allowed us an extra person in the room — me — on Obama's request. (When relations later deteriorated, the Russians stopped being so flexible.) At that first meeting, I felt a little guilty attending this small group meeting while our ambassador, a few cabinet members, and several very senior White House officials were in the room next to us, sipping tea, eating cookies, and waiting. But it was Obama's decision. Over the next three years at the White House, I would attend every Obama meeting with Medvedev and Putin, and listen in on every phone call.

The setting for our meeting was spectacular: a majestic, imperial room adorned with gold leaf. The contrast with the small, earth-toned, living-room feel of the Oval Office could not have been starker. As we exited the Kremlin later that day, Obama remarked to me that his digs back home seemed very modest compared with Medvedev's office. I reminded him that the grandeur of the Kremlin had helped trigger a revolution. He flashed me one of those big, toothy smiles at that.

Our small group meeting lasted several hours. Both leaders had prepared thoroughly, wanting to demonstrate their command of their new portfolios and their desire to make the Reset work. In this more intimate setting, they covered the most important topics in U.S.-Russia relations, which back then meant Iran, North Korea, Afghanistan, Georgia, arms control, and missile defense. The two presidents spent a good chunk of time affirming their shared desire to complete a follow-up treaty to START as soon as possible. Medvedev pressed Obama hard on the need to include limits on missile defense as a condition for deeper cuts in offensive systems. Obama listened carefully to Medvedev's logic but pushed back, claiming that Medvedev was exaggerating the capabilities of American missile-defense systems. I could tell, though, that Obama did not want to dismiss Russian concerns out of hand.

Obama's most emphatic argument concerned the growing Iranian threat, specifically Iran's nuclear weapons program. I could sense that Obama understood that we would need to pursue additional sanctions against Iran to bring its leaders to the negotiating table about limits on their nuclear program. Obama was getting Medvedev ready for this pivot toward "pressure" — our euphemism for sanctions. They agreed that our two governments should conduct a joint assessment of ballistic missile threats in the twenty-first century, focusing in particular on Iran and North Korea, even if the document itself did not single out those two countries.

Most remarkable about Obama's extensive meeting with Medvedev was not the results that day but the rapport. Clearly the leaders were listening closely to each other. They went back and forth several times on the many items on their agenda, going well beyond the talking points on their note cards. The exchanges were rational, unemotional, and analytic, with both leaders trying to find solutions, not just restate problems. All summit hosts want these events to be successful. But Medvedev seemed to want more than a "productive" meeting. He was embracing the Reset as an opportunity to establish a fundamentally new kind of relationship with his young, newly elected, post–Cold War counterpart. As I chatted with Obama on the way back to the room where the rest of our delegation was waiting, I got the sense that the president had also picked up on Medvedev's openness, flexibility, and genuine desire to deepen Russian relations with the United States, as well as his intent to establish a personal bond with Obama.

The next meeting with our full delegations at the table was pro forma, with both presidents running through a list of talking points mostly related to trade and investment. Obama secured a commitment from Medvedev to lift some nontariff restrictions on American poultry imports. This was a breakthrough that would produce real dollars for American farmers. The two presidents also discussed and praised the newly formed U.S.-Russia Bilateral Presidential Commission. It was all very jolly. Obama was cracking jokes. Medvedev also tried his hand at humor. I spent most of the meeting scribbling madly, as I was the note taker for the memcon — the memorandum of conversation — a dreaded assignment I would have to perform for all Obama meetings with Medvedev and Putin.

After the meeting ended, we marched through the grand halls of the Kremlin, piled into our many vehicles, and raced back to the Ritz-Carlton to prep Obama for his press conference at the Kremlin just an hour later. At the Ritz, our security team had assembled what looked like a minisubmarine inside one of the suites. In London, blue tents and weird music were enough to allow for secure conversations, but Moscow required more: a completely sealed, thick-walled space, constructed entirely by our team. A handful of us squeezed inside to have a conversation with the president that the Russians couldn't hear. It was hot in there; the tall David Axelrod looked particularly uncomfortable, crouched over so that his head didn't hit the top of the secure space. I thought it was a lot of trouble to go through to keep our press conference pre-brief secret. Wasn't the president going to say publicly in just a few minutes what we were discussing inside the submarine? But this was Russia, so we took extraordinary precautions.

After an hour in the submarine at the Ritz-Carlton, our motorcade sped back to the Kremlin for a signing ceremony for the new START framework document, the lethal-transit agreement, and the resumption of military cooperation. The signing event seemed a bit over the top to me, but others assured me that the formalities would bring attention to our accomplishments. Only a few months into the job, we were already getting things done on foreign policy: that's what Gibbs, Axelrod, Rhodes, and McDonough wanted the press to write.

At the press conference after the signing ceremony, Obama and Medvedev both trumpeted the Reset. Medvedev described the meetings that day as "very open and sincere," and expressed his and Obama's commitment to "communicate in this mode" moving forward. Medvedev said that he and Obama "came to the conclusion that Russian-American relations and the level achieved today does not correspond to their potential, to the other possibilities of our countries,"[8] and emphasized that the world wanted and needed leadership from the two countries. That bold statement impressed me. If the United States and Russia developed a more cooperative relationship, Medvedev suggested, the whole world would be better off. He was outlining an ambitious vision. Obama echoed his counterpart's sentiments, underscoring that the relationship between the United States and Russia had "suffered from a sense of drift" but was now on track, producing progress "on a range of issues, while paving the way for more progress in the future." He added, "I think it's particularly notable that we've addressed the top priorities — these are not second-tier issues, they are fundamental to the security and the prosperity of both countries."[9]

Obama also underscored some of the differences between the two countries, stating bluntly, "I won't pretend that the United States and Russia agree on every issue . . . For instance, we had a frank discussion on Russia — on Georgia, and I reiterated my firm belief that Georgia's sovereignty and territorial integrity must be respected." But he ended on an optimistic note: "I believe that all of us have an interest in forging a future in which the United States and Russia partner effectively on behalf of our security and prosperity."

Sitting in the front row of this press conference with hundreds of journalists in the room, knowing that millions more would soon watch or read the positive messages delivered by Obama and Medvedev, I felt real pride in our accomplishments. In just a few months, we had managed to turn around a vital relationship. And it wasn't just happy talk: the two leaders were describing a set of concrete, albeit modest achievements that we had accomplished since their first meeting in April.

In the evening of our first day in Moscow, Obama attended an informal dinner at Medvedev's residence outside town. No ties, no staff, only spouses. I liked the idea, but others on our team worried that Medvedev might try to press Obama on issues with which the Russian president was more conversant. Obama was juggling a hundred other issues at the time that were more important than missile-defense configurations or the intricacies of Russia's WTO negotiations. My one real worry was that Obama was exhausted after an overnight flight and a full day of meetings. Of course, Obama had slept in his bedroom on *Air Force One,* but the jet-lagged negotiator is always at a disadvantage. As it turned out, Obama and the First Lady were just fine. The informal dinner gave the president and Medvedev an opportunity to get to know each other on a personal level, talking about their kids, law, internet reading habits, and the like. The only ones who suffered were Robert Gibbs and Marvin Nicholson, the president's trip director. They drove out to Medvedev's residence with the Obamas and then waited for them in a separate building. Their Russian hosts fed them, but also served up some wicked moonshine. Both came back to the hotel completely smashed. Theirs was perhaps the most authentic cultural experience of our stay.

Our second day in Moscow began with a drive back out of town to meet with Prime Minister Putin at his country residence in Novo-Ogaryovo, a Moscow suburb where many senior Russian government officials live. It can be reached in thirty minutes with a presidential police escort or two hours without. To refer to the homes in this region as dachas, or country houses, is a gross understatement. Residents there all live in mansions that are part of giant compounds situated on acres of land. Putin rarely met with anyone important at his prime minister's office in the White House, the building where the Russian government was located. He preferred that guests come to him at his compound. So we did.

After a quick press spray, we spent the first hour of the meeting indoors over breakfast. With President Obama were Jones, Burns, and myself. Putin's protocol office limited participation to "POTUS +3"; his team, as I would come to learn, was militant about keeping these meetings small.

We were scheduled to have sixty minutes with Putin. After pleasantries were exchanged, Obama opened the conversation by expressing his optimism for U.S.-Russia relations. Putin interrupted him early to express a different view. For the next hour, Putin walked through the complete history of U.S.-Russian relations during his time as president. He punctuated his narrative with several instances of disrespect from the Bush administration. He liked President George W. Bush as a person, he told Obama, but loathed his administration. As Putin explained, he had reached out to Bush after September 11, believing that the United States and Russia should unite to fight terrorists as a common enemy. He had helped persuade leaders in Kyrgyzstan and Uzbekistan to allow the U.S. to open air bases in their countries to help fight the war in Afghanistan. But in return, so he claimed, the Bush administration had snubbed him. Putin even suggested that Russia and the United States could have cooperated on Iraq had the Bush administration treated Russia as an equal partner. But it did not, and that's why U.S.-Russia relations deteriorated so dramatically while Bush was president. The Bush team had supported color revolutions in Georgia and Ukraine—a blatant threat to Russia's national interests. In Putin's view, Russia had done nothing wrong; America was to be blamed for the poor relations between the two countries.

Putin knew how to tell a dramatic story. For each vignette of disrespect or confrontation, he told the president the date, the place, and who was at the meeting. During one story, he pointed to a chair he recalled Condoleezza Rice sitting in at the time, right next to Sergei Ivanov, then Putin's defense minister. He must have rehearsed all these details beforehand. For one story about counterterrorism cooperation, Putin told Obama how the Russians had benefited from some information shared with them by American officials. Dramatically, he waved away the waiters serving us tea, leaned in, and told Obama that they had used this information to "liquidate" the terrorists.

As I remember it, Putin spoke uninterrupted for nearly the entire time scheduled for the meeting, documenting the injustices of the Bush administration. This was a guy with a chip on his shoulder. Obama listened patiently, maybe too patiently. I was amazed. There was no way I could have sat for a full hour without saying something. I was also nervous. The meeting was scheduled for sixty minutes, and by minute fifty-five the U.S. president had not said a thing. It was my assignment to read out this meeting to our press corps later that day. I couldn't tell them that Obama had merely listened the entire time!

My worries were misplaced. In the end the meeting went well beyond three hours, and Obama had plenty to say. His main message was again about Reset. He asked Putin to have an open mind about resuming engagement with the United States on issues of common interest. He explained to Putin that he was different, representing a break with many of the policies of the Bush administration. Obama avoided flowery language about friendships and strategic partnerships. Instead, he pledged to always be straight with Putin and to respect Russia.

The two most contentious subjects that morning were missile defense and Iran. Putin explained to Obama why planned American missile deployments in Europe threatened "strategic stability"—otherwise known as mutual assured destruction (MAD)—between our two countries. Putin seemed annoyed—

irrationally annoyed — with the Bush administration's plan for missile-defense deployments so close to Russia's borders. Obama pledged to review America's missile-defense plans and get back to Putin on his decisions. Putin expressed less concern about the Iranian threat than Medvedev had. He talked more generally about the strategic importance of Russia's bilateral relationship with Iran as its most significant partner in the Middle East. Obama urged Putin to leverage his influence with the Iranian leaders to dissuade them from developing nuclear weapons. Curiously, Putin responded by saying that he was not responsible for foreign policy anymore: you need to talk to "Dmitry" about that, he advised. However, when Obama warned about the destabilizing effects of shipping Russia's S-300 antiaircraft system to Iran, Putin engaged. The S-300 was a defensive system, Putin argued. This Russian weapon only threatened those who intended to attack Iran — Israel and the United States. In addition, Putin argued, Russia already had signed a contract with the Iranian government that was worth billions. To renege on the agreement would damage Russia's reputation and hurt Russian industry. Weren't Americans for the rule of law, for the fulfillment of contracts?

Putin's worldview was different from Medvedev's. The day before, Medvedev had implied that stronger ties with the United States would help him advance his interests abroad and at home, and that U.S.-Russia cooperation could help international stability more generally. After three hours of listening to Putin discuss various international issues, it was clear to me that he saw the world in more zero-sum terms. At this moment, he was not *against* cooperation with the United States, but he most certainly did not see closer relations with the United States as necessary for pursuing his definition of Russian national interests. Medvedev was president because Putin had selected him for the job, but the two men did not approach foreign policy the same way, especially regarding relations with the U.S. The KGB shaped Putin's understanding of the world. By contrast, Medvedev was a lawyer, born a decade later — ten fewer years of life in the USSR. That made a difference.

Putin and Medvedev also had very different communication styles. Medvedev organized his thoughts in paragraphs, used eloquent sentence structures, and spoke in a soft voice. Putin had a high voice, but it was sharp, not soft. As our translators over the years explained to me, Putin's language was relatively blunt and more direct; some have even called his manner of speaking vulgar.

Putin also spoke with the confidence conveyed by his years in power; Medvedev had just finished his first year as president. Putin went out of his way to convey his worldly experience to the new young American president, at times even hinting at his own indifference to what we Americans sought to accomplish with U.S.-Russia relations. He did not once utter the word "reset."

At the same time, it was clear that Putin wanted to impress Obama. The breakfast spread was elaborate, featuring several kinds of caviar and exotic eggs. A server in traditional nineteenth-century peasant dress took off his tall leather boot, using it to stoke the fire in the samovar warming water for our tea. Even members of our usually cynical press corps made note of the impressive show.

By the end of the marathon session, I thought Putin had warmed somewhat to the possibility of cooperation with us. When Putin ranted about the stupidity of invading Iraq, Obama responded calmly that he agreed, reminding Putin that he too opposed that war well before it began. That seemed to surprise the prime minister: all Americans did not think alike. I sensed that Putin was considering the possibility that Obama might be different. Putin also liked what he heard from Obama about the need to develop economic ties as part of our diplomatic mission, rather than focus on a limited Cold War agenda of security issues. As Putin escorted us to our cars, I got the impression that he felt betrayed in his attempt to reset Russian-American relations in 2001, and therefore didn't want to be too involved this time around. Been there, done that. At the same time, he was not opposed to us trying again with Medvedev. He made a point of reminding Obama again that Medvedev as president was in charge of foreign policy. Medvedev was our guy, not Putin.

Driving back to the city, my knees awkwardly bumping up against those of the president of the United States as we talked face-to-face in the back of the Beast, I could tell that Putin had made an impression on Obama. Putin had been on the world stage for a decade. He knew all the global players. He had developed firm, clear views about most issues. He wasn't doing policy reviews like us newbies. We all agreed that the meeting had gone well enough. Obama was polite, but held his ground. Putin communicated clearly his disappointment with the Bush administration, but also seemed willing to consider the possibility of a more productive relationship with the new guy in the White House. We also agreed that figuring out ways to engage Putin was going to be a central challenge. Putin clearly had signaled that engagement with us was not in his portfolio. Yet we understood from that session that he remained the primary decision maker in Russia.

Amazingly, Obama and Putin would not meet again until Putin became president in 2012. They would never have a formal summit together, in Moscow or Washington.

Now several hours behind schedule, we headed from Putin's country estate to the New Economic School, a new, private university where Obama would give the commencement address. Compared with his Cairo speech just a few months earlier, Obama's Moscow remarks generated much less attention, but had a similar scope of strategic vision extending well beyond U.S.-Russia relations (or at least that was my view, as one of the authors).[10] In the speech, Obama discussed his approach to five major global issues — nuclear disarmament, countering extremists, promoting global prosperity, advancing democracy and human rights, and strengthening the international order. At the end of his explanation on policy regarding each of these issues, he suggested that Russia might share the same perspective and interests. The speech did not put Russia at the center of the narrative. Instead, Obama outlined his priorities and suggested that this same list might be important for Russia as well, a perfect expression of our new approach to U.S.-Russian relations. As one Russian press account commented, "President Obama did not promote, blame, or flirt with us, but just handed us a proposal to start building a new relationship as partners and not as enemies."[11] That seemed exactly right to me.

From the graduation ceremony, we headed back to the Kremlin for an elaborate lunch — a menu originally planned for a three-hour dinner squeezed into forty-five minutes. Obama and Medvedev then attended a parallel summit with several hundred business leaders from both countries to underscore the message that Reset was not only about increasing engagement between our governments, but also about deepening economic and business connections. But this message seemed only to be putting the jet-lagged leader of the free world to sleep.

I was sitting next to Deputy National Security Advisor Denis McDonough at the business summit when he received an email from Obama's personal assistant, Reggie Love, that the boss was tired and might have to skip the next event: a parallel civil society summit, hosted by American and Russian NGOs. At this third summit (in the government-to-government, business-to-business, and society-to-society triad), NGO leaders from both countries were meeting as peers, sharing their experiences on issues such as community development in urban areas and public health. In cooperation with Russian and American nongovernmental partners, we were experimenting with a fresh way to engage the Russians on issues of democracy and human rights: a departure from the standard practice of Americans telling Russians what to do.[12] If Obama was a no-show, our plan was to have McDonough address the conference on his behalf. But when Denis and I arrived at the hotel and informed the organizers — Russian and American alike — of the possible change, their disappointment was evident. McDonough made a few phone calls and got the event back on Obama's schedule.

As Obama arrived, a Russian civil society leader, Yuri Dzhibladze, was at the podium, speaking about human rights violations at the U.S. detention center at Guantánamo Bay. One of our advance team members urged me to interrupt Dzhibladze, but the president waved off the idea. He sat near the podium and listened. Russians are not known for their brevity. He listened for a while.

By the time he got his chance to speak, Obama was energized. These were his people. Obama was a community organizer himself, an NGO activist. In fact, the cochair of the working group on community development was his former colleague Calvin Holmes, an African American leader from the Chicago Community Loan Fund.

At events earlier in the day, Obama had stuck closely to his scripts. At this event, he did not. He started with some Russian and a joke: "*Dobriy den.* I apologize that I think I'm running late and I'm leaving early. This is a good reason why civil society is so important — because you can't always count on politicians." Then he signaled why the engagement of civil society had to be a central component of the Reset:

> Through the work that you do, you underscore what I believe is a fundamental truth in the 21st century: that strong, vibrant nations include strong, vibrant civil societies . . . We not only need a "reset" button between the American and Russian government, but we need a fresh start between our societies — more dialogue, more listening, more cooperation in confronting common challenges. For history teaches us that real progress — whether it's economic or social or political — doesn't come

> from the top-down, it typically comes from the bottom-up. It comes from people, it comes from the grassroots — it comes from you.

Obama reminded the audience of his work as a community organizer in Chicago, describing that he too "had a lot of setbacks — in fact, we had more failures than successes." It was a tough-love message, and it had special meaning for the Russian participants at the summit. You had to listen to people's concerns, Obama recommended, and then act upon them. That was a perfect message for this audience. For too long, Russian NGOs had focused too much on abstract concepts of democracy and human rights, and not enough on the more immediate needs of their citizens. Obama was pushing them to connect with their communities.

But he also applauded their courage, arguing that there could be no progress for Russian citizens' concerns without the fight for stronger democracy: "Meeting these challenges, in turn, requires what many of you have dedicated your lives to sustaining — a vibrant civil society; the freedom of people to live as they choose, to speak their minds, to organize peacefully and to have a say in how they are governed; a free press to report the truth; confidence in the rule of law and the equal administration of justice; a government that's accountable and transparent."

He stressed that the values those in the audience were striving for are universal, and that "the United States of America will support them everywhere. That is our commitment. And that is our promise."[13] He ended his remarks with exuberant praise for their work; the place erupted in raucous applause.

Sitting in the front row, I turned a few times to see the crowd's reaction to Obama's remarks. I saw a sea of smiles, Russian and American alike. Some in the room were courageous activists whom I had known for twenty years. I was elated to have delivered the president to them, especially with this powerful message of solidarity.

The last official event on the president's schedule that day was a roundtable with Russian political opposition leaders. We had the representation of communists, liberals, and centrists. We invited no one from the "party of power," United Russia, since Obama had spent most of the previous day meeting with its leaders, Medvedev and Putin. We also decided not to include the nationalist Liberal

Democratic Party of Russia (LDPR), headed by Vladimir Zhirinovsky, out of concern that the extreme and flamboyant Zhirinovsky might dominate the entire discussion.

The roundtable discussion was lively, covering everything from the sorry state of democracy in Russia to missile defense. Obama listened intently to the participants' speeches, and ended the session with a blunt statement that they —not we—were the drivers of political change in Russia. Obama echoed his speech from earlier that day, stressing that the United States had an interest in seeing Russia become more democratic. But ultimately, he argued, we were going to be observers of that process, not players. I assumed that the room was sufficiently fitted out with listening devices that the Kremlin would hear his message as well.

My friend Boris Nemtsov assured me as we left the meeting that Obama had hit the right note: urging Russians to take charge of their own destiny and not wait for the Americans. Just two years later, Nemtsov and his allies would do exactly that, organizing the largest demonstrations in support of democracy in Russia since 1991.

After Obama's personal assistant, Reggie Love, entered the room for a third time with a note to the president to signal that time was up, the president thanked the attendees, stood for some photos on his way out, and ended the formal events of the Russia summit. As we left the venue, he thanked me for a great two days, and then hurried out, knowing he was late for dinner with his family. I was thrilled. If POTUS was pleased, I was pleased.

That night, the president dined on the rooftop of the Ritz-Carlton with the First Lady, his two daughters, and his mother-in-law, the scene perfectly captured in a photograph of the first family looking out onto the Kremlin and Red Square. I admired him for spending this time with his family, even if he got some heat from critics at home and in Russia for not spending more time with Russians.

A group of us from the White House celebrated at Tiflis, one of my favorite Georgian restaurants in Moscow. Someone worried that our presence at a Georgian restaurant could be perceived as an unfriendly political statement. If we could be so lucky! This was Moscow, not Washington. No one there knew who we were. No one took our photo.

But we did celebrate. The summit felt like a huge success. We'd got some agreements done. Obama had spent meaningful time with Medvedev, enough for us to assess that we could do real business with the Russian president. The long session with Putin also seemed to have gone as well as could be expected. And we practiced dual-track diplomacy, executing the Reset strategy just as it had been scripted a few months earlier. Obama had spent nearly an entire day meeting with nongovernmental groups and leaders. No American president, not even Ronald Reagan, had ever devoted so much time in Russia to students, businesspeople, civil society activists, and opposition leaders. Importantly, all of this presidential face time with nongovernmental leaders did not generate negative news stories in the Russian government-controlled press. The loudest critics, as best I could tell, were those within our own government, as well as a few voices in the think tank world back home, who worried that we were disparaging Medvedev in our outreach efforts. But the Russian president clearly was not offended. He wanted Reset to work. In 2009 it was possible to practice dual-track engagement.

McDonough, Rhodes, and I finished the evening with beers on the rooftop of our hotel, looking out at the lit-up Kremlin. We all agreed that the Reset was one of the biggest successes of our young administration. Of course, that was in part because we had little to show for ourselves in other foreign policy areas at the time! Nonetheless, we three shared a sense of real possibilities, real change, and real outcomes.

There was an added facet to my sense of satisfaction. On this trip, I was the lead for a delegation whose members I was still just getting to know, including the president. As an academic, I had spent most of my time sitting alone in front of a computer screen, writing. I'd been a solo artist. That night in Moscow, I felt firmly on a team, a winning team, and one that I really liked. Although it was clear, in the summer of 2009, that we had a lot of work to do moving forward, momentum seemed on our side.

9

NEW START

After the July summit, the race was on to get a new arms control treaty done before the existing treaty (START I) expired on December 5, 2009. Over the next several months, negotiations over the replacement treaty dominated my days at the White House.

Rose Gottemoeller, who had been sworn in as assistant secretary of state for the Bureau of Arms Control, Verification, and Compliance in April 2009, assembled a negotiating team that met frequently in Geneva with its Russian counterpart, headed by Deputy Foreign Minister Anatoly Antonov. Gottemoeller was an excellent choice to lead the U.S. team: knowledgeable about the issues, calm in demeanor, persistently optimistic, and a team player. She was also a true believer in arms control, so this was an assignment of a lifetime for her.

Antonov shared Rose's expertise on all things nuclear, but that was where their similarities ended. He was gruff, emotional, and difficult, never giving up anything without a fight or an in-kind trade. He seemed to be always worried about the hardliners back in Moscow, and therefore was extremely cautious. To make progress on any major issue in the treaty, we frequently had to go over his head, often to President Medvedev himself.

Back in Washington, my NSC colleague Gary Samore chaired the interagency team responsible for policy decisions on the treaty. The core of our Washington team consisted of Samore; myself; Jim Miller, then the Pentagon's principal deputy under secretary of defense for policy; Sandy Winnefeld, then vice chairman of the Joint Chiefs of Staff; and George Look, a senior director at the White

House charged with treaty-based nonproliferation efforts. Officials from the Department of Energy and the intelligence community also provided needed technical input and advice. Our team's job was to give "guidance" to Rose and her team in Geneva, which we did frequently via secure video teleconference. During crunch time for negotiations, we had two SVTCs a day, one at 8:00 a.m. Washington time and another at 2:00 or 3:00 p.m. Washington time, the end of the day in Geneva. It was an exhausting pace.

It often felt like we in Washington were holding Rose back, but maybe that was our job. As the one sitting across from the Russians, she was reaching out, trying to find solutions. On more than one occasion, Rose offered Antonov fallback positions that the interagency team had intended to be used only in exchange for Russian compromises. We in Washington were the conservative naysayers, always pulling back on the reins of our creative Geneva team. I may have been the worst offender, though my Pentagon colleagues did their share of saying *nyet* as well.

The Washington team's proximity to the Capitol several blocks down Pennsylvania Avenue partly explained our behavior. Always looming in the background of our deliberations was the U.S. Senate, and especially Arizona senator Jon Kyl. At the time, Kyl was the de facto leader of Senate Republicans on nuclear weapons matters. Whenever we considered a potential concession to the Russians, someone would say, "Kyl will never support that." Throughout the negotiations, we always assumed that we could not get the sixty-seven votes needed for Senate ratification without convincing Kyl of the treaty's merits. So anyone who wanted to kill an idea but did not want to confront Gottemoeller and her team directly would invoke Kyl's name. It often felt like he was sitting in the White House Situation Room with us.

I was not a "high priest of arms control" — a phrase that Tom Donilon coined to describe the experts on our team who knew the details of arms control and believed in the mission. But I quickly realized that the only way I could participate in the substance of these negotiations was to aspire to become one. As an undergraduate at Stanford, I had taken several courses on nuclear weapons and arms control, including one on U.S.-Soviet relations taught by Chip Blacker and Condi Rice. I remembered the basics, had all the books, and knew all the acronyms — ICBM, SLBM, MIRV, et cetera. But that was not enough. I didn't want to sit in these meetings every day only to chime in when some question about "the Russians" came up. Samore and Look patiently tutored me. George was especially gracious with his time. He *was* a high priest, having been part of the team that negotiated the original START treaty. Through their mentorship, I got to the point where I could express my own views on such obscure arms control topics as telemetry, type 1 and type 2 inspections, and "unique identifiers."

The main focus of negotiations concerned limits on nuclear warheads, specifically deployed nuclear warheads. In conversations with our team, Obama indicated he wanted to get to as low a number as possible without undermining our strategic deterrent. In Prague on April 5, 2009, Obama established an ambitious objective: "I state clearly and with conviction America's commitment to seek the peace and security of a world without nuclear weapons."[1] The Prague Agenda, as we came to call it, gave clear presidential guidance for our goals in the START deliberations: go as low as possible.

The second major issue in the treaty negotiations involved limits on delivery vehicles — that is, intercontinental ballistic missiles (ICBMs), submarine-launched ballistic missiles (SLBMs), and heavy bombers. We wanted two limits in the treaty, one for deployed delivery vehicles and one for nondeployed delivery vehicles. Until the very end, the Russians refused to include nondeployed delivery vehicles in the treaty text. We were not limiting nondeployed nuclear warheads, which both countries stockpiled by the thousands. So why try to limit nondeployed delivery vehicles, which would be very difficult to count and inspect? I didn't see the point of continuing to push for a limit on nondeployed delivery vehicles, but Rose was committed to the idea.

The third tough issue in these negotiations was missile defense. Securing some limit on missile-defense deployments was of the highest priority for the Russians. Conceptually, they had a point. Defenses and offenses are intertwined. We were way ahead of the Russians in developing accurate interceptors of warheads — though our Russian colleagues continued to grossly overestimate our capabilities. At the time of the new START negotiations, we had only thirty-four ground-based interceptors (GBIs) — our only missile capable of hitting a Russian nuclear warhead — deployed, in Alaska and California. (Interceptors target warheads, not missiles.) But the Russians worried that over time our technology would get better and our missile-defense deployments would expand. We could never protect ourselves from a Russian first strike — on that, we agreed — but the Russian military feared that we might have the capacity to defend the United States from a Russian second strike after we already had destroyed most of their missiles and bombers in a retaliatory first strike. In other words, we could "win" a nuclear war.

When I discussed this logic with President Obama on one occasion, I remember him rolling his eyes incredulously before declaring that no one can win a nuclear war. But such wild scenarios punctuated our negotiations with the Russians. In negotiations like these, both sides must focus on capabilities rather than intentions, and consider the worst-case assessments of each other's technological abilities. From the Russian vantage point, we were ahead not only in missile defense but in offensive capability, since most of our nuclear assets were deployed in submarines rather than in vulnerable fixed land silos. The Russians had not yet caught up with our missile-defense advances, so they wanted to use the treaty to constrain our advantage. It was a reasonable strategy —I would have adopted the same approach were I in their position. But politically, there was no way Obama could sign a treaty with missile-defense limits: it would never pass the Senate. So we argued and argued until the final days of negotiations.

Hundreds of other issues had to be wrestled to the ground before the treaty could be signed, but in addition to the limits on warheads, limits on delivery vehicles, and missile defense, three other lesser issues haunted our negotiations until the very end: the number of inspections, unique identifiers for each warhead, and telemetry.

The debate over inspections could be summed up as follows: we wanted more; they wanted less. Since the United States was not deploying any new delivery vehicles or nuclear warheads, Russian estimates regarding our stockpile were fairly accurate. The Russians, on the other hand, were developing new systems. Establishing an effective inspections regime was therefore our top priority. Ronald Reagan's famous quip, "Trust, but verify," became scripture for American arms controllers. George Shultz once gave me a magnifying glass inscribed with that phrase—I kept it on my desks at the White House and the U.S. Embassy in Moscow for my entire five years in government.

We argued equally intensely over telemetry provisions. The START I treaty mandated that our two countries exchange telemetry—the technical data generated during missile tests — so that both countries could be sure the other was not developing new types of missile systems capable of carrying more warheads. Telemetry was an essential component of START I's verification regime, but not critical to verifying the new treaty, since our new obtrusive inspections regime permitted each country to inspect actual warheads and missiles.[2] But again there was the U.S. Senate to consider. Even our supporters, Senator John Kerry and Senator Dick Lugar — chairman and ranking member of the Senate Foreign Relations Committee — advised President Obama that the new treaty could not be ratified unless it contained the same telemetry provisions as the existing treaty. I listened to the discussion skeptically. How many senators even knew what telemetry was? Could we not explain to them why we really didn't need it anymore? Surely Kyl would understand? The answer was no. Giving up on obtaining this data would be perceived as a huge concession.

We also pushed for unique identifiers, alphanumeric serial numbers that would help us keep track of ICBMs, SLBMs, and heavy bombers.[3] The Russians didn't feel the need for them; again, they already had a good sense of our arsenal. We needed them especially to track their mobile land missiles. We argued for a long time about that technical detail.

Gottemoeller and her team chipped away on the treaty details, consulting our interagency team on all the big moves. When her team reached an impasse with Antonov, we escalated. Sometimes that meant a phone call between Clinton and Lavrov. But over time, this channel helped only marginally, since Lavrov never really seemed to be engaged in the details of the treaty. I also realized that Lavrov did not have the ear of Medvedev, who decided early on to be personally involved in key decisions on the treaty. Eventually, I came to the conclusion that Lavrov did not respect Medvedev's foreign policy instincts. Medvedev was too accommodating of U.S. interests for Lavrov's tastes. I'm sure Lavrov was delighted when Putin returned to the Kremlin.

A more promising high-level channel for us on START emerged between my immediate boss, National Security Advisor Jim Jones, and Medvedev's foreign policy advisor, Sergei Prikhodko. Whenever we encountered gridlock in the interagency process about what to do next on START or any other Russia-related matter, Deputy National Security Advisor Ben Rhodes would joke: "Jones calls Prikhodko." General Jones did not seem to fit in well with the rest of the NSC team. He was twenty years older than most people working there, including Lippert, McDonough, and Rhodes, the troika from the campaign that remained close to Obama. (McDonough and Rhodes ended up staying the entire eight years with Obama at the White House.) Jones hadn't worked on the Obama campaign; he may have even voted for McCain.

Jones might have been a bad fit for Obama's White House, but he was a real asset for the Reset. He never really embraced his role as an advisor or staffer to the president, but thought of his position more as what's known in government as a "principal," a central policymaker and policy implementer. Other principals throughout our national security team fiercely guarded their AORs — areas of responsibility — and kept Jones out of their operations. But things were different with Russia, and with the START negotiations in particular. Clinton recognized early on that her counterpart would not engage on the details of the treaty, and Gates generally wanted nothing to do with the Russians, but Jones was itching to be in the arena. He liked dealing with the Russians and wanted to be useful to Obama on one of his top foreign policy priorities. And for their part, the Russians respected the six-foot-six former Marine Corps general and commander of U.S. forces in Europe as a serious interlocutor.

Whom Jones should call was at first tricky to say, since the Russians divided the equivalent of Jones's responsibilities between two positions. Sergei Prikhodko was Medvedev's advisor on foreign policy, but General Nikolai Patrushev was the secretary of the Security Council. When I outlined the pros and cons of interfacing with either Prikhodko or Patrushev — noting that the Bush team had had only minimal contact with the hardliner and former FSB head, General Patrushev — Jones recommended engaging both. Jones believed in the importance of developing personal relationships, and would not shy away from any phone call that we teed up for him or any trip we proposed. Prikhodko eventually emerged as our more reliable contact, but was not as fond of the telephone. He would sometimes avoid taking these calls, especially if he knew we were using them to move the ball on a difficult issue. And Prikhodko was always a man of few words. Yet, strangely, given their very different backgrounds and personalities, Jones and Prikhodko developed a rapport that helped us get START and other things done. Jones also gave Samore and me near complete autonomy in writing the talking points that Jones would deliver in these calls. That was a huge asset for our diplomacy.

For all the big moves, however, our closer was Obama. He became our lead negotiator on the New START treaty, speaking with Medvedev both on the phone and in meetings on the sidelines of multilateral gatherings. Obama wanted the assignment. Since college, he had developed a keen interest in nuclear weapons, and was therefore ready to invest the time to learn the issues and negotiate the difficult elements of the treaty. Over the next year, Samore and I submitted dozens of requests to the NSC's executive secretary to get time on Obama's schedule for calls with Medvedev; almost all were approved. We also pushed for an Obama-Medvedev meeting any time the two leaders were in the same city. Over time, White House Chief of Staff Rahm Emanuel would get tired of seeing Samore and me in the Oval Office or its waiting area, competing with the Obamacare negotiating team for the president's time. Emanuel even gave us nicknames — Cheech and Chong — though what a couple of national security guys had in common with those comedians, I'll never know. Ultimately, however, it was Obama who wanted to see us, Obama who wanted to make the necessary calls and take the meetings, Obama who wanted the leading roll in these negotiations.

In the fall of 2009 Obama and Medvedev discussed a variety of technical issues related to the strategic arms reduction treaty: during the United Nations General Assembly meeting in New York in September, the APEC summit in Singapore in November, and the Climate Change Conference in Copenhagen in December. I attended all three. We made progress in every session, but Copenhagen was most dramatic. Obama's primary mission in Copenhagen was to close a deal on climate change, but knowing that Medvedev and his top foreign policy advisors would be there, I suggested we bring both our White House and Geneva-based arms control teams along to tack on some treaty negotiations. Led by Jones and Prikhodko, our two delegations would spend a day hammering out the difficult issues, and then have our two presidents join for the last hour or two to finish the heavy lifting on any matters that remained unresolved. By the time we traveled to Copenhagen, it was already December; the old treaty had expired. I had high hopes for this meeting.

A month earlier, Jones and our core START team had traveled to Moscow to float our final offers on some of the big limits on warheads and delivery vehicles. The trip was an adventure. At Jones's request, our team flew to Moscow in a C-17 military cargo plane to which they added a trailer-like pod nicknamed the Silver Bullet. It slept four comfortably, but with five senior officials in our delegation, we were above capacity. So after chatting and drinking for a while inside the pod, I volunteered to exit the warm, comfy, quiet trailer to sleep outside, stuffed in a sleeping bag in an uncomfortable chair, wearing earplugs and surrounded by soldiers. I froze, and got little sleep, but I got to witness an in-air refueling. Ah, the glories of White House work!

The results in Moscow made the difficult trip worthwhile. During an elegant dinner at Spaso House hosted by Ambassador John Beyrle, we dug into the details of our comprehensive proposal with our Russian counterparts. I felt like we were making real progress, and had positioned ourselves to secure agreement on some of the more contentious issues in Copenhagen the next month. At the VIP airport lounge en route back, we were all in a festive mood; Samore offered to buy a round of vodka to toast our team's successes. Then he got the $270 bill for nine shots and quickly became a little less festive. This was the new Russia.

A few weeks later, we flew overnight on *Air Force One* to Copenhagen for a single day of negotiations. As we were landing, the plane's medical staff offered us some kind of stimulant to help keep us awake for the day, while also warning us not to take the drug if we planned on competing in a marathon anytime soon. I had no big races on my agenda, so I took the drug, and was amazed at how well it worked. I also marveled at the fact that I could get this drug on *Air Force One,* but probably not from my doctor back home.

Our negotiating room, tucked within the stadium where the climate-change summit was being held, was a clothing store that had been modified for our use. Naked mannequins observed our deliberations from behind a makeshift curtain. Prikhodko, Deputy Foreign Minister Ryabkov, and Antonov represented the Russian side, and Jones, Samore, Gottemoeller, and myself, the American side. Our aim for the daylong session was to resolve the technical issues such as unique identifiers, telemetry exchanges, and the number of inspections. We wanted to save the discussion regarding limits on warheads and delivery vehicles for the Obama-Medvedev meeting later that afternoon.

We made progress, but not nearly as much as I had hoped. My goal had been to fly home that night with all the big issues resolved — a draft treaty in hand. Our Russian counterparts clearly did not have the authority to make that happen. At one point, I grew impatient, pushing hard on our Russian colleagues to agree to unique identifiers, especially on mobile land missiles, which they had and we did not. Giving a serial number to each missile and heavy bomber seemed like an obvious, trivial move. Why were they resisting so defiantly?

Antonov in turn grew impatient with me. He was used to dealing with the more composed Gottemoeller. My academic style — an appeal for rationality, and little tolerance for illogical arguments — grated on him, so much so that at one point, he stood up dramatically and threatened to leave the negotiations. Jones talked him down, and I decided to keep quiet for the rest of the day. I knew we weren't going to get this treaty done that afternoon with these negotiators. We were going to have to go to Medvedev.

Obama and Medvedev did make some real progress in their meeting, including on unique identifiers and the number of inspections. We also got very close to agreeing to a limit on warheads. However, we left Copenhagen with lots of work remaining, and no draft treaty in hand. As we boarded *Air Force One* for home that evening, I felt like we had failed. The year was ending, and we did not have a new treaty ready to replace the now expired START. Mike Froman, then deputy national security advisor for international economic affairs, comforted me by joking that others — specifically the climate team on which he served at the time — had also not achieved all that they wanted in Copenhagen, but in their case the whole world had been watching. He had a point. Only the mannequins had witnessed our setbacks.

Once we'd reached cruising altitude, Obama drifted to the midsection of the plane, where staff were seated, and paused when he got to me. I started to get up, but he gestured for me to stay seated and kneeled in the aisle. When I apologized for our lack of progress that day, he brushed it off, said we were getting closer, and let me know he felt comfortable living with a lapse between treaties if it meant a better deal in the end. In negotiations, Obama knew, patience is a virtue.

In the first days of 2010, Samore and I proposed a new scheme to get to yes on the strategic arms treaty: send half of the U.S. national security team to Moscow in January for a final push on all outstanding issues. We knew that the Russian Ministry of Defense was a key decision maker on the treaty; specifically, we thought that the chief of the general staff, General Makarov, was the pivotal player, rather than the minister of defense, Anatoliy Serdyukov. Our idea was to get everyone who mattered in the same room. Our delegation was so big that we flew two "Mil Air" planes (shorthand for government aircraft) to Moscow — one for the chairman of the Joint Chiefs of Staff, Admiral Mullen, and his entourage, and another for General Jones and our NSC posse.

The meeting took place at the Ministry of Defense (MOD), not the Ministry of Foreign Affairs (MFA). Makarov and Mullen were the cochairs of this meeting, relegating Jones to a secondary role in the deliberations. Jones's counterpart, Prikhodko, was not even there. Makarov was clearly the most important official on the Russian side. I noticed that Russia's chief negotiator from the MFA, Antonov, sat at the far end of the table, several soldiers down from Makarov. Both the meeting's location and the seating chart confirmed that the MOD was in charge of this event

We got off to a bad start, arriving quickly at an awkward standoff on the number of on-site inspections. I reminded the group that our presidents already had agreed to the number of inspections during their meeting in Copenhagen; Antonov remembered the same exchange differently, and immediately challenged my recollection. There I was, clashing with Antonov again. We agreed to table the issue; Obama and Medvedev would have to resolve it. We then turned to unique identifiers. The Russians again provided what I thought was a revisionist interpretation of the discussion in Copenhagen.[4] At one point, I demanded to speak even though it was not my place, and some of Mullen's staff let me know it later on. I hoped appealing to logic would help move this point forward. It didn't seem to work.

During a break, however, Makarov motioned that he wanted to speak to me directly about unique identifiers. I knew I'd be breaking protocol in speaking directly to him without others in our negotiation team present, but I also believed the formalities at the table had complicated our negotiations and that I might be able to explain our position more clearly and directly in Russian. He listened intently, and seemed to respond to my arguments. Makarov was a straight shooter; no posturing needed. That conversation restored my faith in our ability to get a deal. Most importantly, he appeared to have the same instructions from his commander in chief that we had from ours: get it done.

Later that day, we ceded some ground on inspections, and the Russian team agreed to unique identifiers. We circled around a satisfactory compromise on telemetry and crept closer to mutually agreeable numbers on the limits on warheads and delivery vehicles. Gottemoeller was still pushing for a limit on nondeployed delivery vehicles, but I believed that we could live without it. We were so satisfied with our progress at the end of the day that Jones decided to call Obama to let him know we had reached agreement on the basic parameters of a new treaty. We left Russia the next day thinking we had made some history.

Our optimism evaporated when one major issue we thought had been resolved surfaced again: missile defense. From the very earliest days of our administration, our national security team had been telling our Russian colleagues that we would not allow any limits on missile defense in the new treaty. When Obama and Medvedev signed the framework agreement for the new treaty in Moscow in July 2009, we considered the issue settled. We had succeeded in keeping substantive language about missile defense out of all joint statements, apart from the banal, factual observation that a relationship exists between offensive and defensive weapons. At our minisummit in Moscow in January, Makarov did not broach the issue. But after Moscow, Antonov kept pushing hard for some constraints on U.S. missile-defense deployments with our team in Geneva. We held firm. Even the slightest hint of constraints would torpedo Senate ratification, our legislative team at the White House counseled. We thought it was a done deal.

It wasn't. On the morning of February 24, 2010, Samore and I drifted over to the West Wing to prep the president for a phone call with Medvedev. On my way to the Oval Office, I stopped by the White House Situation Room to pick up cables, but didn't have time to read them. I was focused on what I was going to say to the president, rehearsing my arguments in my head. I considered these pre-briefs with him to be my most important responsibility. The cables could wait.

The pre-brief went smoothly and the call began positively, like most Obama-Medvedev calls back then, until suddenly Medvedev began thanking Obama for the concession that we'd made on missile defense. Obama looked puzzled. He moved the phone away from his mouth and shrugged, as if to say, "What is he talking about?" I shrugged back. We didn't know.

But as the conversation proceeded, I felt that I was witnessing the development of a giant misunderstanding with no easy path for resolution. I finally looked at the cables in my lap. Right there in the pile was a cable from our team in Geneva, warning us that the Russian side might have misinterpreted an informal exchange on missile defense as a U.S. policy change. Based on this new information, I scribbled out in my third-grade handwriting a few new talking points for the president, rushed over to his desk, and showed him my notepad. Obama had to push back.

The call got tougher from there. No matter what the Russians thought they had heard about new concessions from someone else, Obama's position had not changed, he affirmed: we were not going to allow missile defense to creep into this treaty at this late stage. Obama told Medvedev that if our position on missile defense were a deal breaker, he'd rather scrap the treaty. Medvedev sounded surprised, disappointed, and even a little angry — an emotion I had rarely heard from him. Obama stood firm, stressing that the U.S. position on missile defense was nonnegotiable. Medvedev ended the call ambiguously, hinting that we might indeed have reached an impasse; that maybe our positions could not be reconciled. A treaty that an hour earlier had seemed a done deal was now in serious jeopardy.

After Obama hung up the phone, I saw him genuinely angry for the first time. Already late to his next meeting, Obama marched out, but not before turning to Donilon, Samore, and me to say as I recall, "People have to be held responsible for their actions." In other words, we had screwed this up; we were accountable. Later, when I mused to Donilon that the kind of outburst we'd witnessed must happen often, Tom informed me otherwise. That was the first time he had ever seen the president so worked up.

As an added sting, Obama's terrifically talented photographer, Pete Souza, had snapped a shot of me leaning over the president while I was showing him my newly scribbled talking points. As he did with all his favorite photos, Souza later blew it up into a "jumbo" and hung it on a West Wing wall for several weeks. It was a cool photo, and I got all sorts of compliments on it, but few knew the full story. My worst day of work at the White House was forever memorialized in this jumbo.

After that tense phone call, we waited. Medvedev was probably waiting as well, hoping that we might reconsider. Obama, however, was clear in his instructions to our team. We were not moving. Eventually, the Russians reengaged. Medvedev realized that Obama was not bluffing, and decided to move forward with the treaty without the missile-defense constraints. They knew that our missile-defense capabilities were not going to expand significantly enough over the duration of the New START treaty to undermine mutual assured destruction. Antonov received new instructions from Moscow, and we resolved the missile-defense standoff. They would postpone that fight for another day.

In New START's preamble, we allowed the following innocuous phrase: "Recognizing the existence of the interrelationship between strategic offensive arms and strategic defensive arms, that this interrelationship will become more important as strategic nuclear arms are reduced, and that current strategic defensive arms do not undermine the viability and effectiveness of the strategic offensive arms of the Parties."[5] And then we agreed to disagree. Each government issued nonbinding unilateral statements on the relationship between missile-defense deployments and the treaty. The Russians warned that future U.S. deployments could undermine the treaty, while we issued a statement in response recording the U.S. position that there was no relationship between our planned missile defenses and the integrity of the treaty.

And we left it at that. This issuing of unilateral contradictory statements seemed like a strange practice to me, but our resident arms control high priest, George Look, reminded our team that such side notes had been essential to getting to yes on the original START treaty two decades earlier.

With the missile-defense standoff resolved, other elements of the treaty fell into place. Following months of prodding from Obama, Medvedev finally agreed to add a chapter to the treaty on telemetry. Obama raised the issue so many times that it became a kind of inside joke between the two leaders: both were sure that none of their predecessors had uttered that obscure word so many times. We also got a provision on unique identifiers into the treaty, and Obama made one final push for twenty annual on-site inspections, but ultimately settled for eighteen. On one of his last phone calls with Medvedev on the treaty, Obama tried to reduce the upper limit on each side's deployed warheads by 50, chiding Medvedev in a friendly way to try to explain the difference between 1,550 and 1,500 warheads, since both numbers were capable of blowing up the planet. But Medvedev held firm, and we acquiesced to 1,550. That number was still 30 percent lower than the limit in the 2002 Moscow Treaty, and 74 percent lower than the limit in the original START treaty — no small achievement.[6] The final treaty also secured 700 as the limit for deployed strategic delivery vehicles:

the Pentagon was not ready to go lower. And finally, amazingly to me, Gottemoeller managed to get an upper limit of 800 on launchers, both deployed and nondeployed, to accompany the treaty's limit on strategic delivery vehicles. She was patient and persistent. I was impressed.

On March 26, 2010, we arranged a congratulatory phone call between Obama and Medvedev, and then announced to the world that New START was done and ready for signing a few days later, on April 8, at an official ceremony. After congratulating our team, Obama headed to the press briefing, flanked by Secretary of State Clinton, Secretary of Defense Gates, and Joint Chiefs chairman Mullen, to begin the long and arduous process of getting the treaty ratified by the U.S. Senate.

In my mind, Senate ratification was a battle for another day. That afternoon, in my office, we were going to celebrate. General Jones ventured over from the West Wing to join our party in the EEOB. We had just finished a big piece of work, closed out a big account. It felt great — as a concrete achievement, and even more, for me, as a team effort.

At the time of New START's completion, our administration had very few foreign policy wins, so we purposely played up the importance of the accord. We highlighted the dramatic reductions in deployed weapons, and trumpeted the more obtrusive kind of inspections that this new treaty allowed — echoing Reagan's "Trust, but verify" dictum. We reminded our press corps that this was the first arms control treaty in a decade — the first *comprehensive* treaty in two decades (the 2002 Moscow Treaty was just a three-page document). If ratified, our treaty would be the first major arms control agreement negotiated and entered into force during a Democratic administration. We also argued that completion of this treaty would aid our nonproliferation work in other areas of the world. Since we were now making progress toward our commitment in the Nuclear Non-Proliferation Treaty (NPT) to reduce our nuclear stockpile, we could press more successfully for other nonnuclear signatories to meet their commitments, including first and foremost Iran.

Our efforts generated some good press. The *Washington Post* reported, "Experts from the right and the left agree the treaty extends a verification plan that has allowed the world's two nuclear giants to maintain stability that has existed for the past 20 years."[7] The *Financial Times* dubbed Obama "a president with an endgame," someone committed to getting things done.[8]

To underscore the importance of the treaty and generate momentum for Senate ratification, we organized a fancy signing ceremony. At the White House, we debated the venue. I suggested Reykjavik, the Icelandic capital where Reagan and Gorbachev in 1986 had discussed radical cuts in the U.S. and Russian nuclear arsenals. Others worried, however, that the memory of the failure in Reykjavik would not send the right message. Someone else pushed for Geneva, which I opposed — too boring, and way too reminiscent of the Cold War. We settled on Prague, Rhodes's idea. In addition to underscoring Obama's Prague Agenda, the Czech capital was a testament to the changes in the international system that had occurred since the end of the Cold War. Prague was no longer the capital of a Warsaw Pact country, but a NATO ally. For this reason, some Russian government officials scorned the choice, but Medvedev ultimately agreed with our proposal, and so the president returned to Prague for the second time in two years. I was delighted to accompany him, along with Gary Samore.

At the signing ceremony at Prague Castle, it was clear that the U.S.-Russia Reset had borne fruit. Obama said he was honored to be in Prague to "mark this historic completion of the New START treaty," while Medvedev called it "a truly historic event." Stealing a phrase often used by Obama in their private talks, Medvedev ended his remarks in English, describing the treaty as a "win-win" outcome.

The end of negotiations with the Russians on New START marked the beginning of negotiations on the treaty with the U.S. Senate. Throughout the deliberations with the Russians, we had kept key senators apprised of our progress. Senators Lugar and Kerry came to the White House a few times to receive updates on the status of negotiations, and a congressional delegation had traveled to Geneva. But now the preseason was over. It was time to focus our complete attention on getting the sixty-seven signatures needed for ratification.

I assumed ratification would be easy. Who wouldn't support parallel cuts in our nuclear arsenals? Who wouldn't see the value of our new inspections regime; that is, who wouldn't want to gather intelligence on Russia's nuclear arsenal? Who wouldn't understand how the ratification of this treaty would help us address the nuclear threat gaining steam in Iran? I was wrong. We had to fight hard to get the votes.

Our first strategy for winning ratification focused on Senator Kyl. Senator Lugar had been more deeply involved in arms control over the decades, but as a moderate Republican, he had no pull within the party. Kyl was key. If we got him, everyone else would fall in line. Initially our administration engaged Kyl in an implicit trade: his support for our treaty in return for our support for modernizing our nuclear weapons. Specifically, Kyl wanted us to add an additional $10 billion to the $80 billion appropriation for nuclear weapons modernization already included in Obama's 2011 budget proposal.[9] These numbers seemed outrageous to me, but we eventually gave Kyl what he wanted (although we could not guarantee that Congress would support these budgetary allocations for ten years). At the time, analysts thought a deal was in the works. A November 2010 *New Republic* article described Kyl as the agreement's "unlikely savior."[10]

In conversations with the president on our ratification strategy, I could tell Obama was skeptical. Our approach was too rational. It left out the politics. "Kyl's never going to back me on anything," I remember Obama saying during one conversation in the Oval Office. In the early stages, however, our approach seemed to be working. We kept allowing the price tag for modernization to inflate. And on the substance of the treaty, we seemed to be winning the argument. We emphasized that the new limits allowed us to keep our triad of sea-, land-, and air-based delivery vehicles, which was important to some senators. We assured senators in Montana and North Dakota that our planned cuts would not lead to the closure of one of the ICBM bases in their states. We hammered home the importance of ratifying the treaty as soon as possible to get inspections going again, and trumpeted how we had successfully avoided limits on missile defense. We had a good story to tell.

On November 23, 2010, however, Senator Kyl circulated a memo to his colleagues in the Senate, explaining why he could not support ratification that year.[11] It was a bombshell. "There remain a few substantial concerns about the adequacy of the proposed budget," the memo read. I happened to be in McDonough's office when the news popped up on his computer screen. He had a two-word response: "Those ratfuckers."

We had to change our strategy. Rather than focusing quietly on one key senator capable of bringing others along, we shifted toward a more public campaign as a means for putting pressure on and winning the support of individual senators. And so, we courted validators. We invited George Shultz, Henry Kissinger, Bill Perry, and Sam Nunn to the White House to secure their backing. This group of former statesmen had earned the nickname "the Four Horsemen" after publishing a seminal op-ed — and op-eds are rarely seminal — in the *Wall Street Journal* in 2007 calling for the abolition of nuclear weapons.[12] After meeting with the president and then watching a film about their disarmament efforts with Obama sitting in the audience in the White House theater, they made a very strong statement in support of the treaty. Shultz was particularly forceful.

The endorsement of the treaty by Ronald Reagan's secretary of state was not enough to get us to sixty-seven. So we recruited more validators. Former secretaries of state James Baker and Madeleine Albright, former national security advisor Brent Scowcroft, former secretary of defense William Cohen, and several others joined the president for a meeting in the Roosevelt Room, after which they all came out in favor of the treaty. Biden and Clinton then went to work on individual senators. Because the vice president had logged three decades in the Senate before moving over to the White House, Obama asked him personally to lead the final push.[13] Aware of the weight of military endorsements, Biden involved our military colleagues in our lobbying offensive, knowing that they might have more sway with conservative Republican senators than we arms-control-loving liberals at the White House.

To secure the last votes, President Obama acquiesced to send one more letter of assurance to the U.S. Senate on missile defense. In his December 19, 2010, letter, Obama reaffirmed that the new treaty would not prevent the United States from developing a missile-defense system in Europe.[14] I could tell that the president did not want to send this letter, but our legislative experts advised that this final act was required in order to ratify the treaty before the end of the year. I was surprised at how hard this work turned out to be; our politics had become very polarized even on issues of national security.

On December 22, the last day of the session of that U.S. Congress, I joined the vice president's motorcade for a drive down to the Capitol. I wanted to witness this piece of history. From the balcony, we watched every senator vote. By that day, we knew we had sixty-seven votes, but hoped for more, as a few senators' decisions remained unknown to us. For instance, we still thought McCain might vote in favor. After his name was called, he walked down the aisle to the front of the chamber, lifted his hand with his thumb up, and then dramatically turned it down. But McCain couldn't stop us. Thirteen Republicans joined all the Senate Democrats in voting for the treaty, giving us seventy-one votes.[15] It was not the ninety-five that we might have hoped for, but it was enough for ratification. Biden and our negotiating team exchanged high fives. We had just closed out an account — a really big account.

Rhodes and I had promised each other that we would celebrate New START's ratification over vodka at the Russia House on Dupont Circle, where Washington Capitals hockey star Alex Ovechkin was said to hang out. That afternoon, however, Rhodes called me to say he couldn't go because he was traveling with the president to Hawaii that night for the Christmas holiday. So instead, we did a few symbolic shots in his office that afternoon — my first and only time drinking hard liquor in the White House.

An hour later, I got a call from the president's secretary asking me to come see the president. At first, I panicked. I had just knocked back two vodka shots with Rhodes and now I had to go brief the president? But then I learned it was a presidential request to do more drinking. The New START team assembled in the Oval Office to sip champagne and celebrate ratification. When Obama praised us for our work with the Russians, I jumped in to add, "You, Mr. President, were the real lead negotiator for the New START treaty." He replied, as I recall, "Yeah, I did a little staff work on that one." We all laughed as he flashed his famous broad smile. At that point in his tenure as president, the new treaty might have been his greatest foreign policy achievement.

After our Oval celebration, he climbed aboard *Marine One*, and headed out for his annual two-week Christmas vacation in Hawaii. I went back to my office and packed up all the New START files. No more discussions of telemetry; no more deliberations about type 1 versus type 2 inspections; no more arguments about hidden clauses in the treaty constraining our missile-defense deployments. We were done.

I hoped the new treaty would give us momentum heading into the new year. With respect to issues in U.S.-Russia relations, 2011 was going to be even harder than 2010, in part because of our accomplishments that year. When you check off something on your list, you can usually bank on the next item on the list being harder; otherwise, you would have tackled it sooner. For U.S.-Russia relations in 2011, that meant missile-defense cooperation and WTO accession. Or at least that's what I thought in December 2010. I had no idea that the biggest issues in 2011 would be popular uprisings in Egypt, Libya, Syria, and, eventually, Russia.

10

DENYING IRAN THE BOMB

Soon after starting work at the White House, I came to understand that preventing Iran from obtaining a nuclear weapon ranked as President Obama's top foreign policy priority. That meant it was also my number one priority. To achieve that objective, we quickly developed a strategy for reaching out to Iran's leaders (the diplomatic track), to be followed by sanctions (the pressure track) if engagement failed. Russia would play a critical role in the pursuit of both strategies.

In the early months of 2009 we vigorously but quietly pursued engagement with the ayatollahs, including Ali Khamenei, the supreme leader himself. During the 2008 presidential campaign, Senator Obama asserted boldly that he would be ready to meet and talk with anyone, including the theocrats in Tehran, if engagement advanced U.S. national interests, adding, "The notion that somehow not talking to countries is punishment to them . . . is ridiculous."[1] In his inaugural address, President Obama reiterated his commitment to engage adversarial, autocratic regimes.

Over time, this approach to diplomacy would produce the Reset with Russia, rapprochement with Burma, and the reestablishment of diplomatic relations with Cuba. But when I heard Obama deliver that line at his inauguration, I knew the new president had in mind Iran, first and foremost. Only weeks after his inauguration, Obama sent a letter to Ayatollah Khamenei outlining his new vision for relations between the U.S. and Iran.[2] In March, Obama taped a respectful message to the Iranian people on the occasion of Nowruz, the Persian New Year, something none of his predecessors had done. He was serious about extending an open hand to Iranians.

To translate aspirational rhetoric into policy, the U.S. administration — together with the International Atomic Energy Agency — developed the Tehran Research Reactor proposal. The idea was straightforward: Iran had a research reactor in Tehran that was running out of fuel. Back in 2006 the Iranian government also had started to enrich uranium.[3] The Iranians claimed that they needed to enrich uranium to replenish the TRR and, eventually, to fuel their Russian-built nuclear reactor under construction at Natanz. This same low-enriched uranium, however, could also be converted into weapons-grade uranium for use in a bomb. So our nonproliferation team — Gary Samore and Rexon Ryu at the National Security Council and Bob Einhorn at the State Department — together with IAEA director General Mohamed ElBaradei and his associates, came up with a creative scheme. Under the deal they proposed, the Iranian government would ship to Russia its low-enriched uranium. Russia would enrich this uranium to higher levels before sending it to France to fabricate into fuel rods for use in the TRR.[4] These fuel elements could not be repurposed for use in a nuclear weapon.

I thought the proposal was brilliant. The deal would reduce the stockpile of enriched uranium held by Iran, thereby increasing the time it would take Iran to produce a nuclear weapon. Suspending enrichment processes already under way was not part of the discussions at this stage; we recognized that Iran would eventually be able to replenish its stockpiles. Our objective at this point was to buy time for a more comprehensive agreement. We also were testing Iran's intentions, assessing whether the regime would commit to comprehensive talks with us regarding its nuclear weapons program.

Under Secretary of State Bill Burns first discussed the TRR idea quietly with Russian deputy foreign minister Sergey Ryabkov. Once we knew the Russians were interested, Obama pressed Medvedev hard to own this proposal with us. We believed the Iranian government would be more likely to accept the scheme if it came from Russia. Moscow had close ties with Tehran; we didn't even have an embassy there. Russia also had the technical capabilities to make the proposal work. Medvedev told Obama that Russia had no interest in Iran obtaining a nuclear weapon. The Russians did, however, have a real interest in Iran developing its nuclear energy program, as Rosatom, Russia's nuclear energy agency, wanted to build additional nuclear power plants in Iran. We could live with that. Our shared interest in preventing Iran from acquiring a bomb allowed us to cooperate as negotiating partners.

Our brilliant plan left out one key player, however: the Iranians. They were not eager to negotiate. The Iranians had a reputation in our government for stretching out every negotiation, even on the tiniest of issues. True to form, they dragged out the negotiations on this fuel-swap proposal for months. The supreme leader, Khamenei, rebuffed Obama's overtures as insincere. He warned that pursuing negotiations with the United States would be "naive and perverted," and declared that Iranian politicians should not be "deceived" into starting such talks.[5] With that opinion from the supreme leader, our team viewed hardline president Mahmoud Ahmadinejad as the Iranian leader most likely to support our talks, which underscores just how long the odds of success were.

Watching from the sidelines at my NSC perch, I was amazed at the patience of our negotiators. Diplomats are trained to engage and deliberate — that's what they do. I thought we needed to end these talks and pivot to the economic sanctions more quickly. But I eventually came to embrace the logic of playing out the engagement track first as a means to gain international support — most importantly, support from Russia — for future robust sanctions later on. We wanted to give Russia a prominent role in the authorship of the TRR proposal, so that if the Iranians rejected it, they would be rebuking not just the evil United States but one of their closest allies. With time, our TRR fuel-swap idea became known in Tehran as the *Russian* proposal. Our strategy worked.

This "Russia in the lead" approach was not always well received by some members of our administration and some allies. But Obama supported the strategy, and tried to cultivate Medvedev as his close ally in dealing with the Iranian problem. China, for all of its alleged aspirations to become a world superpower, was happy to let Russia play the role of our lead partner. We also convinced our European allies that this Russia-first approach served all of our interests, even if some of our friends in Paris and London were annoyed that we sometimes talked to the Russians about Iranian matters in bilateral channels. Together, in the era of the Reset, we believed that the United States and Russia could take on even this most difficult security issue of our time. We understood the low probabilities for success, but reasoned that we were more likely to achieve our objectives regarding Iran with Russia on our side.

Our strategy for preventing Iran from developing nuclear weapons focused solely on government-to-government channels; our TRR proposal was a pitch to the ayatollahs ruling Iran. Then, out of the blue, another player appeared on the scene during the summer of 2009 and disrupted our diplomacy. That player was the Iranian people. In fact, over the next five years, protesters pushing for democracy — in Iran, Egypt, Libya, Syria, Russia, and Ukraine — would complicate and eventually erode our reset with the Russian government.

On June 12, 2009, Iranians went to the polls to select a president. Iranian elections are not free and fair, but they can sometimes be competitive, and this election was competitive. The official tally asserted that the incumbent, Mahmoud Ahmadinejad, won, but the main opposition leader, Mir-Hossein Mousavi, and his supporters claimed to uncover massive voter fraud. Two days later, Ahmadinejad gave a victory speech, and Iran's supreme leader blessed the outcome of the election. In response, massive public protests erupted in what became known as the Green Movement, for the color associated with the Mousavi campaign. The Ahmadinejad regime responded violently, killing several people, which in turn sparked even bigger protests and then more violence from the regime, leading to the deaths of dozens of innocent protesters and the arrest of hundreds more.[6] Mousavi and other leaders were placed under house arrest.[7]

Our administration did not cause or sponsor the Green Movement in Iran, though some critics suggested as much. On the contrary, we were trying to secure a deal with the Iranian regime on nuclear weapons, and the Green Movement protesters were distracting the Iranian government officials we sought to engage. Some analysts, both inside and outside the U.S. government, predicted that the protest movement would strengthen the hardliners inside the Iranian regime — the same group most hostile to a nuclear deal.

I had a different view. As I had written with my Stanford colleague Abbas Milani, I believed that only the emergence of a democratic regime in Iran would eliminate for good the Iranian threat, including the nuclear threat.[8] Arms controllers didn't end the Cold War with the Soviet Union; democrats inside Russia and the other Soviet republics did. I maintained the same prediction about Iran.

Few of my colleagues, however, were interested in my views on this subject. I was the Russia guy; Iran was not my AOR. I pushed in the margins when possible, advocating for public statements of solidarity with the Iranian protesters. Ben Rhodes was an ally, and added supportive, albeit cautious, phrases into Obama's public remarks.

Although I pushed for us to say more at the time, I also appreciated what some State Department colleagues, Iran experts, and many Iranian American friends told me at the time: that too much engagement, too much association with the United States, would discredit the Iranian opposition movement. Members of the opposition did not want visits to the White House or photo ops with American diplomats; such contacts would taint them. Yet I still felt we were saying and doing too little. I defended our policy with journalists and critics, one night engaging in a "frank exchange of views"—to use the diplospeak I would later master as a U.S. ambassador—with *Washington Post* columnist Bob Kagan and *New York Times* columnist David Brooks on the merits of our approach. I lost that exchange, in part because I did not truly believe we had the correct approach. That was a first for me: defending a policy I did not fully support. But that's part of working in government. I didn't like the feeling.

After the regime killed innocent protesters and arrested opposition leaders, I felt even more strongly that we had to condemn these acts of violence. As Ronald Reagan had done with the Soviets, we could work on an arms control agreement with Iran without checking our values at the door. Obama agreed and issued a statement along these lines at a June 15 news conference: "I am deeply troubled by the violence that I've been seeing on television. I think that the democratic process—free speech, the ability of people to peacefully dissent—all those are universal values and need to be respected."[9] The next week, Obama used even stronger language while discussing the events in Iran: "The United States and the international community have been appalled and outraged by the threats, beatings and imprisonments of the last few days."[10] At the same press conference, he lamented the tragic murder of Neda Agha-Soltan, a young protester whose assassination was captured on video and viewed by millions of people around the world. These statements aside, American security interests and values were clashing; and in this round, security trumped values. No senior foreign policy official within our government believed that we had an effective way of influencing political events inside Iran. Because no plan for democracy promotion was ever advanced, our arms controllers could remain focused on their negotiations on the fuel swap. Stopping Iran from acquiring a nuclear weapon was the higher goal.

By the end of the summer of 2009, a consensus emerged within our government around the need to change our strategy from engagement to pressure. The Bush administration had achieved a major diplomatic victory in 2006 by securing support for UN Security Council Resolution 1737, which imposed sanctions on Iran in response to its continued uranium-enrichment program. We now had to up the ante. Russian support was going to be key, since new sanctions against Iran would mean losses in revenue from Russian exports. In particular, we wanted new sanctions prohibiting the sale of heavy weapons to Iran. Those weapons all came from Moscow.

Our courtship of Medvedev for support for new sanctions began with a meeting between Obama and the Russian president on September 23, 2009, at the Waldorf Astoria Hotel in New York during the UN General Assembly meeting. At that meeting, Obama informed Medvedev about Iran's construction of a second nuclear-enrichment facility near the holy city of Qom. Obama was sharing this intelligence with Medvedev for the first time. Obama argued that Iran's secrecy about this underground plant clearly proved that the Iranians were not enriching uranium for peaceful purposes. They were lying to us and to the entire world; we had to respond. Obama did not insist right then that we cease negotiations with the Iranians over the TRR deal, but he did recommend that the U.S. and Russia begin preparations to implement new sanctions.

As had become our tradition with Medvedev, our delegations at this meeting were small, just three per side. While Obama and Medvedev were talking one-on-one following the meeting, Lavrov came up to me and asked why we hadn't told them about Qom before. I replied, "I thought you knew." Throughout the interaction, Lavrov kept an absolutely straight face, always a talent of his. To this day, I don't know if the Russians really were surprised about the secret nuclear facility, or if they knew but didn't want us to know they knew. But sharing this news with Medvedev did create the desired effect. He expressed disappointment with the Iranians. As we had hoped, Medvedev felt like their Iranian partners had deceived and disrespected the Kremlin. Medvedev did not defend Iran, nor did he assume the role of lawyer, arguing that Iran had a legal right to this and that (as other Russians had done in the past and would do again). He was not ready to discuss sanctions in detail that day, but hinted that he was open to the issue going forward.

After the meeting, Obama stated before the press that "serious additional sanctions" remained a possibility if Iran did not respond adequately to our offer of negotiations.[11] In his remarks, Medvedev clarified the Russian position on sanctions: "Sanctions rarely lead to productive results. But in some cases sanctions are inevitable."[12] We were as thrilled with this statement as Tehran must have been threatened by it. On the most important security challenge of our time, we were now working closely with Russia. Perhaps most importantly, Medvedev's remark demonstrated his desire to work with us, even on an issue that required him to take actions unpopular at home.

Satisfied we were on the right track, I hastened to make my train back to Washington so I could be home in time for my son's birthday party. I was already a day late — his birthday fell on the same day as the Obama-Medvedev meeting. His grandparents had traveled from Montana, and the celebration had been delayed so that I could be there. I already felt bad that I had missed his actual birthday, but I knew that one day late was not the end of the world. And then I got a call on my BlackBerry from the White House Situation Room. It was General Jones, informing me that news was about to break about the Qom facility — that is, we were breaking it — and now was the time to make a major push to get the Russians on board with Iran sanctions. Jones wanted a group of us from the NSC to travel to the G-20 summit in Pittsburgh in order to draft a joint statement with the UN Security Council's five permanent member states (the P5) about next steps on Iran. Since Russia's participation was critical, I was needed in Pittsburgh.

I was upset. I wanted to get home for my son's birthday party. Couldn't it wait? Do these statements really matter? I understood the gravity of the moment, but I also knew that we would be working this diplomacy for several months. But I was staff, so I turned around, headed back to the Waldorf Astoria, got on a White House bus to the airport, climbed aboard *Air Force One,* and flew to Pittsburgh. Usually, I liked being on the team — the Obama team, the White House team, the U.S. government team. That day, I did not.

That evening, Samore, Einhorn, and a few other experts briefed David Sanger from the *New York Times* on the details of the Qom facility. Still new to the White House, I was struck by the strangeness of this practice. In essence, the president had authorized the declassification of intelligence that we were now providing to the *New York Times.* Why were we giving this story to Sanger hours before we shared our intelligence with the world? Later, however, I would use this technique many times, as it allowed members of our team to go into real depth with someone who could write smartly on the details. And Sanger did. He was an experienced reporter with expertise in the subject, and already had obtained some key details through his own reporting. The next morning, the story about Iran's secret site would startle the world.[13] Sanger later wrote about his government briefing on the Qom facility in his book *Confront and Conceal:*

> As they laid it out on a coffee table in the hotel suite, it was clear that this new site was relatively small: it had enough room, they estimated, for three thousand centrifuges. That is not enough to make fuel for a nuclear power reactor, but plenty, as Samore put it later, "for a bomb or two a year." The fact that it was built on a military base said everything you needed to know; it reopened the question of why the military was so involved in a program that the Iranian government has said is entirely civilian.[14]

Before Sanger's bombshell went up on the *New York Times* website, we launched a full-court press to get all five permanent members of the UN Security Council, plus Germany — a grouping known as the P5+1 — to endorse a single statement denouncing Iran. My assignment, of course, was the Russians. Medvedev had flown to Pittsburgh to attend the G-20 with several of his key aides, and, as luck would have it, his delegation was staying at the same hotel as ours. That evening, General Jones had dinner with several of his counterparts from the P5+1, including Prikhodko, Medvedev's foreign policy advisor. Over dinner, the group made progress toward getting a single agreement. Over the course of the evening, however, it became clear that the positions of the Russians and French were too far apart to reconcile in a single day.

Sometime well after midnight, McDonough, Rhodes, and I woke Ambassador Kislyak to tell him that we just couldn't sign on to the draft agreement that was floating around. Whatever had been signaled earlier in the evening, we were now walking it back. The statement wasn't tough enough. Instead, we agreed that the United States, the United Kingdom, and France would issue a joint statement the next morning. Medvedev, we hoped, would then issue his own statement, similar to ours, though perhaps not quite as robust. We hoped the Chinese would do the same. Kislyak respected our need to backtrack, and didn't try to embarrass us. He agreed that this way forward was a good, second-best option.

The next morning, as the president's caravan sped toward the G-20 convention site, General Jones got word that French president Nicolas Sarkozy wanted to issue a few follow-on remarks after Obama had read the joint statement from the United States, the United Kingdom, and France. This proposal sounded dangerous to me. Sarkozy liked to make news, and I feared he might deviate from the message contained in our joint statement, thereby undermining Obama. Moreover, although we had failed to secure Russian and Chinese support for our joint statement, we still wanted both Medvedev and China's President Hu Jintao to express general support for our pivot to a more coercive strategy with Iran. If we sounded too bombastic, we worried that Medvedev might shy away from his own tough statement. Sarkozy, however, was not to be denied.

Before a crowd of journalists from around the world, Obama read the statement on behalf of the three Western allies in the P5. Obama made clear that "Iran's decision to build yet another nuclear facility without notifying the IAEA represents a direct challenge to the basic compact at the center of the non-proliferation regime."[15] Obama then explained in detail why Iran's refusal to declare the Qom facility violated the Nuclear Non-Proliferation Treaty (NPT), which Tehran had signed. The evening before, we had decided to give Iran one last opportunity to come clean, so the joint statement did not call explicitly for new sanctions, but instead hinted at their possibility: "The Iranian government must now demonstrate through deeds its peaceful intentions or be held accountable to international standards and international law."[16]

Sarkozy then stepped up to the podium to deliver brief remarks that were tougher than the joint statement, and in my view, embarrassed President Obama. Specifically, he called for sanctions by December even after we had agreed the night before to save that step for later in the negotiations.[17] I could not believe that the French president was upstaging my guy! And then, not to be left standing on the stage muted, Prime Minister Brown added his two cents, using equally tough rhetoric.

During the subsequent press conference with our nonproliferation team, Samore and Einhorn provided additional details about Qom, and gently but effectively paved over the differences between the joint statement read by Obama and the tougher remarks offered by Sarkozy and Brown. I watched this performance from the sidelines standing next to Robert Gibbs, our press secretary, and David Axelrod, one of Obama's closet advisors. After forty-five minutes of back-and-forth on the minutiae of Iran's nuclear program, Axe became nervous. It was almost axiomatic that the longer a press briefing went, the more likely the president or our briefers would make a mistake, or say something newsworthy that we didn't want to be newsworthy. Finally, Axe turned to me and said, "This is not open mic night at the Brookings Institution. We gotta shut this down." Two minutes later, the press conference was over.

Thankfully, Sarkozy's statement did not alienate Medvedev. Later that day, the Russian president issued a statement echoing his lament that new sanctions were becoming inevitable.[18] That was good enough for me.

On the short flight home that evening, everyone on *Air Force One* was feeling pretty good about what we had accomplished with the Russians on Iran over the last three days. Medvedev was upset at the Iranians. In hiding the Qom facility, they had deceived not only us, but also their friends in Moscow. Our discussion on Iran with Medvedev and his team revealed a more troubled relationship between Iran and Russia than was often assumed in public discussions. That gave me hope.

That evening, however, I was preoccupied with getting back to Washington for my son's much-delayed birthday celebration. My family met me downtown at an expensive steak and lobster place near the White House, a location I had selected in a meager attempt to beg forgiveness from my son, wife, and parents. I did bring Luke a signed card from the president, one of Obama's White House note cards with his signature message, "Dream Big Dreams!" He still has the card; I think he has forgotten my tardiness to his birthday. I'm not sure my mother has.

We left Pittsburgh prepared to pursue deep, comprehensive sanctions, not just incremental additions. We couldn't achieve that outcome without a Russian vote in the Security Council. On Iran, China would follow Russia's lead. How far the Russians were prepared to go would therefore determine how far the entire Security Council would go. Over the next several months, we developed a multipronged strategy aimed at nudging Russia toward new, crippling, and comprehensive sanctions against Iran. Just weeks after completing the new strategic arms agreement, we were on to our next big round of negotiations with the Russians.

Deliberations with the Russians on a new Security Council resolution shared some features with the negotiations on New START. In New York, our ambassador to the United Nations, Susan Rice, took the lead in negotiating every section of the resolution. It was hard, frustrating diplomacy. She had to engage her Russian counterpart, Vitaly Churkin, who, like Antonov in the arms treaty negotiations, was kept on a tight leash by Moscow. Unlike the bilateral START negotiations, however, negotiations for the UN resolution required Susan to keep our allies happy, together, and moving in the right direction at the right pace.

To move the ball in New York, others in our administration engaged their counterparts in Moscow. Secretary of State Clinton pressed Lavrov, playing a more active role in these negotiations than she had on New START. Jones also made calls to Prikhodko. But for all the heavy lifts, we again turned to the president. Obama called Medvedev several times to discuss sanctions. During one particularly long phone call — an hour and forty-five minutes — the president glanced over at Donilon and me three different times while reporting to Medvedev, as I remember it, that "our staffs don't do anything." He made these pronouncements with a grin on his face; I think he actually liked negotiating with the Russian president. And we had no other option but to deploy the president; Medvedev was the key interlocutor on all matters regarding the Reset, and he would only talk to Obama. Medvedev also willingly played the role of chief negotiator. Later that year, at his joint press conference with Obama at the White House, Medvedev jokingly complained that "the ear starts to get stiff" after two hours on the phone, but the results made it worth the while.[19] Medvedev was right. Only the top two guys could get the really hard stuff done.

Of course, there was one other senior Russian official to engage on Iran whom we all believed was *the* key decision maker: Prime Minister Putin. Protocol, however, made it almost impossible for us to talk to him directly. President Obama's counterpart was President Medvedev. As discussed earlier, we floated the idea of trying to establish a line of communication between Biden and Putin, but Putin rejected that proposal. During this cooperative period in U.S.-Russia relations, Prime Minister Putin did meet with Secretary Clinton when she traveled to Moscow in 2010, and with Vice President Biden when he visited Moscow in 2011. But we still wanted to figure out a way to have our top decision maker engage with Russia's top decision maker on Iran. The stakes were too high not to try.

I had an idea. In the fall of 2009, Chicago was one of the finalists to host the 2016 Summer Olympics. I suggested that Obama call Putin to ask him for advice on how to win an Olympic bid, since Putin had successfully secured the 2014 Winter Olympics for Sochi. Putin, not Medvedev, was the expert on all Olympic matters, so we had the perfect excuse to call him and not the president. After getting Putin on the phone and chatting a while about the politics of winning Olympic bids, Obama could pivot to a discussion about Iran. That was our plan — that was my plan — and the president agreed to give it a shot.

During the call, Putin became animated when discussing the ins and outs of securing votes from members of the International Olympic Committee. He was indeed an expert on the subject. He seemed to enjoy sharing his insights with Obama, though he ended that part of the conversation by telling the president bluntly that the United States had no chance. Brazil had it locked up. And Putin was right. Even after Obama personally flew to Copenhagen to make a final push for Chicago's bid, we came in fourth.

When Obama gently turned the conversation to Iran, Putin offered the president a brief lesson on the Russian constitution: the Russian president, not the prime minister, was responsible for foreign policy, he explained. But it just so happened that President Medvedev was sitting with him, having lunch (even though it was 5:00 p.m. in Moscow). So he handed the phone to Medvedev, who proceeded to discuss Iran with Obama. Putin loved dramatic gestures, and this was a clear reminder that we had to deal with Medvedev on all foreign policy matters. Iran was no exception.

Over the next three years, I devised various proposals to get Obama and Putin together. For instance, we floated the idea of Putin leading a trade delegation to the United States and then building a stop in Washington into the itinerary. We encouraged various private actors — oil companies and universities — to invite Putin to give a speech, so that we could tag on a stop to Washington. I even pushed for a quick trip to Moscow in May 2010 to have Obama attend the fifty-fifth anniversary of the end of the Great Patriotic War. In Moscow, Obama could see Putin without breaking protocol. The president liked the idea and wanted to go, even if everyone else at the White House thought it was a kooky proposal, since the trip would come less than a year after Obama's first trip to Moscow and only a month before Medvedev's visit to Washington. I lost that one. The result of all these unrealized schemes, however, was that Obama and Putin did not speak again for the next three years, until Putin became president again. No matter what the Russian constitution said, that lack of contact was not good for U.S.-Russian relations.

Although the formal objective of our April 2010 trip to Prague was to sign and celebrate the New Strategic Arms Reduction Treaty, we used the meeting to make progress on the Iran account. In fact, Prague was all about Iran. Before joining the larger delegations, Obama and Medvedev spent several hours in our usual small-group format hammering out the details of new sanctions. At that meeting, Medvedev explained why this negotiation, as distinct from the START negotiations, would not result in win-win outcomes for both countries. Russia would suffer much more than the United States if more sanctions were implemented: Russia would lose billions in trade with Iran, while the United States would lose next to nothing, since our trade with Iran was already minuscule.[20] Medvedev complained in particular about our desire to ban the export of heavy weapons, as these sales produced serious profits for Russian companies. As Medvedev spoke, I had the impression that some powerful interest groups within the Russian military-industrial complex — many of whom were close to Putin, not Medvedev — were pressing the Russian president hard to reject new sanctions. Medvedev also argued that new sanctions would damage Moscow's relationship with Iran, one of Moscow's closest partners in the Middle East. Again, the United States had little to lose, since we did not even have diplomatic relations with Iran. All the diplomatic and economic costs of new sanctions fell on Russia, not the United States.

In response, Obama pressed Medvedev to think strategically. Obama explained that over time he aimed to make economic relations between the United States and Russia much more lucrative for Russia than Russia's trade with Iran.

Obama also argued that he wanted to make U.S.-Russia relations more important diplomatically to Russia than Iranian-Russian relations. We were trying to develop a fundamentally deeper relationship with Russia, one that over the course of years and decades would be far more valuable to Moscow than its bilateral ties with Iran.

To signal his seriousness, Obama brought to Prague some sweeteners. He promised to lift sanctions on Rosoboronexport, Russia's state arms export corporation, which had been on the U.S. sanctions list since 2008 for selling arms to Iran, as well as on three other Russian entities that had first been sanctioned in 1999.[21] Obama also told Medvedev that he would resubmit the 123 Agreement with Russia to the U.S. Congress. The 123 Agreement, an accord that would allow for cooperation between American and Russian nuclear energy firms, had been withdrawn from congressional consideration by the George W. Bush administration after Russia invaded Georgia in August 2008. Rosatom, Russia's state nuclear energy company, might be hurt by new sanctions against Iran, but would benefit directly from this agreement. Obama also pledged to push hard for Russia's accession into the WTO, an outcome that would facilitate greater trade and investment between our two countries.

In the short term, Obama's package would not compensate for Russia's economic losses if new sanctions were introduced. But Medvedev seemed impressed by Obama's reasoning and commitment to strengthening U.S.-Russian economic ties. I left Prague confident in our ability to pass a new resolution.

A lot of diplomatic maneuvering took place between Prague and the vote to adopt UN Security Council Resolution 1929 on June 9, 2010. Clinton pushed Lavrov. Jones called Prikhodko. And our ambassador at the United Nations, Susan Rice, did the heavy lifting of drafting and negotiating a new resolution. She seemed to be charming, cajoling, pressing her Russian counterpart, Vitaly Churkin, daily. Eventually, Russia voted in favor of the most robust set of sanctions ever implemented by the UN Security Council against Iran. The new resolution moved well beyond the previous rounds of sanctions. Beyond imposing more extensive military and cargo sanctions, UNSCR 1929 called upon states to suspend financial activities that could contribute to Iran's banned nuclear activities and established a framework for sanctioning Iranian banks. The resolution also banned the import of many categories of heavy weapons, including tanks, attack helicopters, combat aircraft, warships, armored combat vehicles, and missiles capable of delivering nuclear weapons.[22] It also banned participating countries from allowing Iran to invest in those countries' nuclear-related technology, and it placed forty companies and organizations under a travel ban and asset freeze.[23]

In New York, our efforts to include language banning the transfer of the S-300 antiaircraft system, which Moscow had already committed to sell Tehran, ultimately failed. Russian officials argued that the S-300 was a defensive system, and that Iran should have the right to defend itself. Israeli officials pushed hard for the inclusion of a restriction on this system in the resolution. But we could not get an S-300 ban into the resolution.

And then Medvedev made a startling move. On the day the resolution passed, he announced that the Kremlin had decided unilaterally to cancel the S-300 contract. We were both surprised and delighted. Medvedev again was leaning forward, in a manner that seemed to contradict what Putin himself had told us in 2009 about the importance of fulfilling weapons contracts with Tehran. This bold action seemed to prove that Medvedev wanted to deepen ties with Obama and the United States, even when it meant parting ways with Putin. (Years later, after Putin had returned to the Kremlin and the Reset was over, Putin reinstated the contract.)

We kept our end of the bargain. We lifted sanctions on the four Russian organizations. We resubmitted the 123 Agreement to Congress. It sat there for ninety days — the mandatory congressional review period for any nuclear-cooperation agreement — without any action. That meant it entered into force at the end of the year. And most importantly, we accelerated and completed negotiations on Russia's WTO accession. These negotiations proved even more complicated than the Iran sanctions resolution, forcing us to do battle with the Russians on all sorts of new topics, including tariff rate quotas (TRQs) for pork, ractopamine, complex subsidies for production of American cars in Russia, and codification of a new method to regulate trade on the Russian-Georgian border. But on November 10, 2011, eighteen years after being constituted, the Working Party on the Accession of the Russian Federation "completed its mandate." The WTO's Ministerial Conference formally approved the accession on December 16, 2011, and Russia became the 156th member of the WTO on August 22, 2012.

Both Clinton and Bush had tried to close this account and failed, but we finally did it: one more major achievement of the Reset completed. As I had done with my NEW START and IRAN SANCTIONS files, I packed up all of my WTO folders at the end of 2011 and moved them into cabinets destined for the archives.

After reaching agreement with Russia on UNSCR 1929, Obama signed legislation authorizing new unilateral financial sanctions by the U.S. targeting Iranian individuals and companies. According to Obama, these sanctions would "make it harder for the Iranian government to purchase refined petroleum and the goods, services and materials to modernize Iran's oil and natural gas sector."[24] Our European allies followed suit. The degree and precision of sanctions on Iranian banks were extraordinary. In coordination with our European Union allies, we froze hundreds of billions of dollars of Iranian money in foreign bank accounts. As a final blow, in March 2012, Iranian banks were disconnected from the Society for Worldwide Interbank Financial Telecommunication (SWIFT), the global provider of secure financial messaging services to banks. This act, making Iran the only country to have been excluded from SWIFT due to sanctions, damaged Iran's ability to process international payments, and as a result limited its foreign trade capabilities.[25] Sanctions were having an impact.

Moscow was upset with these additional sanctions. In a letter to Obama dated March 5, 2011, Medvedev complained that these unilateral sanctions might adversely affect Russian companies. Later that spring, Lavrov became especially animated in one of his meetings with Secretary Clinton that I attended, asserting that we had double-crossed them. He warned publicly that unilateral sanctions "go against the principle of the supremacy of international law as laid down in the U.N. Charter."[26] Lavrov had a point. We had not briefed the Russian government on our intentions to go beyond UNSCR 1929. We reasoned that if we did, the Russians would back away from the Security Council resolution.

Despite this new friction with the Russian government and business community, I supported the additional sanctions. The financial sanctions in particular may have been more important than the Security Council–imposed sanctions in increasing pressure on the Iranian regime, as Iran eventually could not get payment for its oil exports. We paid a price in our relations with Russia, and yet, for the next five years, the Russian government never tried to subvert the UN sanctions regime. Most individual Russian companies even complied with our unilateral sanctions. As Andrey Kostin, the CEO of VTB, Russia's second-largest bank, explained to me when I visited him with a request to stop doing business with a particular Iranian bank, his bank valued its reputation in the United States more than its small accounts with Iranian companies.

Sanctions took time to bite. But three years after their implementation — on November 23, 2013 — the P5+1 reached an agreement on the "First Step Understandings Regarding the Islamic Republic of Iran's Nuclear Program." This agreement established the basic parameters of a grand bargain: the lifting of sanctions in return for Iran stopping its nuclear weapons program. Two years later, on July 14, 2015, the Iranians signed the Joint Comprehensive Plan of Action (JCPOA) with the P5+1. In doing so, Iran agreed to stop producing fissile materials (either separated plutonium or enriched uranium) at its declared nuclear facilities for at least ten years, while also complying with an elaborate international inspections regime. In return, we agreed to lift the sanctions we had worked so hard to implement in 2010.[27] The deal was controversial; hardliners in both Tehran and Washington criticized the agreement as a sellout. I held a different view. It was the best agreement that could be achieved at the time. While assessing the agreement's ultimate success would have to wait several years, the JCPOA ranked at the time as one of the biggest foreign policy achievements of Obama's presidency.

Many factors combined to produce the deal. In June 2013 Iranians elected a new president, Hassan Rouhani, who proved much more willing to negotiate with us. Rouhani, in turn, named Javad Zarif as his foreign minister and chief negotiator. Zarif was a very accomplished diplomat fluent in English who enjoyed a reputation in the West as a pragmatic interlocutor, and proved instrumental in getting to yes on the JCPOA. Had a more conservative candidate won the Iranian presidential election, the deal might not have happened. Effective diplomacy also played a crucial role. Obama asked Deputy Secretary of State Bill Burns to lead a secret mission to establish bilateral contacts with Iran prior to the June election. These initial contacts laid the groundwork for the final round of P5+1 negotiations, at which Secretary of State John Kerry and Under Secretary of State Wendy Sherman played instrumental roles in negotiating an agreement.

But none of these factors would have had any impact without sanctions. The mullahs ruling Iran needed to feel economic pressure before they engaged seriously in negotiations. And Russia played a critical role in putting these sanctions in place, even when their implementation cost Russia dearly. Even after Putin returned to the Kremlin, Russian diplomats continued to play a positive role in negotiating the JCPOA agreement. On this most important national security issue for the United States, the Reset delivered.

11

HARD ACCOUNTS: RUSSIA'S NEIGHBORHOOD AND MISSILE DEFENSE

In his speech to the Munich Security Conference in February 2009, Vice President Biden stated that we would not reset relations with Russia at the expense of relationships with other countries in Russia's neighborhood. Later that summer in Moscow, Obama stressed his commitment to defending sovereignty, with Georgia on his mind: "In 2009, a great power does not show strength by dominating or demonizing other countries. The days when empires could treat sovereign states as pieces on a chessboard are over . . . Given our interdependence, any world order that tries to elevate one nation or one group of people over another will inevitably fail. The pursuit of power is no longer a zero-sum game — progress must be shared."[1] That was our policy, and our approach to the Reset.

Few in Tbilisi, Vilnius, or Warsaw believed us. As we developed positive momentum in our relations with Russia, our critics in Central and Eastern Europe, as well as their friends in Washington, grew nervous. Analytically, I thought that our approach was logical, moral, and grounded in historical experience: closer cooperation between Washington and Moscow could make Russia less threatening to its neighbors. If Russia was trading more with Europe and the United States, and developing mutually beneficial projects with our European allies and us, then Russia would have more to lose from aggression against its neighbors. Historically, better relations between the United States and Russia had correlated with greater security in Eastern Europe. For example, Moscow had invaded Hungary in 1956 and Czechoslovakia in 1968, and ordered the crackdown on Solidarity in Poland in 1981—all times when tensions with the United States were high. Conversely, improved relations between Gorbachev and Reagan in the 1980s made Soviet intervention in Eastern Europe in 1989 less likely. Warmer relations between the United States and the Soviet Union in the Gorbachev era had fostered permissive conditions for political reform, communist collapse, and democratic breakthrough in Warsaw Pact countries.

I hoped the same dynamics could eventually apply in the Medvedev era, to the benefit of both our bilateral relations with Russia and European security as a whole. Chief among our hopes were that Medvedev (or Putin) would not take further military action against Georgia, and that we could work toward developing a common agenda between NATO and Russia on issues such as counterterrorism and missile defense. If these projects were successful, we hoped that Moscow would stop defining European security in zero-sum terms.

At the same time, I believed that for the Reset to succeed, we had to do more to enhance the security of our NATO allies in closest proximity to Russia. Any appearance of weakening alliance relations in the pursuit of closer ties with Moscow was both bad policy and bad politics. Translating this aspiration into policy proved challenging. NATO expansion was not an option. Croatia and Albania joined in 2009, but expansion farther east lost momentum at the NATO summit in Bucharest in May 2008. Despite Bush's prodding, Germany, France, and several other alliance members were not even prepared to offer a Membership Action Plan (MAP), the usual path by which countries joined NATO, to Georgia and Ukraine. As a compromise, NATO issued a confusing statement, which welcomed "Ukraine's and Georgia's Euro-Atlantic aspirations for membership in NATO" and insisted "that these countries will become members of NATO," but also kicked the start of their MAPs to the distant future.[2]

A few months later, in August 2008, Russia invaded Georgia, scaring further NATO allies away from commitments to Georgian membership. Despite all the cheap, public talk about support of Georgia from European and American politicians, few NATO members wanted to extend security guarantees to a country partially occupied by Russian soldiers. This majority opinion within NATO emerged well before Obama became president. Regarding Ukraine, there was neither push nor pull for membership. Even under President Yushchenko, the leader of the Orange Revolution in 2004, a majority of Ukrainians opposed NATO membership. After the pro-Russian Yanukovych was elected Ukraine's new president in 2010, the idea faded completely. We spent little time debating Ukraine membership in NATO because it was a nonstarter both within the alliance and within Ukraine.

Enhancing NATO, however, was an option, and one we pursued. In preparation for NATO secretary-general Jaap de Hoop Scheffer's White House visit on March 25, 2009, we discussed contingency plans for defending Estonia, Latvia, and Lithuania. At the time, there were no such plans for these new allies, only for our older friends in the West. Obama listened to the reasoning, and questioned it. All allies had to be treated equally. The divide between "New Europe" and "Old Europe" may have meant something to the old Cold Warriors, but it had little resonance with our new post–Cold War president, who was looking for ways to reassure our NATO allies on Russia's borders. I was thrilled to hear his reasoning. Any perception of collusion between the United States and Russia at the expense of small countries and NATO allies in between would derail the Reset.

Missile defense offered another way for us to enhance our commitments to our European allies and partners, but to defend against Iran, not Russia. As with most other important policies, Obama ordered a major review of our missile-defense policy. Regarding the European dimension of the review, our new national security team quickly agreed that the George W. Bush administration proposal to deploy ten ground-based interceptors (GBIs) in Poland only offered protection for the United States, and then only in the very distant future, when and if Iran developed an intercontinental ballistic missile.[3] Well before that might happen, however, Iran's short-range and medium-range ballistic missiles could attack our allies in Europe and our armed forces stationed in Europe and the Middle East,[4] and our ten GBIs in Poland would not be capable of defending Europe from Iranian missiles. So we refocused our discussion on developing new missile-defense architectures that could deter or counter an Iranian threat to Europe.

The Standard Missile 3, or SM-3, was our interceptor of choice. Although smaller and slower than a GBI, SM-3s could be deployed on the Navy's Aegis ships as well as land sites. They also were cheaper, so we could deploy more of them — eventually hundreds more — and thereby protect the entire European landmass from an Iranian missile attack. We eventually agreed, first among ourselves in the U.S. government and then with our allies, that we would deploy SM-3s on land in Romania and Poland, as well as on Aegis ships around the European continent.

SM-3s based in Europe would have no capability against Russian warheads.[5] They were too slow, and could not catch up to Russian ICBMs. Interceptors hit warheads, not missiles, a fact frequently forgotten in the missile-defense debate. Consequently, to protect the United States, we were better off launching interceptors from Alaska or elsewhere in the United States rather than Europe to protect us from a Russian missile attack. We reasoned, therefore, that our deployment of dozens of SM-3s at several locations in and off the coasts of Europe greatly enhanced European defense from an Iranian attack without threatening Russia.[6]

We also built flexibility into the program, calling our new deployment plan the European Phased Adaptive Approach (granted, not the most graceful name, but we quickly reverted to its acronym, EPAA) to allow us to expand or contract deployments depending on the evolving threat. As Obama said in his Prague speech on April 3, 2009: "As long as the threat from Iran persists, we will go forward with a missile defense system that is cost-effective and proven. If the Iranian threat is eliminated, we will have a stronger basis for security, and the driving force for missile defense construction in Europe will be removed."[7] If the threat expanded or accelerated, EPAA could be ramped up.

Finally, we decided to change the type and location of the radars in the system. The Bush administration proposed building a radar called the European Midcourse Radar (EMR) in the Czech Republic. Czech citizens didn't want the radar, and the Russians complained with good reason that such a radar in this location was capable of tracking Russian ICBMs.[8] We came up with a different configuration: two radars in Turkey and Romania. By placing these radars closer to Iran, we had greater capability to track the Iranian threat, and in locations less threatening to the Russians.

The missile-defense review process was both contentious and confusing given the technical issues involved. I spent hours in tutorials with outside experts from Stanford to educate myself sufficiently so that I could participate effectively in these debates. Our recommendations also were going to be viewed as political, both because the Bush administration had made its own decisions about missile-defense deployments in Europe that we were now questioning, and because anything we changed would be interpreted as a concession to Russia. Nonetheless, by the time we finished the review, it was clear to me that we were enhancing, not degrading, the security of our European allies with our new plan. Secretary Gates, Bush's last secretary of defense, came to the same conclusion: "I sincerely believed the new program was better — more in accord with the political realities in Europe and more effective against the emerging Iranian threat."[9] President Obama expressed the same sentiment when he announced our new missile-defense architecture for Europe on September 17, 2009, emphasizing that "it ensures and enhances the protection of all our NATO Allies."[10] That was something no other American president had done.

We thought that our European allies would be appreciative — especially since the United States was going to foot the bill. Initial reactions, however, were anything but appreciative, especially among our Eastern European allies. Our rollout of EPAA was a complete disaster. In diplomacy, sometimes form does trump substance.

We planned to discuss the new proposal with our allies, but as it happened *Wall Street Journal* reporter Peter Spiegel learned of the details of EPAA before they did, disrupting our diplomatic timetable. As Gates recalled, the leak "made us look like a bunch of bumbling fools, oblivious to the sensitivities of our allies."[11] Spiegel's reporting forced us to accelerate the announcement of our new program. An interagency team flew hastily to Poland and the Czech Republic — the two countries that had agreed to host the Bush administration's missile defense — to explain our future plans, only hours before we announced our ideas to the world. That was not smart diplomacy. The Pentagon leak to Spiegel also compelled us to announce EPAA on a most inopportune date: September 17, the anniversary of the Soviet invasion of Poland in 1939. That coincidence produced all sorts of bad press for our decision well before anyone had read the technical details of our plan.

Once we learned that Spiegel was working on the story, our press team engaged him. There was a plan for Secretary Gates to talk to the journalist. But that interview never happened, meaning that Spiegel had only partial information about what was going to be released when he wrote his story. The *Journal* ran his story on page 1, under the large-point headline "U.S. to Shelve Nuclear-Missile Shield."[12]

At 5:45 that morning, McDonough emailed me to ask if I thought we'd screwed this one up (he used stronger language!). I replied, "Yep." For the rest of the day, the rest of the fall, and the rest of our administration, we were playing defense.

The *Journal*'s headline was misleading. We were not "shelving" missile defense, but replacing Bush's plan with our new and improved program. Over time, we were going to deploy more missiles in more places in Europe than Bush had planned to do. And unlike the Bush plan, we were going to protect all of Europe, not just the United States, from a potential Iranian missile attack. Nor was our proposal a gift to the Russians, as many stories that day claimed, especially since the old Bush plan had almost no capability against the Russian arsenal. The only slight gain for Russia was our decision not to move forward with the radar in the Czech Republic, which would have helped us detect Russian ICBMs. But of course, we had many other ways to track those same Russian weapons. And there was one detail completely ignored in the press coverage: the Czech people did not want the radar built in their country.

That morning, Secretary Gates did damage control from the Pentagon, and General James Cartwright joined us at the White House to continue the counteroffensive for a good chunk of the day. At the time, Cartwright was deputy chairman of the Joint Chiefs, a four-star Marine general, a former commander of the U.S. Strategic Command, which is responsible for strategic weapons, and a genuine grand strategist on all matters concerning intercontinental rockets and nuclear warheads. In addition to the four stars on his lapel and his 1950s crew cut, he had a commanding speaking style. We needed him badly that day. I joined him for some of these briefings, always making the case that our changes in missile-defense deployments in Europe had nothing to do with the Reset. At one of these briefings, *New York Times* reporter Peter Baker slyly asked me, as I remember, why I was participating in the briefings if the changes were not about Russia. He had a point. I peeled away after that one was finished.

With time, we made progress in explaining the virtues of our program. Our allies eventually appreciated that our proposed deployments were designed to protect them, not us, and the subsequent debate inside Russia underscored that

EPAA did not constitute a gift to the Kremlin. Gates later pointed out, "For the first time since before Reagan's 'Star Wars' speech, building a limited American missile defense had broad bipartisan support in Congress."[13] Still, we never got much credit for this achievement in Washington or Warsaw.

Our fumbled rollout of the EPAA only reconfirmed suspicions about the Reset among our critics in Central and Eastern Europe. Particularly demoralizing for me was an open letter criticizing the Reset signed by two dozen Central and Eastern European intellectuals, think tank members, and former policymakers. I lamented that they'd gone public with their attack instead of first engaging with our new team privately, but most importantly, I believed we already were implementing, or planning to implement, their recommendations as laid out in the letter. That obviously was not their perception. In diplomacy, perceptions matter, so we had work to do. The letter also offended me personally, given my long track record as a champion of Eastern European democracy and sovereignty and my regular criticisms of belligerent Russian foreign policy actions. My friend Ron Asmus, the American who had worked behind the scenes to help produce the letter, assured me that it was nothing personal, but published as a political act to get the Obama administration's attention. That was an important lesson for me. I no longer spoke for myself or had an individual reputation. I was now part of the Obama team.

In July 2009, just a few weeks after Obama's trip to Moscow, I wiggled my way onto Vice President Biden's trip to Ukraine and Georgia so that I could help explain the logic of our Russia policy. As we drove down George W. Bush Boulevard after landing in Tbilisi, I realized what an uphill battle we were fighting. Biden was our best messenger, but he was bringing a tough-love message. Hardest of all, he had to explain to President Saakashvili our convoluted, multistage policy for developing military cooperation. After Russia's invasion in August 2008, Saakashvili had argued quite logically that he needed defensive weapons to repel a future Russian attack. President Bush and his administration had provided strong rhetorical support, but no new guns. We continued that policy, but we also provided a road map—albeit a long and winding one—for when we would begin to provide lethal assistance. Consistent with NATO policy, Biden also kicked the can down the road on NATO membership. We hinted at our aspiration to offer a Membership Action Plan to Georgia, but insisted on all sorts of preliminary reforms and actions by the Georgians before they could pursue this preliminary agreement.

No one in Tbilisi liked our message, but it was good that Biden had delivered it in person. Biden knew how to talk with Saakashvili. He could grow animated at times without alienating the Georgian president. And Saakashvili understood perfectly well that Biden was his best ally within the Obama administration.

During our stop in Kyiv, we hardly discussed security issues at all. No one, not even the fiercest Ukrainian critics of Russia, was worried about a Russian annexation of Crimea at the time. Soon after Russia's invasion of Georgia, Putin said publicly that Russia had no legitimate claims on the peninsula: "The Crimea is not a disputed territory . . . Russia has long recognized the borders of today's Ukraine."[14] Furthermore, the Ukrainian government was not pushing hard for NATO membership in the summer of 2009, but instead focused on joining the European Union. Given those priorities, Biden's main message, both in private and in public, was about fighting corruption, reforming the energy sector, and reducing Ukrainian dependence on Russia. As he said in his public address, "Ultimately, democracy and free markets will flourish when they deliver on what people want most — honesty, the elimination of corruption, a decent job, the ability to care for their parents and educate their children, physical security and economic opportunity, a chance to build a better life."[15] Ukrainian president Yushchenko, and everyone else we met on that trip, nodded enthusiastically in agreement with Biden about the need to tackle corruption and end energy dependence on Russia, yet the Ukrainians seemed unprepared to do much about either issue. It was a depressing stop.

Back in Washington, I made a special effort to engage with Central and Eastern Europeans. Czech ambassador Petr Kolar hosted a dinner for me and all the other ambassadors from the Central and Eastern European region to give me a chance to explain the Reset. Our exchanges sometimes became heated, but at the end of a long evening, we all admitted that we enjoyed the meeting and agreed to do it again, which we did. I also accepted any invitation to visit the Georgian Embassy. Often, I was the only U.S. government official at these events, outnumbered by think tank members and former Bush administration officials. Still, I reasoned that I was the best messenger for our Russia policy, even if it meant listening patiently to a lot of criticism.

Sometimes I even spoke publicly, an unconventional act for an NSC senior director at the time. For instance, at the Peterson Institute in D.C. in the summer of 2010, I reminded the audience that we were constantly fighting off Russian attempts to link Georgian issues to other areas: "We're not throwing Georgia under a bus in the name of a U.N. Security Council resolution."[16] I also won some minor but, for me, symbolically important battles regarding our policy of supporting Georgia. For instance, on June 24, 2010, the day President Medvedev visited the White House for the first and only time, I fought hard and ultimately won my fight to keep the word "occupation" in our "U.S.-Russia Relations: 'Reset' Fact Sheet," which stated, "We continue to call for Russia to end its occupation of the Georgian territories of Abkhazia and South Ossetia."[17] But I remained frustrated by our poor policy options when it came to Georgia. We wanted to end Russian occupation, but did not have a strategy for achieving that objective. Neither more engagement nor more confrontation with the Kremlin was going to compel Russia to withdraw. That cold, hard fact bothered me, but it was a fact, no matter what our critics asserted. Our administration and our NATO allies were simply not prepared to risk war with Russia to liberate Abkhazia and South Ossetia. I thought that we should state our policy honestly, and not fuel false expectations, but others explained to me that sometimes ambiguity served U.S. national interests. To this day, I'm not sure.

Georgia and Ukraine were not the only post-Soviet countries that demanded our attention in the first years of the Obama administration. I also devoted serious attention to the five countries of Central Asia — Kazakhstan, Kyrgyzstan, Tajikistan, Turkmenistan, and Uzbekistan — also in my portfolio at the National Security Council. For all these countries, we adopted a similar strategy as the Reset: engagement on a broad agenda, not just military issues but also trade, human rights, and culture. Under Secretary Burns and I first traveled to the region soon after Obama's trip to Moscow in July 2009, both to brief Central Asian leaders on the Reset with Russia and to begin the development of our own engagement with each of these countries.

Our stated goals of expanding the dimensions of our relations with these countries clashed directly with our central objective of expanding our military supply routes through Central Asia. That's why Burns and I met with the cantankerous president of Uzbekistan, Islam Karimov, who lectured us about the evils of both Russian *and* Western values. That's why we praised Kazakhstan's President Nursultan Nazarbayev as a wise and strategic leader. And that's why I sat down for a meal that included goat eyeballs with the corrupt Kyrgyz president Kurmanbek Bakiyev. We also met with civil society leaders, and mentioned the importance of human rights in our statements. But compared to our outreach efforts with civil society and opposition leaders in Moscow with Obama, the effort was muted.

As we worked on expanding our Central Asian relationships, I was troubled by the centrality of security issues in all of these engagements. A U.S. Central Command general had greater sway than any ambassador in that region. The Pentagon was fighting a difficult war in Afghanistan and the military needed all the help it could get. There were no transportation routes through liberal democracies that got you to Afghanistan, and so democracy in that part of the world was seen as an admirable but distant goal. At the time, I worried about what I had called in my academic writings the "false promise of autocratic stability."[18] In addition to the creepiness of engaging with some of these leaders —Karimov, for instance, had allegedly boiled political opponents to death—I was worried that these autocracies, and Tajikistan and Kyrgyzstan in particular, were not stable. I also feared how quickly autocrats could change course, since they had no parliament or voters to hold them accountable. Karimov invited the United States to open a military base in 2001, but then abruptly closed our access in 2005 after blaming us for provoking a popular protest against his regime. My anxieties about autocratic allies, however, did not produce fundamental changes in policy. Within the Interagency Policy Committee on Central Asia, which I chaired, I pushed, and people politely listened, but continuity won out.

On April 7, 2010, my academic worries became real-life policy challenges. That day, protesters in Kyrgyzstan stormed the president's office, parliament, the internal security headquarters, and KTR, Kyrgyzstan's main television broadcaster. Dozens of people were killed.[19] President Bakiyev fled the country, eventually ending up in Belarus, but his departure did not defuse the conflict. By June the crisis in Bishkek had metastasized into an ethnic clash between Kyrgyz and Uzbeks throughout southern Kyrgyzstan, especially in the cities of Osh and Jalal-Abad. That summer, roughly four hundred thousand ethnic Uzbeks were displaced, including over one hundred thousand who had crossed the border into neighboring Uzbekistan, seeking refuge.[20] Uzbek president Karimov was making threatening statements about the need for his country to protect ethnic Uzbeks still being attacked in Kyrgyzstan. War seemed likely.

As the senior director for Central Asia at the National Security Council, I coordinated our effort to respond to the crisis. Our immediate concern was the safety of Americans living in Kyrgyzstan, both our embassy staff and a handful of others, including Peace Corps volunteers, scattered in small towns in the south. Our next concern was maintaining our operation at the Manas Air Base, including jet fuel supplies for our war effort in Afghanistan. But my biggest worry was avoiding a full-blown war, whether an ethnic civil war within Kyrgyzstan or an interstate clash between Kyrgyzstan and Uzbekistan. The word "genocide" lingered uncomfortably in our conversations.

Thankfully, we had a powerful partner who also did not want to stand by and witness large-scale conflict in that part of the world: Russia. The role of Russia's intelligence services in supporting the overthrow of Bakiyev remains murky. Russian leaders despised Bakiyev. He was a corrupt double-dipper, taking money from them and us.[21] Some analysts have suggested that Moscow wanted to exact a price against Bakiyev for agreeing to keep the Manas Air Base open just weeks after Russia had pledged $2 billion in assistance to Kyrgyzstan. Nonetheless, whatever the involvement of the Russian intelligence services in Bakiyev's overthrow, Medvedev personally did not welcome the spread of violence, especially along ethnic lines. Obama and Medvedev agreed that our two countries had a mutual interest in working all our diplomatic, intelligence, and military channels in the region to contain the fighting if at all possible. At the Nuclear Security Summit in Washington in April 2010, Obama, Medvedev, and Nazarbayev huddled to coordinate our cooperation on defusing the Kyrgyz crisis, and a few months later, during Medvedev's visit to Washington in June 2010, Medvedev and Obama further discussed the need to work together.

Obama even agreed to consider bringing in international peacekeepers—including Russian peacekeepers—as long as they were deployed with a UN mandate. As Obama stated during a press conference with Medvedev, "One of the things that we discussed is creating a mechanism so that the international community can ensure that we have a peaceful resolution of the situation there, and that any actions that are taken to protect civilians are done so not under the flag of any particular country, but that the international community is stepping in."[22] Medvedev did not object. That was a huge step forward in U.S.-Russian relations.

Regime change — a "color revolution" — was occurring in a former Soviet republic. But this time, in the era of the Reset, the United States and Russia were cooperating to achieve a win-win outcome, not competing against each other. We leaned heavily on the Uzbeks not to invade. Russia pushed hard on the new leadership in Bishkek to restore order in the southern cities, where large Uzbek communities resided. I traveled to Astana to plead with the Kazakh government to reopen the Kazakh-Kyrgyz border, which they had abruptly closed when the fighting in Kyrgyzstan erupted, thus triggering a huge hit to the Kyrgyz economy. We eventually succeeded in avoiding worst-case scenarios. This instance of regime change in a former Soviet republic did not precipitate a major clash between the United States and Russia and did not erupt into war, as we would witness tragically in Ukraine just four years later.

In May and again in July 2010, I also traveled to Bishkek to try to mend our relationship with the new government, and keep the Manas Air Base open. Privately, I felt vindicated by my warning about the false promise of autocratic stability, but you don't win any points in government for bragging about how your theories proved correct. Instead, we needed a new strategy to get us out of the huge hole created by our past approach. We had embraced the fallen dictator whom the new regime in Bishkek hated. With good reason, the new government distrusted us.

It was a stroke of good fortune for us that Roza Otunbayeva became the interim president of Kyrgyzstan. I had known Roza for decades. In the 1990s she had served as Kyrgyzstan's foreign minister, but had become an opposition leader as autocratic rule strengthened under Bakiyev. Because of my personal ties with her, my White House bosses sent me out to Bishkek as the senior Obama administration official to make contact with the new revolutionary regime. In my bilat with President Otunbayeva — my first meeting with a head of state in which I was the head of the American delegation — Roza was very friendly toward me, but very tough on our government. Why had we backed a dictator for so long? she asked. Whatever happened to American values? I had no rebuttal, only a pledge to operate differently in the future. I promised to provide more transparency regarding U.S. government contracts and assistance in Kyrgyzstan. Specifically, we planned to publish information about all U.S.

government transfers, including rental payments for the base, on a government website—which we eventually did, much to the dismay of some of my Pentagon colleagues. We also promised to allow competitive bidding on the lucrative fuel contract that supplied the Manas Air Base with jet fuel, which everyone in Kyrgyzstan believed was used to provide side payments to the Bakiyev family. As I dug into the issue, I agreed that the arrangement seemed fishy.[23] After much debate within our government, we also delivered on this promise.

My meeting with Prime Minister Almazbek Atambayev—who became president of Kyrgyzstan the following year—was much tougher. He also beat up on the United States, but this time with cameras rolling. Unlike with Roza, I had no personal relationship with him. In our meeting, he expressed his strong desire to develop close relations with Russia. When I tiptoed into a question about the future of our air base, his voice grew louder and his contempt stronger. And then he pulled back, saying we needed to work on a long-term solution.

For a while, we succeeded. We increased our economic assistance to this poor, tiny, remote Central Asian country, paid a higher rent for the air base, and provided more transparency about our transfers. We also provided new resources for assisting economic growth, peace and security, and democratic reforms. And I secured a meeting for President Otunbayeva with President Obama in New York in September 2010. We managed the transition from supporting a dictator to engaging with these new leaders in part because Medvedev and his team allowed us to do so. We were proving that U.S.-Russia relations need not be a zero-sum game; that both of our countries could pursue legitimate interests in Russia's neighborhood without undermining Russian foreign policy objectives.

After breakthroughs in cooperation on arms control, trade, Afghanistan, Iran, and Kyrgyzstan, we had the courage to tackle with Moscow one of our most audacious aims: missile-defense cooperation. We hoped to transform this issue from a point of contention into an area of cooperation. The aim was extremely ambitious. But in the heyday of the Reset, ambition was not shunned but celebrated.

In discussing missile-defense cooperation with Moscow, our objective was to enhance the security of Europe—all of Europe—from a missile attack launched from outside the continent. The most optimistic advocates for missile-defense cooperation, both inside and outside government, pushed for an integrated system between Russia and the United States or between NATO and Russia. That seemed crazy to me. We were not military allies. Russia was still spying on us. Too much integration would allow the Russians to obtain information about our defenses, which they might share with our enemies if Russian-American relations soured. But I did support more modest steps, which might create the conditions for deeper cooperation in the distant future.

As negotiations developed, Medvedev proposed a division of labor, or what he called a "sectoral" plan. We would protect a "Western" European sector and Russia would protect an "Eastern" European sector. It was a bad idea. We were not going to leave the defense of NATO members such as Estonia, Latvia, and Lithuania in Russia's hands. Moreover, Russia's missile-defense capabilities were years behind ours. They did not have the hardware to carry out the mission they were proposing.

Still, somewhere between these radical proposals and the option of doing nothing, I thought there was room for cooperation. Russia did have radars positioned in locations close to Iran that could provide us with valuable, early-warning information about an Iranian attack on Europe.[24] In turn, we collected information from our radars that should have been useful for Russia, and not only in the Middle East. I also believed that it was in our interests to have the Russians take a first shot at an Iranian missile flying through their territory. We still would be ready to take a second and third shot if they missed. And though it was taboo to discuss in public, long-term planners had to worry about new, future threats from the region. For instance, if Islamic fundamentalists one day seized control of Pakistan, then we would want Russian missile-defense systems capable of effective action against this new threat.

These kinds of concerns eventually created modest agreement in our government about the prudence of exploring missile-defense cooperation with Russia. Obama wanted to try, so we tried. The first step was to disabuse Medvedev of his plan for a "sectoral defense" in which the U.S. would have to cede responsibility for protecting its own allies. Obama took on that assignment himself at the NATO summit in Lisbon in November 2010. At this summit, the NATO-Russia Council (NRC) met at the head-of-state level for the first time in two years. The substance of the NRC discussion was unprecedented: around the table, heads of state from NATO countries discussed in earnest missile-defense cooperation with a Russian head of state! In the official statement, Russia and the NATO alliance "agreed to cooperate on missile defense against shared threats. The NATO-Russia Council will resume theater ballistic missile defense exercises and leaders tasked it to identify opportunities for Russia to cooperate with NATO's new territorial missile defense capability by June 2011."[25]

By the way, the issue of NATO expansion was not discussed at all. In fact, in five years in government, I cannot remember a single serious conversation about NATO expansion between Obama and a Russian official.

With the cameras off and the press escorted out of the room, Medvedev outlined his "sector-based missile defense system" to all the other leaders of the NATO alliance. Obama spoke next on behalf of the alliance to explain why we could not agree to the Russian proposal. No matter how logical geographically, NATO was not going to leave the defense of NATO allies to Russia.

Medvedev's proposal did raise the question about the defense of European countries in between Russia and NATO. Who would defend Ukraine, for instance, against an Iranian attack? As I listened to these discussions, both at the NATO summit and at other venues, I would occasionally wonder about the probability of an Iranian attack against Ukraine or Latvia. We seemed to be committing a lot of diplomatic attention and resources to addressing very low-probability events. But within our government and the NATO alliance, that kind of question was taboo.

While diplomatically dismissing Medvedev's sectoral plan, Obama also enthusiastically endorsed the idea of missile-defense cooperation. It would take years, maybe decades, to implement, but the very existence of such a joint project would advance American relations with Russia and help remove missile defense as an obstacle to further reductions in nuclear forces.

As we walked out of the NATO-Russia Council meeting, Obama asked me to set up an impromptu one-on-one meeting with Medvedev. He was worried that his rebuttal of Medvedev's proposal might have been too blunt. It was a smart move. In one of the more bizarre settings for a presidential meeting that I remember, the two of them sat down together in a small, windowless room, which had been hastily cleared by our Secret Service just seconds before, together with just one translator (theirs, not ours). That was their first meeting or phone call that I was not in on, and I didn't like it, as I always worried about miscommunications. Lavrov, Prikhodko, Donilon, and I hovered about outside, hoping our two leaders would not commit us to ideas on which we could not deliver. Immediately afterward, however, Obama debriefed me on the exchange, reporting that Medvedev still wanted to try to find a way to work together on missile defense in Europe even after we had rejected his proposal. That was an excellent sign, which underscored for me the willingness of these two leaders to keep engaged even on the hardest of issues.

As we strolled down the corridor, I promised the president that our government would work assiduously with our Russian counterparts to have something concrete ready for the two presidents to discuss the next time they met, which would be in Deauville, France, on the sidelines of the G-8 summit in May the following year. Somewhat artificially, Deauville became the new deadline for getting something serious completed with the Russians on missile defense.

In the run-up to the Obama-Medvedev meeting in Deauville, our diplomatic teams, headed by Under Secretary of State Ellen Tauscher and Deputy Foreign Minister Sergey Ryabkov, worked furiously to complete a modest statement on cooperation. They hoped that at Deauville, a joint presidential statement would direct officials and experts from both the United States and Russia to focus on certain priority issues and endorse further work in both political and military channels. As Obama touched down in London the day before Deauville, we got a draft text of the joint political understanding of missile-defense cooperation upon which Tauscher and Ryabkov had agreed. As I learned subsequently from Lavrov in Deauville, Medvedev had presented the statement to the entire Security Council, which included Prime Minister Putin. It had been a tough deliberation, Lavrov reported to me, but they approved the text.

Our process, however, was more chaotic. Because the president was already traveling, our national security team was divided between London and Washington. National Security Advisor Tom Donilon organized a secure conference call from London with our interagency team on missile defense. The State Department pushed for release of the statement at the end of the Medvedev-Obama meeting. The Defense Department disapproved, arguing that we needed more time. At issue was whether the language on assurances about our European missile-defense deployments not adversely affecting Russia might be used by Russia to limit our systems in the future. The language was convoluted, maybe intentionally by the Russians.

Donilon eventually decided that the joint statement was not ready. Sitting together in a blue tent with weird sounds swirling in our secure space at our London hotel, he told me that he couldn't recommend that the president agree to a statement that he himself did not understand. Tom was a lawyer, and he sensed entrapment language in the statement. Tom recommended postponement to the president; Obama concurred.

At the time, I also did not understand the rush for releasing a joint statement the following day. There was always going to be another "Deauville"—some G-8, G-20, or APEC summit in the near future. Diplomats rarely stop trying to negotiate. There's always the possibility to schedule another meeting!

In his meeting with Obama in Deauville, however, I was surprised to hear Medvedev express real disappointment with our decision. Evidently, there was some urgency for the Russian president. I had never seen him so animated. It seemed to me that he had been led to believe that we already had signed off on the statement.

After the formal meeting, Obama and Medvedev stayed in the room to talk one-on-one for a few moments, while both Lavrov and Prikhodko cornered me to ask what had happened. They both were furious. As Lavrov explained, Medvedev had expended a lot of political capital to convince Putin and other hardliners in the Security Council to approve the statement. Now we were pulling back, and thereby undermining Medvedev internally. Although Obama, Tom, and I felt no urgency for an agreement at this meeting, maybe we were not paying enough attention to Medvedev's internal politics. At the time, he was trying to prove to Putin his value as president. To win reelection, Medvedev had to win one vote: Putin's. To win that vote, Medvedev above all else had to demonstrate his unique abilities to work with Obama to achieve results that were good for Russia. We had just handed the Russian president a defeat at the very moment when Medvedev believed Putin was deciding his fate. It played into the narrative of Medvedev's critics back home that he was weak and susceptible to being pushed around by the Americans.

The momentum for missile-defense cooperation died in Deauville. Our working groups continued to meet. Donilon and Prikhodko met on the sidelines of the G-20 summit in Cannes, France, to discuss again the parameters of a new statement on missile defense. But without the impetus from the presidents that the Deauville draft statement was designed to provide, no further progress was made. And in retrospect, that was a good thing. Had we signed on to that agreement and then Putin returned to the Kremlin several months later, eager to pivot in a more hostile way against us, we would have looked weak, if not naive. Putin would have tried to use this agreement to constrain our missile-defense deployments, not work together to enhance our mutual capabilities. As discussed later, there could be no missile-defense cooperation with Russia while Putin was in power. And he came back to power just months after Deauville.

12

BURGERS AND SPIES

Soon after President Obama's trip to Moscow in July 2009, we'd begun planning a reciprocal summit in Washington for the following year. My hope was that the two presidents would meet every year for the length of their time in office. Little did I know that Medvedev's visit in June 2010 would be the last full-fledged Russian-American summit of the Obama era.

In planning for Medvedev's visit, we had three objectives: show results from the Reset, emphasize our engagement with both the Russian government and society, and highlight our expanding agenda, with a particular focus on economic ties between our two countries. We also wanted to be creative, to do some new things that would reflect the new, post–Cold War era.

One innovation was to have Medvedev begin his visit in California, not Washington, D.C. While meeting in Moscow with Medvedev on a trip with Secretary Clinton, I personally encouraged the Russian president to travel to my part of America, including Stanford and Silicon Valley. At the time, Medvedev was trying to jump-start Skolkovo, a high-tech business area just outside Moscow. Most Americans rolled their eyes at the idea of a state-led Silicon Valley in Russia, but I was a supporter. I knew the odds were long for the project, but reasoned that it could be in the United States' interest to have the Russian government invest in high tech and education. Having lived in the Valley for most of my life, I also knew that U.S. government investment in research and development, including through grants to Stanford, had played a central role in the

Valley's growth. Done smartly, the government could be a good partner. I also saw business opportunities in Russia for our high-tech firms. Cheerleading for Skolkovo seemed like an easy call.

With our help, Medvedev had a terrific trip to California. George Shultz and his wife, Charlotte, hosted the Russian president for dinner with some of the "founding fathers" of Silicon Valley, so that Medvedev could hear directly from them how it all began. The next day he met Steve Jobs at Apple, John Chambers at Cisco, and Eric Schmidt at Google, and opened an account at Twitter. He then gave a speech at Stanford, dressed in jeans and a blue blazer (no tie) — very much the Valley uniform — reading his remarks from an iPad. In an informal session organized by my former students, the Russian president also met with Russians working in the Valley at a coffee shop near campus. The optics of the trip was so very un-Russian, and most certainly un-Soviet. When Obama asked Medvedev about his trip, Russia's young, tech-savvy president went on and on about the Valley's energy. He even showed Obama his new iPhone, given to him by Jobs. I was thrilled. American "soft power" was having a clear and positive effect on the head of a state ruled for decades by leaders who despised America.

At the White House, Obama and Medvedev discussed frankly and cooperatively a long list of issues, but Kyrgyzstan, Iran, and Russia's WTO bid topped the agenda. In the small group session in the Oval Office, Obama and Medvedev focused especially on how to coordinate our efforts on Kyrgyzstan, even dancing around the idea of a Russian peacekeeping mission. I was scared to death of this proposal, worried that Russian intervention would produce more harm than good, and relieved to hear Medvedev express caution. But if ethnic fighting did escalate, and some military force had to come in to stop a civil war and keep the peace, it was not going to be American soldiers. Medvedev and Obama disagreed about some particulars regarding the Kyrgyz crisis, but the tone of cooperation was striking. The American and Russian presidents were working together to defuse a conflict in a country that Russians only a few years earlier claimed to be in their sphere of influence. That was progress.

On Iran, Medvedev complained about new sanctions that the United States and our European partners had imposed, which went beyond those stipulated in UN Security Council Resolution 1929. He suggested we had done a bait and switch, and he had a point. We would hear this complaint often for the next several years. In June 2010, however, Medvedev deliberately tried to downplay our disagreements. He wanted a successful summit.

On Russia's accession to the World Trade Organization, Obama that day committed to a deadline of September 30, 2010, to complete all of our bilateral negotiations. After eighteen years of deliberations, we were now going to close out this account. We committed to this date by releasing a joint statement, despite the misgivings of our trade negotiators. But Obama sensed how important this statement was to Medvedev and instructed our teams to get it done during a lunch break. We did.

Most interesting for me was the two presidents' extended exchange on the relationship between democracy and development. In my presence, Obama never lectured anyone on democracy or human rights. But with Medvedev, he was not afraid to have an analytic discussion about what we in academia would call the relationship between political institutions and economic growth. The president basically outlined a theory of development that would have been familiar to my Stanford political science students. He used more diplomatic language, but his argument was that autocracies could steer economic growth in countries with low levels of development while still industrializing, but could not do so in richer countries as they entered the postindustrial phase of their economic development. There is empirical data, by the way, supporting this hypothesis.[1] His message about Russia was indirect but clear — Russia was going to need to liberalize politically to sustain growth over the long term. Medvedev seemed to agree. It was not a contentious interaction, but an intellectual exchange. My sense was that Medvedev believed he was doing what Obama was suggesting, albeit perhaps on a slower timeline than his critics wanted. But Medvedev always perceived himself as a modernizer and reformer. He was just working in a very constrained context — one shaped and dominated by Putin. Temperamentally, he was also cautious, playing a long, patient game. He did not seem to disagree with Obama's analysis and even embraced the same ultimate objective: liberal democracy. He just had his own strategy for achieving that goal. Ultimately, his strategy did not succeed.

In planning for Medvedev's visit, we debated about what kind of social event to organize. His trip was deliberately billed as a "working visit," parallel in protocol to Obama's trip to Moscow the year before. We toyed with elevating the summit to a "state visit," which would have included a formal state dinner, with tuxedos and ball gowns. I was all for that idea, as I wanted to take my wife to one state dinner at the White House before I left. (It never happened.) Ben Rhodes had a different idea — he proposed that Obama and Medvedev go for burgers at Ray's Hell Burger in Arlington, Virginia, just outside D.C. Ray's was a classic American grease joint, complete with rolls of paper towels on the tables. I worried that Ben's idea might seem vulgar to the Russians, in contrast to the very elaborate lunch that Medvedev had hosted for us at the Kremlin. Did Medvedev even eat hamburgers? But I was wrong; Rhodes was right. The outing was a success, both as a bonding experience for the two presidents and as a symbol of the new era in U.S.-Russia relations. Medvedev hopped in Obama's car and the presidents drove to Virginia together, had the chance to speak one-on-one for a couple of hours, with only translators present, and produced an iconic photo for the Reset era. Of the hundreds of meals that Obama had with heads of state over the course of his eight-year administration, his "burger summit" with Medvedev at Ray's Hell Burger was one of the most memorable.

After lunch, Obama and Medvedev walked through Lafayette Square in D.C. to the offices of the American Chamber of Commerce to meet with several dozen Russian and American CEOs of some of the largest companies on the planet. We were following the same script from the Moscow summit the summer before, moving from government-to-government interaction to engagement with business and civil society. It was a hot, humid day, so the two presidents took off their jackets and strolled casually. Even with the dozens of bodyguards surrounding them, the mood was relaxed and friendly.

The substance of the business summit was unremarkable, but the symbolism was important. Obama and Medvedev both highlighted their commitment to build a fundamentally different kind of relationship, focused just as much on economic ties as security issues. Obama started his remarks by using the dreaded "*f* word," saying that it was "a pleasure to be here with my friend and partner, President Medvedev, and I want to thank him again for his leadership, especially his vision for an innovative Russia that's modernizing its economy, including deeper economic ties between our two countries."[2] I was always wary about using the word "friend" in diplomacy, but I could not stop the president's reference in relation to Medvedev. Obama then praised Medvedev for his modern ways, joking about his new Twitter account and the need to retire the red "hotline" phones of the Cold War. He noted the benefits of trade with Russia for the United States, including the tens of thousands of jobs created from it. Obama even dared to hint at the importance of deeper democratic reforms in Russia, noting that he and Medvedev "discussed this — issues of transparency and accountability and rule of law remain absolutely critical. This is the foundation on which investments and economic growth depends. And I very much appreciate and applaud President Medvedev's efforts in this area."[3] Obama was summarizing the longer conversation the two presidents had in the Oval Office earlier that day. Medvedev in turn reported on his trip to Silicon Valley, underscoring that he was deeply impressed by what he had seen there. He also outlined a package of incentives to stimulate American investment in the Skolkovo project.

Around the table at this meeting were some of Russia's and America's most important business leaders, and there was high praise for the two presidents and their focus on economic relations and the desire to create incentives in support of trade and investment. At one point, Medvedev affirmed that "President Obama is doing great work," referring to the president's support for legislation that could stimulate greater economic cooperation between the two countries. The American Chamber of Commerce did not normally lead cheers for Obama. On this day, however, its members embraced him with glee.

Knowing of the president's love for basketball, one of the meeting's participants, Russian billionaire Mikhail Prokhorov, owner of the Brooklyn Nets, tried to give Obama a team jersey. The Secret Service intercepted him, however, so he passed the jersey on to me; I assured him I would get it to Obama, and as we parted ways Prokhorov gave me a big thumbs-up on the session. I walked back to the White House, pondering the idea that maybe we really could make economics, not nukes, the focal point of U.S.-Russia relations in the future. On that day, such a fundamental change in our bilateral agenda seemed not only possible, but probable.

Rounding out this summit, we worked with an American NGO, IREX, to organize a peer-to-peer meeting between civil society leaders from the United States and Russia. Neither Medvedev nor Obama made an appearance at this event, but Secretary Clinton attended, and gave a moving speech. She told the assembled activists, "We need creative, committed, courageous organizations like you and yours to find innovative solutions, to expose corruption, to give voice to the voiceless, to hold governments accountable to their citizens, to keep people informed and engaged on the issues that matter most to them."[4] She also pushed back on the Russian narrative that American NGOs working in Russia were violating Russian sovereignty: "In one of my early discussions with Minister Lavrov, he said, well, you know, we don't like it when you have so many NGOs coming to Russia. And I said, well, send Russian NGOs to the United States. [Laughter.] We'll be happy to have them. And I really mean that. I think the more exchange and the more . . . cross-fertilization the better."[5] In her remarks Clinton gave a shout-out to me, which I appreciated, especially in front of so many Russian civil society leaders, since it signaled that engagement on human rights issues was not just a State Department concern, but a White House priority as well.

At the press conference later that day, Obama and Medvedev were effusive about the results of the Reset. President Obama went out of his way to stress that our new relationship was not just about arms control and trade agreements, but also people-to-people ties:

> The new partnership between our people spans the spectrum, from space to science to sports. I think, President, you're aware that recently I welcomed to the White House a group of young Russian basketball players — both boys and girls — who were visiting the United States. We went on the White House basketball court, and I have to admit some of them out-shot me. [Laughter.] They represented the hope for the future that brings our countries together . . .

Obama then pivoted, as he always liked to do, back to World War II, the last time Russians and Americans called each other allies.

> Those were the same hopes of another generation of Americans and Russians — the generation that stood together as allies in the Second World War — the Great Patriotic War in which the Russian people suffered and sacrificed so much. We recently marked the 65th anniversary

> of our shared victory in that war, including that historic moment when American and Soviet troops came together in friendship at the Elbe River in Germany. A reporter who was there at that time, all those years ago, said: "If there is a fine, splendid world in the future, it will largely be because the United States and Russia get on well together. If it is in trouble, it will be because they don't get on well. It's as simple as that." . . . Mr. President, the decades that followed saw many troubles — too many troubles. But 65 years later, it's still as simple as that. Our countries are more secure and the world is safer when the United States and Russia get on well together. So I thank you for your partnership and your commitment to the future that we can build together, for this and for future generations.[6]

Medvedev echoed the same, promising to explore new areas of cooperation beyond the traditional agenda of U.S.-Russia relations.

At this summit, we took stock of what we had accomplished in just eighteen months.[7] First and foremost, there was the New START treaty, still not ratified, but signed earlier in the year. There was also UNSCR 1929, which put in place the most comprehensive sanctions against Iran ever. We had cooperated on North Korea, including our joint support for UN Security Council Resolution 1874 in response to North Korea's most recent nuclear test. On nonproliferation, there were numerous achievements, including Russia's participation in the Nuclear Security Summit on April 12–13, 2010, and a new protocol to the Plutonium Disposition Agreement, which committed both countries to disposing sixty-eight metric tons of weapons-grade plutonium (enough for seventeen thousand nuclear weapons). There was joint work on Afghanistan, including the expansion of the Northern Distribution Network, which had allowed for thirty-five thousand soldiers to fly to the war through new NDN routes. Russia had provided 30 percent of the fuel for our planes and vehicles in Afghanistan, while Russian companies flew over twelve thousand cargo flights in support of our war effort. American and Russian military cooperation of this magnitude was unprecedented. We were committed to finishing the WTO accession negotiations. We had resubmitted to Congress the 123 Agreement with Russia, which would foster civil nuclear cooperation between our two countries. Trade had expanded. We had worked together in Kyrgyzstan to promote security, even though the situation on the ground remained volatile. There were a dozen other areas of new collaboration, from military cooperation and energy efficiency to civil society engagement and the defense of human rights and democracy in Russia. Russian favorable attitudes toward the United States had increased to 60 percent in May 2010, while negative attitudes had dropped to 26 percent. Positive American attitudes toward Russia paralleled these gains in Russia. The Reset was producing real results.

Both presidents also engaged each other on the hardest security issues of that time in a spirit of cooperation, not confrontation. Obama and Medvedev were even discussing at length democracy and the rule of law, topics that would have been taboo two years earlier. As Medvedev left town, I went back to my office to celebrate with my team what seemed to us meaningful and historic work.

Only a few weeks after this successful summit, the Reset hit a major snag. The problem? Spies.

All countries — all countries with the capability, that is — engage in espionage. The United States and Russia are especially good at it, especially against each other. Yet whenever such activity is uncovered, the targeted nation convulses with shock and anger. The Russian clandestine operation in the United States that triggered the new drama in June 2010, however, was truly extraordinary.[8] The operation had got its start years before the Reset. I only learned about it that summer, but a small group of counterintelligence officials at the FBI, CIA, and elsewhere in U.S. intelligence and law enforcement communities had been tracking the ring for some time. A dozen Russian intelligence officers had been living in the United States for years, pretending to be Americans, just like on the television series *The Americans,* which was inspired, in fact, by this operation. For a decade, these Russian spies with false passports and hidden identities slowly were penetrating American elite society, becoming interns, working the Silicon Valley circuit, and inching closer to decision makers in the U.S. government and business worlds. We called them "the Illegals," because they were not declared as intelligence officers, as is customary with top officials from our respective intelligence agencies, and they were not using government jobs in the Russian Embassy as a cover, which also happens. Our counterintelligence team had discovered the group long before, but events that summer compelled their arrest by the FBI.

Many press reports about these Russian spies characterized the operation as a poorly executed blast from the Cold War past. How were these individuals, living normal lives in the New Jersey suburbs, a threat to U.S. national security? the argument went.[9] As one press report ridiculed, their assignments "were to collect routine political gossip and policy talk that might have been more efficiently gathered by surfing the Web."[10] Many journalists devoted special attention to Anna Chapman, a young, attractive redhead who looked like she could have been cast in a James Bond movie. She would later cash in on her looks and fame, becoming a celebrity back in Moscow. But though she might have looked the part, she didn't act it, as her efforts yielded no American government secrets.

I saw little humor in the matter. Instead, I was impressed with the strategic patience and the enormous resources that the Russian government had devoted to the spy ring. Had we not tracked its operation, it could have done serious damage. For instance, one of the operatives had applied to become an intern at a major foreign policy think tank. As I looked around the White House, I saw lots of young special assistants with the highest levels of security clearances, who just years earlier had been interns at think tanks. Imagine if one of the Russian Illegals landed a job as an executive assistant to the national security advisor? Chapman was developing relationships with some of America's most important companies in Silicon Valley. A decade later, these Russian spies could have secured important jobs in American government and industry.

The FBI's counterintelligence unit made the decision to arrest the eleven sleeper agents (ten Russians and a naturalized citizen from Peru) because of a potential compromise in its surveillance operation.[11] U.S. intelligence operatives worried that the Illegals might get tipped off and flee the country, so they arrested them all one morning and charged them with conspiracy to act as unregistered agents of a foreign government. Some also were charged with conspiracy to commit money laundering. Not surprisingly, senior leaders in the FBI, CIA, and other intelligence organizations wanted to prosecute, convict, and imprison these spies in the United States. As an additional measure, some recommended that we expel dozens of Russian diplomats and other known spies working under the cover of Russian Embassy employees. It was a tense moment in U.S.-Russian relations.

John Brennan, deputy national security advisor for homeland security and terrorism at the time, guided the interagency process for determining our response. It was clear to me that our senior leadership at the White House, including President Obama, did not want to derail our relations with Russia over these arrests. After all, the United States routinely spied on Russia, and many other countries in the world, including at times our allies.[12] This specific Russian operation did violate the informal norms of the spy world. Running longtime sleeper cells was extraordinary. Nonetheless, Obama and his closest advisors decided it was not in the American national interest to dramatically interrupt cooperation with Russia on so many other security and economic issues because of this spy ring.

After some heated days of deliberation, we decided not to prosecute the eleven spies, but instead to trade them for four individuals being held in Russian prisons on charges of espionage for us. Some at the FBI and CIA were not happy with what they considered a soft response. They wanted the court drama; they wanted the world to know about their work against the KGB; they wanted convictions. But Obama did not, and it was his decision to make. So we orchestrated a spy swap, on the tarmac in Geneva, just like in the Cold War days.

My only real contribution to this ordeal was pressing for a Russian academic, Igor Sutyagin, to be included in the detainee swap. Sutyagin's case was complicated. In 1999 he was convicted of espionage and sentenced to fifteen years in prison.[13] Sutyagin claimed that he was innocent, and I believed him. I had met him at Stanford when he was a visiting scholar there. He was a respected researcher on nuclear weapons. He didn't seem like the spying type. Russian prosecutors claimed that he transferred "secret" materials to a British consulting firm that was a CIA front.[14] My suspicion (which I never confirmed) was that he'd agreed to do some research for a foreign think tank that had links to an intelligence organization. How could he have obtained secret information? And how could he have known that this firm had links to a Western intelligence organization, even if Russian counterintelligence agents did? I was glad we were able to arrange for Sutyagin to be included in the prisoner exchange, even though I knew he was not going to like it, since his inclusion would be twisted by Russian officials into implying an admission of guilt.

In London after his release, Sutyagin stated emphatically that he was not a traitor, and planned to return to Russia. As he said at the time, "I haven't asked for political asylum here, that would suggest I am running away. And I don't want to run away from my own country."[15] Several years later, Sutyagin attended a talk I gave in London, and afterward thanked me for helping to secure his freedom. He had not returned to Russia, but had found work at a UK think tank. That was a good day. Sometimes government power can be wielded to do the right thing.

The drama surrounding the Illegals was a sober reminder to me that not everyone in Russia and the United States believed in the Reset. Obama's and Medvedev's commitment to more productive relations hadn't prompted Putin to pull the plug on Russia's espionage activities. In fact, the hardliners—the *siloviki,* as they are called in Russia—were not very present at any of our meetings with Medvedev. As we would learn later, they were watching carefully their new young president as he embraced our new young president, and they were doing so not with enthusiasm but anxiety. If we could not convince this constituency back in Moscow to get behind our ambitious plans for a fundamentally new kind of bilateral relations, we would fail.

Likewise, the 2010 spy drama underscored the skepticism surrounding Reset in the U.S. government. For those involved in American intelligence and counterintelligence, there was no Reset, no change in behavior at all from their Russian counterparts. On the contrary, in the Putin era, Russian spending on espionage had increased dramatically. In the spy-versus-spy world, Russia remained the United States' top enemy, no matter what praise our two presidents might heap on each other at a press conference in the East Wing of the White House. American Cold War thinking and habits were not going to change overnight, especially when Russia's key decision maker was a former officer of the KGB.

13

THE ARAB SPRING, LIBYA, AND THE BEGINNING OF THE END OF THE RESET

I had always planned to return to Stanford after two years, the standard interval of government service for professors. As 2011 began, just as I started making plans to leave D.C. that summer, a new, unexpected "project" appeared in the inbox of every NSC and Obama administration official: how to respond to the tumultuous events in the Middle East that became known as the Arab Spring. The challenge was both invigorating and frustrating; it ended up consuming a major chunk of my time in 2011.

Contrary to what Putin would later insist, the United States government did nothing to spark the Arab Spring. When a young fruit vendor, Mohammed Bouazizi, lit himself on fire in front of a government building in an act of protest against the indignities of Tunisian dictatorship, he was acting independently on his own beliefs. The demonstrations that followed throughout North Africa and the Middle East sprang up organically, with no help from the CIA, the State Department, or the White House.

As both a scholar and activist interested for decades in democratization, I was captivated by the events in the region, and therefore invented reasons to interact with NSC colleagues working on our policy responses. Tunisia had only a marginal impact on American national interests. When the Arab Spring fever spread to Egypt, however, our top foreign policy advisors became engaged. Hundreds of thousands of Egyptian protesters were mobilizing to oust the United States' longtime ally in Egypt, President Hosni Mubarak. The American government had all sorts of interests tied to Egypt, including, first and foremost, maintaining peace with Israel. Between 1979 and 2011 the United States had doled out tens of billions of dollars in bilateral foreign aid to Egypt to maintain stability there.[1] Only Israel had received more U.S. aid over that period. We could no longer just sit by and watch.

The start of these popular uprisings, as well as the speed by which they grew internally and spread throughout the region, surprised senior leadership in the Obama administration. But it's wrong to charge the administration with not having considered the prospects for political change in the Middle East prior to the Arab Spring. The year before, four of my NSC colleagues—Samantha Power, Dennis Ross, Gayle Smith, and Pradeep Ramamurthy—had launched an interagency policy discussion group, Presidential Study Directive 11, to consider exactly this possibility. Jeremy Weinstein, at the time the director for development and democracy, staffed these discussions and eventually authored the study. How likely was political change in the Middle East? What would these possible changes mean for American national interests? How, if at all, could we shape such events? I participated in these meetings, not as a Russia expert but as an academic specialist on democratic transitions. The PSD 11 group focused mostly on prospects for gradual, evolutionary change, meaning that events in 2011 swept away earlier, more conservative assessments. But the existence of this group did establish relationships between people in the U.S. government who might not otherwise have met, which played a positive role in shaping the administration's response to the Arab Spring.

In my experience, historical analogy is the analytic method of choice for senior foreign policy makers trying to get a handle on world events unfolding in real time. When trying to understand a new problem, they rarely use or even read analyses informed by social science methods such as game theory, statistical data, or randomized control trials. And logically, these analogies are made to historical cases with which the individuals are most familiar. I watched this play out dozens of times during my five years in government, and it was particularly striking during our struggles to understand the Arab Spring, and especially events in Egypt in the winter of 2011.

For many on our senior national security team, the Iranian Revolution quickly became the analogy applied to understand the events unfolding in

Egypt. Like the shah of Iran, Mubarak had been a close ally of the United States for several decades. Islamic radicals were threatening to topple Mubarak's dictatorship just as Islamic radicals in Iran had overthrown the shah in 1979, or so it was assumed. If extremists in Egypt succeeded, the result would be as disastrous for our interests in Egypt as the shah's expulsion, the ascension of the mullahs, and the creation of the Islamic Republic had been for American interests in Iran since the 1980s. This analogy also cast Obama as Jimmy Carter. President Carter, as champions of this analogy hinted, had been weak, too focused on human rights, and not worried enough about American security interests. To avoid Carter's fate, therefore, Obama had to stand behind Mubarak. It was a compelling analogy.

In her memoir, Clinton reflects on the internal deliberations in the White House over the U.S. response to developments in Egypt:

> Some of Obama's aides in the White House were swept up in the drama and idealism of the moment as they watched the pictures from Tahrir Square on television . . . I shared that feeling. It was a thrilling moment. But along with Vice President Biden, Secretary of Defense Bob Gates, and National Security Advisor Tom Donilon, I was concerned that we not be seen as pushing a longtime partner out the door, leaving Egypt, Israel, Jordan, and the region to an uncertain, dangerous future.[2]

Clinton and Gates, therefore, advocated for some of the same policy recommendations that followed from the Iran analogy. Back Mubarak. Don't repeat Carter's mistakes. In her memoir, Clinton explicitly invoked the Iranian analogy: "Historically, transitions from dictatorship to democracy are fraught with challenges and can easily go terribly wrong. In Iran in 1979, for instance, extremists hijacked the broad-based popular revolution against the shah and established a brutal theocracy. If something similar happened in Egypt, it would be a catastrophe, both for the people of Egypt as well as for Israeli and U.S. interests."[3]

I shared Clinton's concern about the challenges that arise between the breakdown of an autocratic regime and the consolidation of a democratic system of government. I also shared her analysis of the Iranian Revolution: it could have resulted in a democratic system, had the mullahs not commandeered the process. At the same time, I could think of dozens of historical cases of autocratic breakdown that had ended differently than Iran. I also disagreed with the policy recommendations that followed from the embrace of the Iranian metaphor. Reaffirming our support for Mubarak would lead to more bloodshed, in my view, and therefore was not a viable option.

Eventually I had the chance to inject alternative historical analogies into our policy debate. One evening in the West Wing, I was briefing Tom Donilon about some Russian matter when he asked me what I thought about the events in Egypt. As the resident political science professor, I was a frequent target for Tom's questions about academic theories and American foreign policy history. I looked forward to these conversations, which usually took place late at night when the pressure of Tom's inbox was a little less intense. On Egypt, I told him that I recognized the appeal of the Iranian analogy, but suggested that there were dozens of cases of autocratic breakdown that might illuminate alternative trajectories for the state. The Egyptians were not doomed to follow the Iranian script. Even the outcome of the Iranian Revolution was more contingent than most assumed. That night, I focused in particular on three other analogies: Chile, the Philippines, and South Korea in the 1980s. In all three cases, the dictator in power at the time was a close ally of the United States. In all three countries, our autocratic allies claimed that they needed our backing to prevent anti-American radicals — back then communists, not Islamists — from taking over. Yet, as societal pressures in these countries began to mount, the Reagan administration helped ease all three autocratic rulers out of power in an orderly, evolutionary process. The transition to democratic rule in these instances did not bring radicals to power and did not fundamentally disrupt American bilateral relations with the three countries. Maybe Egypt could follow this path? Importantly, these analogies cast Obama as Reagan, not Carter. And everyone, it seemed, wanted to be compared to Reagan.

Tom was skeptical, but intrigued. He asked me for reading recommendations on these cases. I considered giving him an article I had written on the topic,[4] but instead gave him my copy of George Shultz's memoir, *Turmoil and Triumph.* I knew Shultz's book would make a bigger impression than my academic article! Soon thereafter, Tom asked me to write up dozens of one- to two-page case studies of transitions from authoritarian rule over the last fifty years,

At the end of my sophomore year at Stanford University, I traveled from Bozeman, Montana, to Leningrad, USSR, June 1983. It was my first trip abroad. *Courtesy of the author*

After ratifying the INF Treaty, the first agreement between the US and the USSR to scale down their nuclear arsenals, President Reagan and General Secretary Gorbachev tour Red Square, May 1988. *Sovfoto, UIG / Getty Images*

On March 10, 1991, I joined hundreds of thousands of protesters in Manezh Square. They were demanding that the Communist leadership step down. *AP Photo / Dominique Mollard*

After amicable talks on advancing peace in Bosnia, Presidents Bill Clinton and Boris Yeltsin joke during a joint news conference in Hyde Park, New York, October 23, 1995. They had a special bond. *Wally McNamee, Corbis Historical / Getty Images*

President George W. Bush's first meeting with Vladimir Putin, at Brdo Castle, outside Ljubjana, Slovenia, June 16, 2001, when Bush famously commented on his sense of the Russian president's soul. *Alain Buu, Gamma-Rapho / Getty Images*

Conferring with President Obama on January 26, 2009, before his first call to President Medvedev from the Oval Office. The two leaders, according to our readout, spoke about the "importance of stopping the drift in US-Russia relations and building a serious agenda for their bilateral relationship." *Official White House Photo by Pete Souza*

Deputy National Security Advisor Denis McDonough and I were ready to brief President Obama before his interview with Russian television just days before the July 2009 summit in Moscow, but he had other ideas. We conducted most of this briefing in the Rose Garden, while also throwing a football. *Official White House Photo by Pete Souza*

Briefing President Obama aboard *Air Force One* midflight to Moscow, alongside, from left, NSC Chief of Staff Mark Lippert, Under Secretary of State William J. Burns, Deputy National Security Advisor Ben Rhodes, and Deputy National Security Advisor Denis McDonough, July 5, 2009. *Official White House Photo by Pete Souza*

Presidents Obama and Medvedev meeting at the Kremlin with their respective delegations, July 6, 2009. The ugly brown "lock bag" beside my chair contained our classified documents. *Official White House Photo by Pete Souza*

President Obama's first meeting with Prime Minister Putin. At Putin's residence outside Moscow, National Security Advisor General Jim Jones, Under Secretary for Political Affairs Bill Burns, myself, Foreign Minister Sergey Lavrov, and Deputy Chief of the Government Staff Yuri Ushakov were in attendance, July 7, 2009.
Official White House Photo by Pete Souza

Secretary of State Hillary Clinton presents Foreign Minister Sergey Lavrov with a "reset button" in Geneva, Switzerland, March 6, 2009.
Official State Department Photo

Briefing President Obama on details regarding missile defense issues in the New START negotiations as he talks to President Medvedev. He was not impressed with our staff work that day. Oval Office, February 24, 2010.
Official White House Photo by Pete Souza

Two happy arms control negotiators, Obama and Medvedev, after signing the New START treaty, Prague Castle, Czech Republic, April 8, 2010. *Getty Images*

Left: After ratification of the New START treaty, our NSC staff shares a toast with President Obama in the Oval Office, December 22, 2010. *Official White House Photo by Pete Souza*

Below: Presidents Obama and Medvedev at Ray's Hell Burger in Arlington, Virginia, in what came to be known as the Burger Summit, June 24, 2010. *Official White House Photo by Pete Souza*

Secretary of State Clinton swears me in to my new post as the US ambassador to the Russian Federation, with my family looking on, State Department, Washington, DC, January 10, 2012. *Astrid Riecken / Getty Images*

Above: Weeks before I arrived as ambassador, on December 10, 2011, demonstrators gathered in an unauthorized rally to protest electoral fraud in Moscow's Bolotnaya Square. The Kremlin would later blame me for organizing these demonstrations. *Sovfoto, UIG / Getty Images*

Left: Russian propaganda constantly accused me of fomenting revolution. Here, I was depicted on the cover of a calendar as head of the "McFaul Girls," in which every month featured a different Russian opposition leader. *Courtesy of Prawdzinski Vyacheslav*

When the government imposed bans on my speaking at universities, I invited students for lectures at my residence, Spaso House, like this one on December 16, 2013. *Official Photo of the US Embassy in Moscow*

Introducing my wife, Donna, at the July Fourth celebration at Spaso House in 2013, with my sons looking on. Rebecca Neff of the US Embassy's Economic Section is emceeing. *Official Photo of the US Embassy in Moscow*

Secretary of State John Kerry at his first meeting with President Putin in the Kremlin. The main topic: Syria. May 7, 2013. *Official State Department Photo*

Touring Red Square with Secretary Kerry while waiting to see President Putin. It was just two days before the May 9 parade commemorating the Soviet victory in the Great Patriotic War, so the square was closed to the public and we had the place to ourselves. May 7, 2013. *Mladen Antonov, AFP / Getty Images*

At the end of the G-20 Summit, President Obama and I say goodbye before he boards *Air Force One* in St. Petersburg, September 6, 2013. *Official White House Photo by Pete Souza*

My last speech at Spaso House, February 2014. Note that Boris Nemtsov was there; it was the last time I saw him. He was assassinated a year later. *Official Photo of the US Embassy in Moscow*

both cases that resulted in democracy as well as those that did not.[5] As I got deeper into the project, I began to circulate memos on other aspects of democratic transition, such as the difference between parliamentary and presidential systems and the pluses and minuses of various electoral systems. In what I'm sure was a departure from typical NSC process, I sent around slides from my lecture on democratization that I had been teaching for years at Stanford, as well as short papers summarizing the academic literature on democratic transitions. Jeremy Weinstein, another Stanford professor working at the NSC at the time, also engaged in this project, as he knew this academic literature as well. At one point, fellow NSC member Dennis Ross coined the phrase "nerd directorate" to describe our efforts. McDonough would flag some of my more academic emails "Nerd Alert." One of Stanford's nicknames is Nerd Nation, so I took the new label as a compliment. Our executive secretary at the NSC, Nate Tibbits, set up a special email list—#Egypt—to better connect us nerds with everyone in the building working on Arab Spring issues. In another email exchange with Dennis Ross about a paper on Serbia and Polish transitions written for me by Cornell University professor Valerie Bunce, Dennis wrote, "I am beginning to worry—I did not find it so nerdy." We in the nerd directorate were mainstreaming nerdiness!

Not everyone was as appreciative of our efforts. For many in the Obama administration, the Arab Spring was a giant, unwanted crisis, distracting us from other objectives such as the "Asia pivot," management of the Iranian nuclear program, and even the Reset. But in the early phases, I personally found the events in the Arab world exhilarating. People were organizing peacefully to challenge corrupt, authoritarian leaders. What could be more exciting than that? I also believed that, from the margins, the United States could nudge these events in the right direction. As progress on the Reset tapered, the Arab Spring gave me a new, big project on which to engage.

With time, I got the sense that President Obama was beginning to see these events the same way that our side in this debate did. Obama was leaning toward pressuring Mubarak to step down. Not everyone agreed with this approach. Gates eventually played a critical role in convincing the Egyptian military to ease Mubarak out, but remained skeptical of our strategy. So did Clinton. At lower levels of government, many advocated for maintaining the status quo:

the protests would wane, and we could get back to business with Mubarak. Some senior officials in our government were pressing the president to support Mubarak because he could help us with the Middle East peace process, as if such a process existed.

Along with some others, including Ben Rhodes, Samantha Power, and Susan Rice, I found the arguments for continuity in policy unrealistic. We could not control events in Egypt. The choice was not between keeping Mubarak in power or siding with the revolutionaries. The trade-off was between peaceful evolutionary change and violent revolutionary change. If Mubarak tried to hold on to power by using force, tens of thousands of innocent civilians were going to die. Tragically, we witnessed that exact scenario in Syria a few years later. Moreover, fighting for his political survival, Mubarak was not going to play any constructive role in the Middle East peace process. That was folly.

When violence broke out in Tahrir Square on February 2, 2011, Rhodes told me that several senior members of our national security team were blaming us democracy promoters for provoking violence by calling for a democratic transition. Obama eventually and firmly joined our side of the argument, but in doing so alienated some of his senior foreign policy advisors. On Egypt, our government was divided.

Through military channels, we started to urge others within the Egyptian government to push Mubarak out gently and without arrest as the best and maybe only way to preserve the state. The longer Mubarak stayed, the more likely violence would escalate, and then radical groups would seize power. We argued for a negotiated transition between moderate elements of civil society and those who would remain in the government following Mubarak's retirement.

Those of us advocating for Mubarak's resignation and a negotiated transition worried about the unintended consequences of leadership change. We fully recognized that we didn't know much about who might come after Mubarak. Standing in the Roosevelt Room for a birthday party on February 9, I asked Obama what he thought. His response was tragically prescient. "What I want is for the kids on the street to win and for the Google guy to become president," he said, referring to Wael Ghonim, an internet activist whom the police had arrested during the protests. "What I think is that this is going to be long and hard."[6]

Two days later, on February 11, 2011, Mubarak stepped down. Speaking from the Grand Foyer of the White House that afternoon, President Obama captured the magnitude of what was happening: "There are very few moments in our lives where we have the privilege to witness history taking place. This is one of those moments. This is one of those times. The people of Egypt have spoken, their voices have been heard, and Egypt will never be the same."[7]

Of course, demonstrators in Tahrir Square did most of the work to force Mubarak out. But in the margins, our administration had helped the process, engaging with senior officials in the Egyptian military to urge them to pressure Mubarak to go. In his remarks that day, Obama connected our values to those of the protesters, proclaiming, "Today belongs to the people of Egypt, and the American people are moved by these scenes in Cairo and across Egypt because of who we are as a people and the kind of world that we want our children to grow up in."[8] The moment reminded me of events in Russia in 1991, when Americans and Russians with a shared commitment to democratic values had also stood together.

I was most certainly moved. That day was a special one, made all the more so by messages I received from Egyptians in Tahrir Square praising Obama's remarks. As one activist in Cairo wrote to me that day, "Amazing speech by Obama tonight Mike. Big hit here!" I knew such moments in history were rare and had to be celebrated, marked in some way to help us remember them, since the days that followed would be harder. They always were. To the #Egypt team at the National Security Council, I wrote that day, "No matter what happens moving forward, the successful non-violent mobilization to replace an autocratic leader anywhere, let alone Egypt, is a historic achievement to which your careful but clear policy contributed greatly, Congratulations." I pinged Rhodes and McDonough in an email, asking, Vodka or beer? We gathered with other NSC staffers at my office to watch the festivities in Tahrir Square and celebrate in parallel with them. Inscribed on President Obama's carpet in the Oval Office was a quote credited to Martin Luther King Jr.: "The arc of the moral universe is long, but it bends towards justice."[9] That night, that arc seemed to be bending just a little more sharply, nudged a bit by our own efforts.

But not for very long.

• • •

A well-entrenched autocracy rarely segues smoothly into a new democracy. As Obama rightly emphasized in his remarks on the day Mubarak relinquished power, "By stepping down, President Mubarak responded to the Egyptian people's hunger for change. But this is not the end of Egypt's transition. It's a beginning. I am sure there will be difficult days ahead, and many questions remain unanswered."[10] Egyptians had to take the lead in tackling these issues. I also believed that we had to help them answer these questions of transition, as well as express our firm commitment to a democratic transition, and not just leave them on their own to figure it all out. In retrospect, we did not do enough to help them.

After Mubarak stepped down, Egypt's military leaders — now called the Supreme Council of the Armed Forces (SCAF) — took over as an interim government, pledging to transfer power to civilian leadership as soon as possible. But when? How? To whom? My view, shared by some others, was that we should not be shy about communicating our ideas for crafting a peaceful, evolutionary path toward democratic rule. We were not neutral in this unfolding drama. We had an interest in democracy taking hold. Specifically, I believed the timing, type, and sequence of first elections could influence the prospects for democratic consolidation. On February 6, 2011, shortly before Mubarak's fall, I wrote a detailed memo to Donilon called "Breaking the Stalemate, Crafting a Transition to Democracy in Egypt," in which I argued that we needed to help both sides understand that they had a mutual interest in shifting the drama from the streets to the negotiation table.[11] The regime's interest was in preventing the next large demonstration; for the demonstrators, it was preventing a dissipation of support and the creeping back of the status quo. The military and the opposition had to produce a set of temporary rules — a "pact" — to guide the transition process.[12] At the time, some of us were concerned that a hasty election would advantage the more organized Muslim Brotherhood or Mubarak's old party, the National Democratic Party. We were getting reports from Cairo that the democratic activists, revolutionary groups, and even Islamists were making efforts to unite. But I knew how difficult that process would be; I had watched the liberal opposition fail to unite in Russia after the collapse of the Soviet Union in 1991, and that disunity was not good for democracy.

To get the military out of power as quickly as possible, I pushed for the election or appointment of an interim president. I was familiar with the academic literature on the perils of presidentialism in democratic transitions, yet I still argued that Egypt needed a new, legitimate leader to replace the SCAF. An *interim* president could stay in power until after the transition had been completed, just as had happened in Kyrgyzstan the previous year. Designing the transition rules so that a more neutral figure — for instance, the former head of the International Atomic Energy Agency, Mohamed ElBaradei — could serve as an interim head of state seemed like the highest of priorities. And if the Muslim Brotherhood did succeed in winning that first presidential election, at least it would be to an interim, transitional position, not a fixed four- or five-year term. Above all else, the Egyptians needed time. An interim head of state would give them more time to sort out transitional arrangements.

Our government discussed some of these ideas with General Tantawi, the head of the SCAF. Secretary Gates spoke to Tantawi frequently. National Security Advisor Tom Donilon also called Tantawi on February 15, 2011, to discuss explicitly these lessons learned from other successful democratic transitions, including first and foremost the dangers of quick elections.

Unfortunately, the SCAF moved forward with a referendum on March 19 in which a 77 percent majority of voters approved constitutional changes that cleared the way for swift parliamentary and presidential elections. The Muslim Brotherhood supported the referendum; other liberal groups had opposed it, arguing that the accelerated timetable would not allow for new political parties to organize.

If parliamentary elections had to proceed hastily and precede a presidential election, my recommendation was to craft electoral laws that would prevent a "winner takes all" scenario. The Muslim Brotherhood had organizational capacity that the young, liberal demonstrators in Tahrir Square lacked. I championed the adoption of Morocco's clever electoral law, which at that time allotted two seats in parliament from every electoral district on a proportional basis, making it difficult for the first-place finisher to win enough votes to secure a second seat, thereby guaranteeing that the second-place finisher in each electoral district still won a seat in parliament. This electoral law had suppressed parliamentary representation of Morocco's largest Islamist party, the Party of Justice and Development (PJD). I believed the same could be done in Egypt. I worried that a system of single-mandate electoral districts — the American system for electing representatives to the U.S. Congress — would favor the Muslim

Brotherhood, which enjoyed high visibility in Egyptian communities due to its long track record of social service provision. Most of the liberals were completely unknown, and would therefore be defeated easily, I thought, by Muslim Brotherhood candidates in head-to-head matches.

There were other steps that could be taken to increase the prospects for a successful negotiated transition to democracy. On March 4, I sent Donilon and McDonough another memo: a blueprint for transition that advocated for a slower transition, an interim presidential election followed by parliamentary elections, and a system of proportional representation, among other suggestions. As an incentive for adopting these recommendations, the United States could offer Egypt a onetime economic package — a "democratic dividend" — to demonstrate its commitment to moving the democratic process in Egypt forward.

Of course, I cannot say whether implementation of this road map would have produced a more successful transition in Egypt. But it was never tried. We pressed some of these ideas, albeit not as forcefully as I would have liked, but the Egyptians — both in and out of government — had their own ideas. For instance, the idea of establishing negotiations between the existing autocratic regime and the protesters, as the Poles and South Africans had once done, never gained support within the military, and found little traction with the opposition. The Egyptian military had dominated politics for decades; the asymmetric balance of power between the military and societal challengers changed only slightly even after the revolution. There were two rounds of national dialogue tailored to create momentum for negotiations, but the results were disappointing.[13] The military was not in the habit of talking directly to civil society, and the opposition was not united, making it difficult to determine who would represent society in negotiations with the military.

I also witnessed unease within our government about our involvement in Egypt's transition. Not everyone was on board with my active-engagement approach, arguing that we had to respect the sovereignty of the Egyptian people and let them figure it out. I privately wondered why these sovereignty champions had been so quiet during decades of American subsidization of Egyptian autocracy. We were not very neutral then! But in government meetings, I kept that thought to myself and argued instead that providing ideas about democracy was not an imperial act but an educational exercise, especially when there was a demand from some in Egypt for this advice. I sensed that Obama wanted guidance on how to best assist the Egyptians. I saw the tension explode once in a meeting in the White House Situation Room in early March 2011. The president clearly was frustrated with the conservative advice that he was receiving from some in his senior leadership, now well after the fall of Mubarak. As the president stormed out of the room, I heard several members of his senior national security team whisper unflattering musings about the "young revolutionaries" who had captured the president's ear. Rhodes in particular was singled out. I'm sure some would have said the same about me had I not been in earshot.

Beyond encouraging our government to engage with the Egyptian military, I also urged more discussion with Egypt's opposition leaders about institutional design questions. Lines of communication with the opposition were more difficult for our government as a whole, and for me personally. Our government had little credibility with these democratic activists. For decades, we had supported their enemies. Why should they listen to us now? Besides, they believed they had already won. I had seen this same dynamic before, in the fall of 1991, when Russia's democrats did not want to be bothered with technical details of forming a functioning state. They were revolutionaries, not engineers. Russia's democrats had been wrong to ignore these details in 1991; I worried the Egyptian democrats were making the same mistake in 2011.

We also debated whether our outreach to opposition activists should include the Muslim Brotherhood. Were they the problem, or could they be part of the solution? The discussion reminded me of a similar policy debate we had about engaging the Russian Communist Party after the collapse of the Soviet Union. Eventually, our embassy started to meet with Muslim Brotherhood emissaries, but we knew very little about each other, making complex conversations about democratic transition difficult. And we had no contact with and very little information on the Salafists, Islamists even more extreme than the Muslim Brotherhood who would surprise everyone several months later by winning nearly a quarter of the popular vote in the parliamentary election.

While the Egyptians' interest in engaging with us was fading, I began to sense that, similarly, the Middle East hands in our government were becoming less enthusiastic about engaging with me. I was not an expert on the Middle East, so why was I attending Deputies Committee meetings on the Arab Spring? I understood their point of view. But some of the exchanges during those meetings revealed that several senior government officials responsible for the Middle East did not know how democratic transitions work, and had no comparative historic context to help them understand events in "their" region. And why should they? Transitions to democracy had never occurred in "their" part of the world. I wrote to Rhodes about it in frustration, and said the same to Donilon and McDonough. Denis urged me to keep coming to these Deputies Committee meetings, but I eventually dropped off. Informally and discreetly, I continued to provide input when asked, usually in written memos. But generally, I sensed a growing reluctance within our government to offer specific advice on transition issues to the Egyptians, and in turn less demand for my expertise. And then, Libya began to dominate our policy deliberations. I supported the use of force there, but no one at senior levels in our administration was asking for my opinion on that issue. I was a democratic-transitions guy, not a military intervention expert.

I continued to look for ways to share my views on the Arab Spring, usually through editing and contributing to public statements penned by Rhodes. I still pressed for the idea of managed transitions in Egypt, Bahrain, and Syria when anyone would listen.[14] I also engaged in shaping the larger policy debate about why it was in America's interest to help democracy take hold in the Middle East. Months into the popular uprisings throughout the region, many senior officials in our government remained skeptical. However, one senior official — the most senior official — was on our side, that is, those of us pressing for the United States to remain engaged in promoting democratic reforms in the Middle East. In an interview with a Russian television station on the occasion of his fiftieth birthday that summer, Obama listed the Arab Spring among the most important events in his lifetime, right next to the civil rights movement, the end of the Cold War, and the release of Nelson Mandela. At least in the spring and summer of 2011, Obama understood the magnitude of what was happening in the Middle East. He wanted to be a part of it.

After Mubarak fell, Obama decided to give a second Cairo speech. He could communicate his vision for the Middle East to the world, the American people, and also our own government. I saw such a statement as an opportunity to clarify and codify our policy on the Arab Spring.

In this speech, delivered on May 19 at the State Department, Obama clearly laid out why the United Sates had an interest in seeing democracy take hold in the Middle East.

> Two years ago in Cairo, I began to broaden our engagement based upon mutual interests and mutual respect. I believed then—and I believe now—that we have a stake not just in the stability of nations, but in the self-determination of individuals. The status quo is not sustainable. Societies held together by fear and repression may offer the illusion of stability for a time, but they are built upon fault lines that will eventually tear asunder.

Obama then underscored that the people of these countries, not the United States, were driving the process of reform, but that we would actively support their efforts.

> It will be the policy of the United States to promote reform across the region, and to support transitions to democracy. That effort begins in Egypt and Tunisia, where the stakes are high—as Tunisia was at the vanguard of this democratic wave, and Egypt is both a longstanding partner and the Arab world's largest nation. Both nations can set a strong example through free and fair elections, a vibrant civil society, accountable and effective democratic institutions, and responsible regional leadership. But our support must also extend to nations where transitions have yet to take place.[15]

Never before had a U.S. president stated so explicitly that the United States considered the promotion of democracy in the Middle East a foreign policy objective. Moreover, President Obama made clear that this goal was not secondary to other interests. In the battles under way between autocrats and democrats in Tunisia, Egypt, Libya, Syria, and Yemen, we were taking sides.

Because of my work on the speech, I was invited to travel with President Obama to the State Department for its delivery. After the president finished his remarks, we took the elevator down to the basement to climb back into our cars in the president's motorcade for the five-minute ride back to the White House.

Obama and Clinton were chatting as they walked by me, but I caught the president's eye. After he had concluded his conversation with the secretary, he turned around and shouted back at me, "How'd I do, McFaul?" Of course, I shouted back, "Excellent." I was excited to be present for what I thought could be a major policy pivot for our administration. Obama had made a powerful statement in support of democracy in the Middle East. I allowed myself to speculate that this speech could become a historic document, depending on how things turned out. In government, you get credit for good things that happen during your administration, even if you're not directly responsible for making them happen. If democracy took hold in the Middle East on our watch, this speech would associate the Obama administration with that positive outcome.

Privately, however, I was worried. The speech was excellent as an aspirational statement. It was thin, however, on strategy and implementation, on what concrete policy actions the United States was going to pursue to foster democratic development in the Middle East. Ironically, for an address about supporting democracy, the four concrete policies announced all related to economic reform, including a plan from the IMF and World Bank to stabilize the economies of Egypt and Tunisia, $1 billion in debt relief and $1 billion in credit for Egypt, the creation of enterprise funds in Tunisia and Egypt modeled on those set up in Eastern Europe after the fall of communism, and the launching of "a comprehensive Trade and Investment Partnership Initiative in the Middle East and North Africa." The speech had no ambitious proposals or major new initiatives for supporting democracy, just a pledge from the president to sign a presidential policy directive (PPD) on these issues in the future. Saad Eddin Ibrahim, an Egyptian human rights activist who had recently returned home after years living abroad, wrote to me from Cairo that day with his assessment of the speech: "Grateful for economic aid; but did not do enough to support other beleaguered democracy fighters in the region."[16] Saad was right. We should have been offering more ideas and assistance to Egyptian democrats, and to others fighting for the same throughout the region. At the time, I believed Obama needed some signature new initiative for advancing democratic values in the region, such as Kennedy's Peace Corps or Reagan's National Endowment for Democracy. In a policy paper I drafted for my NSC bosses on August 24, 2011, called "Advancing American Interests with Egypt over the Long Term," I warned that regime change in Egypt could take decades, that "autocratic restoration" remained a real possibility, and that we therefore needed to create new institutions to stay engaged for the long haul. I floated the idea of establishing the "Foundation for Universal Values in the Middle East," or a "Public Policy School for Democracy, Human Rights, and Universal Values," or more modestly a "Virtual Library for the Study of Universal Values." Others, including those drafting and trying to execute the PPD on the Arab Spring, proposed more ideas. But we never succeeded, in part because of internal bureaucratic inertia and in part because events in the Middle East (with the exception of Tunisia) turned quickly away from discussions of democracy and toward military conflict. We were presented with an opportunity to do something big, but we did not deliver.

In retrospect, I still believe that some different institutional and timing choices after Mubarak's fall could have produced higher probabilities for democratic consolidation in Egypt. A managed transition could have slowed down the process and avoided hasty elections, giving less organized groups more time to prepare. Allocating all parliamentary seats through proportional representation might have suppressed the electoral success of the Muslim Brotherhood and the Salafists. Had the liberals supported one candidate for president instead of several, their candidate could have reached the second round and had a much better chance of defeating the Brotherhood candidate, Mohamed Morsi. After all, Morsi won only 51.7 percent of the vote in June 2012, running in the second round against Ahmed Shafik, the last prime minister in Mubarak's government. That was a huge missed opportunity.

And then we stood by passively when the Egyptian military moved to seize power again in June 2013, a move that set back democracy's prospects for years. At the time, President Morsi, the Muslim Brotherhood, and the Salafists were all fading in popularity. A new round of free and fair elections might have produced a genuine, peaceful change of power. Morsi was no democrat, so a free and fair election would have proved difficult to pull off. But we had leverage at the time, and we chose not to use it to push him in a democratic direction. We disengaged at precisely the wrong time.

Perhaps our efforts would not have made a difference. Americans often overestimate their power and importance in these transitions. And other external actors, including Saudi Arabia and the United Arab Emirates, were engaged, pushing the process in an antidemocratic direction.[17] They invested billions and got the result they wanted — restoration of a military dictatorship. In retrospect, perhaps we had few real options available. But at least we should have tried, or tried a little harder, to build upon the opening provided by Mubarak's resignation in February 2011. What at the time seemed to me like one of the greatest achievements of Obama's foreign policy now feels in retrospect like a squandered opportunity.

But before democracy began to crumble in Egypt, things got worse in other countries swept up in the Arab Spring, first in Libya, and then Syria.

We spent the spring of 2011 reacting to events on the ground in the Middle East. Many of our early responses were bilateral, between the United States and Tunisia, Egypt, and Yemen, respectively. Libya (and Syria, discussed in detail in chapter 20), however, required a multilateral response, and eventually a military response. Consequently, to act, we needed the Russians.

Muammar Gadhafi, Libya's autocratic ruler for decades, was crazy — literally so, in my view. At the United Nations General Assembly meeting in September 2009, I sat with Obama and a few other White House officials in our holding room, watching Gadhafi roll through an address featuring all kinds of kooky ideas. In his rant he labeled the UN Security Council a "terror council," argued that swine flu was a weapon invented by the West, offered to host the United Nations headquarters in Libya since New York was no longer safe, and repeated his idea of creating "Isratine" out of the existing state of Israel and the Palestinian territories. I found it shocking that this guy was allowed to propagate these ideas from the same podium from which Obama had just spoken. Someone in our group remarked that our international order must be pretty insane if Obama and Gadhafi were treated as equals.

Obama, however, respected that international order. To borrow a term from international relations theory, the president was a "liberal institutionalist," believing that American support for international rules, norms, and organizations strengthened America's influence in the world over the long run. Generally, I held a similar view, but was more skeptical. In particular, I had written in the past that dictators forfeited their right to international sovereignty when they did not respect the sovereignty of their citizens.[18] Why should a maniacal dictator like Gadhafi get to represent the Libyan people at the United Nations? Who selected him? And shouldn't the UN have the power to censure or throw him out of the organization? But my ideas about international law did not matter.

Obama's did. Throughout his tenure in office, Obama maintained a consistent respect for both using and abiding by UN procedures to tackle even the toughest of issues, including how to stop looming genocide in Libya.

When the Arab Spring tide reached Libya, Gadhafi adopted a different response to peaceful protesters than had Ben Ali in Tunisia or Mubarak in Egypt. Gadhafi made clear his intention to use brutal force against his opponents. Most frighteningly, the Libyan dictator ordered his army to retake the eastern city of Benghazi, a rebel stronghold at the time, and warned that those who did not welcome his soldiers would be wiped out like "rats and dogs." He proclaimed bluntly, "The issue has been decided . . . We are coming tonight . . . We will find you in your closets . . . We will have no mercy and no pity."[19]

Our analysis was that Gadhafi's assault on Benghazi would produce massive civilian casualties. As Secretary Gates wrote in his memoirs, "By March, there was a real danger [Libyan forces] could soon move on Benghazi, and few doubted that the city's capture would lead to a bloodbath."[20] Could the most powerful country in the world just stand by and watch? Gates argued yes. No American strategic interests were at stake, so why get involved in a third war in the Middle East? Gates condescendingly dismissed the opinions of some White House colleagues as "blithely talking about the use of military force as though it were some kind of video game." He asked rhetorically, "Can I just finish the two wars we're already in before you go looking for new ones?"[21] Secretary Clinton, however, parted ways with Gates on Libya and supported the use of force, not as a means for regime change, but as an act of humanitarian intervention. The debate over Libya was even more divisive than our differences over Egypt.[22]

Obama eventually made the decision to intervene in Libya for humanitarian reasons, but not without conditions. First, there would be no boots on the ground, no regime change, no nation building. Obama affirmed this stance in a public announcement about the U.S. military response on March 19, 2011: "As a part of this effort, the United States will contribute our unique capabilities at the front end of the mission to protect Libyan civilians, and enable the enforcement of a no-fly zone that will be led by our international partners. And as I said yesterday, we will not — I repeat — we will not deploy any U.S. troops on the ground."[23] Our mission was defined narrowly — use air power in cooperation with our allies to stop Gadhafi from slaughtering innocents in Benghazi.

Second, Obama instructed that we had to obtain UN Security Council approval. This second condition meant that we had to secure Russian support, and that's where I came in. Early in the Libya conflict, I joined those who advocated for UN involvement. On February 19 I wrote to the #Egypt email group, asking, "Doesn't the level and degree of slaughter in Libya today warrant a UNSC response?" But I was not optimistic about getting the Russians to sign on. As a permanent member of the UN Security Council, the Soviet Union and Russia had never authorized international intervention in a civil war outside their self-defined sphere of influence. Soviet leaders intervened in Hungary in 1956, Czechoslovakia in 1968, and Afghanistan in 1979, but justified these invasions as fulfilling a request made by governments in those countries. Throughout the rest of the world, the USSR had consistently blocked UN approval of American-led interventions.[24] Post-Soviet Russian leaders also remained firm defenders of state sovereignty. Even the pro-Western, pro-American Boris Yeltsin refused to support a UN Security Council resolution in 1999 to authorize the use of force against Serbian dictator Slobodan Milošević to stop the genocidal campaign in Kosovo. In 2011 we were seeking a UN resolution for an almost identical purpose, with Putin, not Yeltsin, as the key decision maker in Moscow. Those seemed liked long odds to me.

At the time, however, Medvedev, not Putin, was Russia's president. That made a difference. In February 2011, Susan Rice secured Russian support for the first resolution on Libya, UNSCR 1970, which called for an arms embargo but stopped short of authorizing force. Aware that only more-drastic measures would slow Gadhafi's offensive, we pressed forward with our allies the following month to adopt a new resolution to authorize a no-fly zone, enforceable through external military intervention. In New York, Rice again leaned hard on Russia's ambassador to the UN, Vitaly Churkin, imploring him to work with us to prevent genocide. Secretary Clinton also engaged Lavrov to obtain Russian support for the Security Council resolution. But the decisive moment in this multidimensional diplomatic effort was Vice President Biden's meeting with President Medvedev in Moscow on March 10. That's when I knew that the Russians would support a military attack on Libya.

It has become conventional wisdom to describe Vice President Biden as a talk first, think later kind of guy. Obviously, he has a record of off-the-cuff remarks to confirm that characterization. And on our trip to Moscow in March 2011, he had one of those moments. While meeting with Russian opposition leaders in the dining room at Spaso House, just a few hours after his meeting with Prime Minister Putin, the vice president undiplomatically commented that he didn't think Putin should run for a third term as president. One of the participants immediately stepped out of the meeting, called a journalist, and reported the vice president's remark. Even before we had finished the meeting, Biden had generated breaking news in Russia. One headline read, "Biden Tried to Persuade Putin Not to Participate in [Presidential] Elections."[25] That was most unfortunate.

His occasional gaffes aside, Vice President Biden struck me as someone who took very seriously the need for preparation heading into important meetings. On that trip to Russia in March, we spent several hours in the "submarine" —probably the same steel-walled, completely contained secure structure that had been used for Obama's trip to Moscow in July 2009—rehearsing his talking points for the Medvedev and Putin meetings. Top on his list of agenda items for Medvedev was Libya.

Unusual for Kremlin protocol back then, Medvedev's schedulers asked for a one-on-one meeting with Biden before the rest of our delegations joined. Biden's national security advisor, Tony Blinken, didn't like that format; he felt it disadvantaged Biden, as Medvedev had primary responsibility in his government for foreign policy and all things American, while Biden was not our point person for Russia. Obama was. In fact, this configuration was unusual even for meetings between Obama and Medvedev; a note taker and two to three other individuals were almost always present in those meetings. To correct the imbalance, Tony instructed me to follow the vice president into the meeting and hope that there would be a seat somewhere in the room. I did as told, even though my move would cause a ruckus. The Kremlin's protocol people freaked out as they watched me march into the meeting room by the vice president's side. I intentionally avoided eye contact with Medvedev's aides in order to avoid interception. Luckily, I did find a seat, albeit one several feet removed from Medvedev and Biden. But at least I was in the room, ready to take notes and to come to the assistance of the vice president as needed. A few minutes later, Medvedev's foreign policy advisor, Sergei Prikhodko, slipped into the room to restore symmetry.

I quickly understood why Medvedev wanted a one-on-one. He hinted that he was ready to support—or, to be more precise, not block with a veto—Western military force against Gadhafi's army. Medvedev shared our analysis about the probabilities of a humanitarian disaster, and the necessity for action. I could tell that this was a tough decision for him. The Russian president carefully presented his reasoning, stressing several times the importance of limiting the military mission. But it was clear to me that he was leaning forward, trying to be cooperative. My own reading at the time was that Medvedev had decided to allow a UN-sanctioned military intervention in Libya not because he had much faith in the mission's success, but rather because he sought to demonstrate to Obama his commitment to deepening cooperation between the United States and Russia.

When our larger delegations met later to discuss other issues in U.S.-Russia relations, Libya was not an agenda item. That omission was striking.

The following day, Biden met with Prime Minister Putin. I heard in Putin's description of the Arab Spring a very different theory of political change than the one Medvedev had articulated the night before. Putin supported modernizing autocrats. He criticized our withdrawal of support from Mubarak in Egypt and affirmed his allegiance to Bashar al-Assad in Syria, not because he had any deep personal affection for the Syrian leader, but because his removal, in Putin's view, would precipitate chaos. We only skirted around the edges of what to do in Libya, but at no time do I remember Putin suggesting that military intervention was wise or necessary. Just as our administration was deeply split over the decision to use force in Libya, so too was the Russian government. Medvedev, however, had a bigger problem. Obama could overrule Gates. Medvedev would have a much harder time overruling Putin.

On March 17, 2011, the United Nations Security Council voted to approve UNSCR 1973, authorizing a no-fly zone in Libya, among other actions.[26] The vote marked a watershed moment for the United Nations Security Council in that it authorized the use of force within a sovereign country for the purpose of preventing genocide. Russia abstained, creating the political cover for China to do the same. Medvedev understood perfectly well that an abstention on this vote meant war. Some of my colleagues in the Obama administration suggested that this achievement marked a major turning point in the evolution of international security norms and institutions. All the major powers were agreeing to take collective action to defend the lives of individuals. That was a major milestone. In a phone call with Medvedev on March 23, Obama praised Medvedev for his support, noting that this level of cooperation reflected a significant, positive upgrade in the level of cooperation between our two countries. In that call, Obama went out of his way to underscore that we were not using this UN Security Council resolution to pursue regime change in Libya.

In my view, though, the Russian vote was a surprising one-off and not the beginning of a new era of consensus in the international system about the use of force for humanitarian missions. Some in our government began to speculate that the UN Security Council could now act in concert in Syria, and help ease Assad out of power. I was skeptical. The only way to gain support from Russia for future actions of a similar nature was to be completely successful in a limited operation in Libya. We were not.

A few days after the UN security resolution passed, the NATO-led military campaign commenced. The United States took out Libya's air defense forces and stopped the Libyan army marching toward Benghazi, but then took a backseat as French and British forces continued the bombing campaign. In a direct rebuke of Medvedev, Putin publicly criticized the UN Security Council resolution and subsequent military action. The Russian prime minister lambasted the operation, saying it "resemble[d] a medieval appeal for a crusade in which somebody calls upon somebody to go to a certain place and liberate it."[27] In Putin's view, Obama was affirming a long-standing American tradition of regime change, similar to President Clinton's war against Milošević and President George W. Bush's attack against Saddam Hussein. Putin maintained the United States had relied upon "a far-fetched and totally false pretext"[28] to invade Iraq; the attack against Gadhafi was no more legitimate. He decried, "And now, it's Libya's turn — under the pretext of protecting civilians. But it's the civilian population who dies during those airstrikes against [Libyan] territory. Where is the logic and the conscience?"[29]

Surprisingly, Medvedev pushed back on Putin in public. On May 21 Medvedev blamed Gadhafi, not us, for events in Libya, stating, "Everything that is happening in Libya is a result of the Libyan leadership's absolutely intolerable behaviour and the crimes that they have committed against their own people. Let's not forget this. Everything else is the consequence of these actions."[30] Medvedev then directly rebuked Putin's analogizing: "It is inadmissible to say anything that could lead to a clash of civilisations, talk of 'crusades' and so on. This is unacceptable."[31] The Russian president made clear that his abstention vote was meant to be supportive of the resolution: "Russia did not exercise [the veto power] for one reason: I do not consider this resolution to be wrong. Moreover, I believe that this resolution generally reflects our understanding of what is going on in Libya."[32]

The public sparring between Medvedev and Putin shocked me. Putin's emotional outburst, followed by Medvedev's on-the-record rebuttal, was unprecedented. Something big was going on internally in Moscow. For years, Putin had told us that Medvedev was in charge of foreign policy. We had detected tensions between Putin and Medvedev on some key policy decisions before, including sanctions against Iran and the continued operation of our Manas Air Base in Kyrgyzstan. But these differences never erupted in public. Putin's remarks confirmed our earlier fears of division within the Russian government over Libya. I worried that Putin's comments signaled the end of his patience with Medvedev, or at least growing reluctance to let Medvedev take the lead in dealing with us. Medvedev had made the case to Putin that his engagement with Obama could produce positive results for Russian national interests. Putin, it seemed, didn't see how intervention in Libya could produce a good outcome for Russia; instead, he viewed Russia's abstention as Medvedev caving in to American influence. In the name of advancing the Reset, Medvedev had just violated a central principle of Russian (and Soviet) foreign policy: resisting American military interventions. For Russian conservatives, Medvedev was becoming too enamored with Obama and his Western ways.

Unplanned, unexpected events in the Libyan war did not help Medvedev's cause. The UN Security Council aimed to stop Gadhafi's advance on Benghazi. But the introduction of NATO air power changed the balance of forces on the ground in favor of the anti-Gadhafi rebels. What started for us as limited military action to prevent Gadhafi from killing tens of thousands of innocent civilians morphed into a civil war. In August, Libyan opposition forces seized control of Tripoli. Two months later, on October 20, anti-regime forces found Gadhafi hiding in a sewer and killed him, without arrest or trial. We had achieved our original mission objective of stopping Gadhafi from slaughtering the innocent citizens of Benghazi. But our intervention also created the permissive conditions for regime change.

At the White House, we understood that events on the ground inside Libya had moved beyond the more limited objectives we had outlined in UNSCR 1973 and promised Medvedev. Though unintended, these developments were still positive. Who could bemoan the fall of a ruthless dictator?[33] Medvedev, however, felt betrayed, not inspired, by events in Libya. When Obama met him in Deauville, France, on the sidelines of the G-20 summit in May 2011, the Russian president was clearly angry about what he perceived to be our bait and switch on Libya. Medvedev again underscored his total disrespect for the Libyan leader, ridiculing his political tract, *The Green Book,* and his erratic ways, but also chided us for going beyond those actions mandated in the UNSC resolution. I had never seen him so upset, sweating as he spoke — in part because the room was warm, but also because he may have felt that his special relationship with Obama was no longer an asset but a liability in his quest to stay on as president for a second term. And he had a point. In retrospect, U.S.-Russian cooperation on Libya may have been both the height of cooperation in the Reset era, as well as the beginning of the end of the Reset. Years later, in defending his annexation of Crimea, Putin said as much, arguing, "You know, it's not that it [the Reset] has ended now over Crimea. I think it ended even earlier, right after the events in Libya."[34] U.S. military intervention in Libya, which helped topple Gadhafi, also inadvertently might have helped remove Medvedev from power in Russia.

14

BECOMING "HIS EXCELLENCY"

At the beginning of my third year at the White House, I began to plot my way back home. I had promised my family that I would only give this government gig two years. My two sons had been troupers, moving to Washington partway through the 2009–10 school year, and I was determined to keep good on my promise of going home in the summer of 2011. I also remembered Condi Rice's observation to me two decades earlier. She had described how people come to Washington with intellectual capital accumulated and then spend it down while working in government, with little ability to renew it while in office. Washington's problem was that people stayed in their jobs well after their intellectual capital had been depleted. I did not want to be one of those intellectually bankrupt bureaucrats! By 2011, I was beginning to feel like I had tried out most of my policy ideas regarding Russia; it was time to go home. The Arab Spring did provide me with a new challenge to apply some of my general expertise and ideas about democratic transitions, but my ability to contribute to that policy challenge quickly began to narrow later in the year.

At the end of 2010, it also began to feel like we were reaching the limits of the Reset. In his *Washington Post* column published just after Senate ratification of the new strategic arms treaty, Bob Kagan warned that progress in the U.S.-Russia bilateral relationship would stall: "Relations with Moscow are about to grow more challenging. This is partly because some of the easy pickings — including this treaty — have already been harvested. The problems that lie ahead are going to be a tougher test of the reset: what to do about Russia's continued illegal occupation of Georgia; how to handle Russia's increasingly authoritarian domestic behavior, its brutal treatment of internal dissent and its squelching of all democratic institutions."[1] New START did not feel like "easy pickings" to me; it had occupied most of my attention for the previous two years! But Kagan's essay resonated with me. I did feel like we had closed out many accounts with the Russians, leaving only the most difficult issues ahead of us. We had declared ending the Russian occupation of Georgia and supporting Russian democracy as U.S. foreign policy goals, but had very weak instruments for pursuing these objectives. Talks on missile-defense cooperation were proceeding, but slowly. A run at deeper cuts in our nuclear arsenals would have to wait until after presidential elections in both Russia and the United States. More generally, Medvedev seemed increasingly focused on his campaign (aimed at one voter, Putin, not the people) for a second presidential term. He was not going to pursue any big initiatives with the Americans until that issue was decided. As I stated bluntly in a public talk in April 2011, we had to have "realistic expectations" moving forward, recognizing that cooperation on some issues would not translate into cooperation on all issues; that some issues were zero-sum; and that a profound values gap remained between our two countries.[2] In an internal policy paper I wrote around this same time called "Resetting Expectations for the Reset during the 2011–2012 Electoral Cycle in Russia and the United States," I warned that any new big breakthroughs in our bilateral relations would have to wait until January 2013. That was a long time for me to tread water at the NSC. In 2011, Russian membership in the World Trade Organization and missile-defense cooperation were the two major goals left from our original 2009 list of Reset objectives. I thought I could stick around to complete those two big tasks — or, in the case of missile defense, at least set that project moving in the right direction — and then head back to California.

When I told my boss, National Security Advisor Tom Donilon, about my plans to leave, he was not supportive. Later, he told me that he had discussed my plans with "the boss," who also thought it was a bad idea. We had too much momentum on the Reset; I couldn't leave now. Denis McDonough, Tom's first deputy at the time, sympathized with my desire to see more of my wife and sons, and also understood the wisdom of getting out of D.C., but he too asked me to think of some way to reconfigure my current job or take a new one that might make it worth my while to stay.

One option was to formalize my work on the Arab Spring by creating a new job, deputy national security advisor for democracy with a special focus on the Middle East. We needed to change our bureaucracy to meet these new challenges and take advantage of new opportunities, or so I thought at the time. My family was not going to stay another year in Washington for me to run out the clock on the Reset, but the enormity of the moment of the Arab Spring might have helped me convince them to hang on a bit longer.

Rather than a new job on the Middle East, however, Donilon and McDonough proposed something unexpected: make me U.S. ambassador to the Russian Federation. They told me that Obama loved the idea. I could stay on the team, still work the Russia account, but have a new experience, and see my family more often. Working at the NSC, I left home early and got home late. Working in Moscow, I would have more control over my time. I would be the "principal," not staff. Tom and Denis gave me the hard sell: in the spring of 2011, maintaining momentum on the Reset remained a top priority for the president. As the author of the Reset, I couldn't leave now. They understood my reluctance to stay a fourth year on the NSC staff, but as Obama's Russia guy in Moscow, I could become a very special kind of ambassador. Because of technology, they argued, I could continue to do my "day job" of making policy back in Washington, but then also assume the role of chief implementer of the policy in Moscow. In addition, I knew everyone at the highest levels in Moscow already, including the president. I would have much better access, both in Washington and Moscow, than any career ambassador. And McDonough pledged to address me as "His Excellency."

I was intrigued, but not sold. Already in the spring of 2011, I had assessed that maintaining momentum on the Reset was going to be difficult. I could work on the Middle East without dragging my family to another country. But we did have a lot of unfinished business regarding Russia. In 2011 we were making progress on WTO negotiations and even missile-defense cooperation. Trade and investment were increasing steadily. Medvedev abstained on a UN Security Council resolution, thereby allowing the use of force in Libya. Demonstrations in Russia against Putin — the issue that would radically disrupt the Reset — were still months in the future.

After several conversations with senior staff at the White House, I wrote a short memo on May 4, 2011, spelling out the pluses and minuses of three different options: (1) ambassador to Russia, (2) deputy national security advisor for democracy with a special focus on the Middle East, and (3) Stanford professor/Obama supporter. In response to my memo, Donilon replied, "Mike — The only one I am clearly against is 3. Let's discuss." Upon further discussion, I could tell that Obama, Donilon, and McDonough valued me more as their Russia guy than as their Arab Spring kibitzer, and they had a point. I knew more about Russia than I did about Egypt or Syria. I eventually decided Moscow was the better fit. I also embraced the new challenge of becoming a diplomat.

My wife, Donna, was excited about the proposal. We had lived in Moscow in 1994–95 and loved it, even though conditions back then were challenging. The idea of living abroad with our children was appealing. Because of the wonders of technology, she would be able to telecommute for her organization, MomsRising. And she was excited by the possibility of augmenting her usual work-at-home wear of yoga pants with some occasional ball gowns.

There was one last hurdle before saying yes — my older son, Cole. In December 2008, the day I told Cole we were moving back to Washington was one of the worst days of my life as a parent. He'd cried uncontrollably for an hour. And he is generally a very happy kid. He had just started to make new friends and settle into his new school in California, and now I was moving him back to Washington. (We had moved back to Palo Alto from Washington in 2006.) In the passion of that moment, I promised him that we would only be gone for two years. Most professors who serve in government only stay two years; some universities require you to give up tenure if you stay longer. Now in 2011, I was thinking of breaking that promise. Donna and I decided we couldn't do that. If I dragged Cole unwillingly to Russia, and then his high school, or college, or other future plans didn't work out, he might always blame me for putting my career ahead of him. That was a burden I was not willing to bear, not even for Obama. So instead we decided that if Cole said no, I'd tell the president no, and we'd go home.

Surprisingly, after checking out the Anglo-American School of Moscow on the internet, Cole said, "I have an open mind about it." On May 4 I emailed McDonough to report that I "had a long talk with Cole about Moscow option last night. He is willing to do [it] if he thinks it's important for the President." Denis responded, "Whoa. Let me work it." From that day on, I knew we eventually would be going to Moscow. For several weeks thereafter, we ruminated on the pluses and minuses, as my family always does. The White House also was "socializing" the idea with the State Department, as my candidacy meant taking an ambassadorship traditionally held by a career diplomat and handing it over to a political appointee.

Unexpectedly, a hard deadline for deciding appeared, one I didn't learn about until I was driving with the president to a G-8 summit after landing in Deauville, France, on May 26, 2011. I was seated in the Beast to do the pre-brief for Obama's meeting with Medvedev later that day. During our drive, Obama told me that he intended to tell Medvedev about his decision to send me to Russia and ask right then and there for *agrément,* a French word for a diplomatic process in which one country agrees to receive the ambassador from another country. We had not quite arrived at yes in our family, but after a quick phone call home, we did so before the president met with Medvedev.

Obama had planned to discuss my proposed new assignment with Medvedev in their one-on-one pull-aside, but forgot. As we were walking through the hotel lobby after their meeting, he remembered, shouted out to Medvedev, and asked to discuss one more thing—his decision to send me to Moscow. Obama and Medvedev had just concluded their toughest meeting yet, yet the Russian president gave me a hearty, congratulatory handshake and said he looked forward to seeing me on a regular basis in Moscow. Lavrov, standing next to Medvedev, looked surprised but pleased. Just a few minutes before, while Obama and Medvedev were doing their one-on-one, Sergey and I had been discussing life after government. He was asking me how Condi's consulting firm was doing, and probing to find out what I planned to do next. I was coy. He sounded like a guy thinking about his next move as well. Had Medvedev stayed on for a second term, my guess is that Lavrov would not have continued as foreign minister. As my life in Moscow became more difficult as ambassador, and I heard secondhand about Lavrov's criticisms of me, I always wondered who the real Sergey was, the person who warmly congratulated me that day in Deauville, or the one who disparaged me behind my back.

Everything in government seems to take longer than it should. My vetting process dragged on. Making sure the subsidies embedded in my Stanford mortgage did not violate conflict-of-interest rules took several weeks to resolve. Even though I already had a top secret security clearance, I had to get a new one to become ambassador. That took time. I knew a lot of foreigners. Finally, on September 14, 2011, Obama formally announced his intent to nominate me as his next ambassador to the Russian Federation. The biggest hurdle — Senate confirmation — was yet to come. Sometimes democracy is a drag.

Strangely, my hearing before the Senate Foreign Relations Committee on October 12 was an exciting, fulfilling event, giving me a chance to sell our policy as well as myself. Unlike some other political appointees, I knew our policy cold. I had helped to formulate and execute Reset, so had little trouble explaining it. I also knew well the country where I was going to be posted. I had lived in Russia several times, spoke the language, and therefore knew the basics that sometimes stumped other political appointees. In my opening statement, I pledged to continue the basic trajectory of the Reset, which for me also meant standing up for human rights and democracy. Knowing that some Republican senators considered our administration too soft on Russia, I devoted special attention to explaining dual-track engagement — engaging both the government and society — and my commitment to execute that policy as ambassador. The hearing lasted several hours, but never was there a moment of confrontation or misstep. I also was thrilled that my family could attend. After it was over, I asked Luke and Cole their opinions on those who'd questioned me. Without knowing their party affiliations, they both named Senator Rubio, a Republican, as their favorite, and Senator Menendez, a Democrat, as the committee member who'd been toughest on me, unfairly so in their assessments.

Coming out of the hearing, I felt like I had momentum. I received many endorsements from prominent Republicans. Bob Kagan and David Kramer, a former Bush administration assistant secretary of state, coauthored an op-ed in support of me. On the day of my testimony, Josh Rogin wrote in his *Foreign Policy* column, "Support in both parties is strong for McFaul's ambassador nomination, despite that he is the key architect of the reset policy that many Republicans oppose. McFaul has a long track record as a democracy advocate, and unlike most other top administration officials, he has maintained close and longstanding relationships with leaders on the GOP side of the aisle."[3] So my confirmation by the Senate should have been a cakewalk, right? Wrong.

The Senate Foreign Relations Committee voted to send my nomination to the full Senate with no drama. Then came the "holds." The U.S. Senate maintains a peculiar tradition that allows a single senator to put a hold on a nomination anonymously and for no reason. As was explained to me at the time, this "courtesy" was extended to all senators two hundred years ago when they had to travel by horseback to get back and forth to Washington to vote. In the twenty-first century, however, the practice was routinely abused, and in my opinion undemocratic. Why should one senator be allowed to block a vote on these important jobs? It was not just a Republican practice. Both parties indulged in the tactic.

As I made the rounds on Capitol Hill, I quickly understood that the several holds on my nomination had nothing to do with me personally. Senator McCain, for instance, was extremely supportive of me, but not of Obama. All of the holds on me, in fact, concerned Obama and his policies. The president's opponents wanted to get something in return for approving *his* candidate. There were lots of quid pro quos kicking around, but eventually the negotiation focused on missile defense. Senator Mark Kirk of Illinois assumed the role as lead negotiator on behalf of several Republican senators blocking my confirmation. In our first meeting, Kirk told me bluntly that he didn't trust Obama. He assigned some personal traits to my boss that I thought were out of bounds. I knew my job was to be polite and listen, but I eventually had to push back and began defending Obama with vigor. He ended the meeting abruptly. I called McDonough, then the person at the White House running point on my confirmation, and told him what had happened. After recapping the meeting, I asked him if I had screwed up. He said yes, reminding me that the goal was to get the holds lifted, not defend Obama's integrity — others at the White House had that assignment. That episode probably earned my family and me a few extra months at the Marriott ExecuStay in Washington, our temporary home for the entire confirmation process.

Kirk agreed to meet with me again. He explained his reluctance to share information about our missile-defense systems with the Russians. He worried that in our haste to get a new arms control deal or cooperate on missile defense, we were going to give the Russians too much information about our interceptors, which in turn they would give to the Iranians. I agreed that it was not in the U.S. national interest to give the Russians any technical information about our missile-defense systems, and assured him that our administration would never do so. That was not enough for him. Kirk wanted assurance in writing. Eventually, we offered a deal. He and his colleagues would lift their holds on me in return for a paragraph in the new National Defense Authorization Act, which obligated the executive branch to inform the Congress in advance of any plan to share information with Russia on our missile-defense systems. That was the ransom for our release from the Marriott!

The process dragged on. We wondered if our kids would have to start their second semesters in D.C. We considered even pulling out and going home. A *Washington Post* editorial published on December 2, 2011, captured my mood at the time: "It's been 2½ months since President Obama nominated the senior Russian director on his National Security Council staff, Michael A. McFaul, to be U.S. ambassador to Moscow. If you've been following how well Washington works these days, it may not surprise you to learn that Mr. McFaul has yet to take up residence in Spaso House. But in the annals of Washington dysfunction, this delay merits a moment of attention, because Mr. McFaul has been acknowledged, pretty well universally, to be an excellent choice for the ambassadorship . . . In these partisan times, he enjoys unusual support from the foreign policy establishments of both parties."[4] Being qualified and having support from both Democrats and Republicans was not enough.

Even after agreeing to the paragraph on missile defense, it took a full-court press on the last day of the Senate session in December 2011 to get me through. Secretary Clinton called Senator Mitch McConnell and pleaded with him to allow my inclusion on the last vote of the year. The majority leader finally relented, and on December 17, I was confirmed by the unanimous consent of the U.S. Senate. At around two in the afternoon, Tommy Vietor—our NSC press spokesperson—pinged me, cc'ing McDonough and Rhodes to say, "Huge win for the good guys." McDonough replied, "Take that, Putin. We are all McFauls today."

If the confirmation process was absolute misery, the swearing-in process was absolute joy. It's like a wedding—all your family and friends come together for a big celebration.

In the several days leading up to the ceremony, I enjoyed lots of feting. I met with all of our senior government officials who had business in Moscow.

Russian ambassador Sergey Kislyak hosted me at his residence for a particularly warm, elaborate, and meaningful dinner. He invited all of my friends in the U.S. government, as well as a few of my important informal outside advisors who had participated in developing the ideas of the Reset. Three years earlier, Kislyak and I had started our relationship with our fists up. But over the years, he had grown on me. We were very different people in so many ways, and yet I felt like we clicked. He made a very thoughtful toast that night, which gave me hope about being able to work closely with other Russian officials in Moscow. If I could win over Kislyak, I could win over anyone.

The afternoon of my swearing-in ceremony was also my last day of work at the White House. There would be no break in between. My family came with me that day to see President Obama, who wished us well in our next adventure. He jawed about basketball with my sons, and thanked my wife for letting me represent him in Russia. That was a memorable Oval session, my last for a long while. My family also met Vice President Biden that day, who, true to form, was effusive in his praise of me and exuberant in his thanks to my sons for their support in allowing me to serve our country in this next assignment. He made a big impression on them both. They were starting to realize that this new job was going to involve them as well, in a much more public way.

At the swearing-in ceremony over at the State Department, Secretary Clinton was gracious in her remarks in front of several hundred of my friends and professional colleagues. She said, "This is a good day for us all — for the United States, which is sending an absolutely top-notch emissary to Moscow, for our partners in Russia. And Ambassador, we're delighted that you are here because we know that this appointment represents the kind of deeper cooperation and closer ties that President Obama stands for." I was an Obama guy. The idea to send me to Moscow had come from the White House, not her building. That she celebrated my appointment so publicly that day went a long way in solidifying my standing within the State Department.

In our last substantive meeting before I left for Moscow, Clinton kept it simple, asking only that I be strong in defending American national interests, including our commitments to democracy and human rights. In her public remarks at the swearing-in ceremony, I was struck by how she made a point of underscoring my reputation as a defender of these values.

> Now of course, as you know, Mike is not only a Russia expert; he's also one of our nation's leading thinkers and writers on democracy. And the coming months and years will be crucial for Russian democracy. Russians from all walks of life and every corner of this great country are making their voices heard, both face-to-face and in cyberspace, expressing their hopes for the future. Few Americans know Russia or know democracy better than Mike McFaul. And I can think of no better representative of our values and our interest in a strong, politically vibrant, open, democratic Russia, as well as a deepening U.S.-Russian partnership.

In my own remarks, I made the case for why the Reset had delivered results in keeping with American national interests, and why I would therefore continue to execute that policy in Moscow. I emphasized my love of being on the team, the Obama team, and now the State Department team. This new assignment gave me a chance to meet some new "teamers," my older son's invented word from his kindergarten basketball days. I also spoke of the challenges that lay ahead.

> The Reset was not completed last summer, nor, however, is it over now, as some are saying. On the contrary, today we're on to the next, more complex phase, when the alignment of our interests and values is not obvious or easy. Cooperating on missile defense, addressing Iran's nuclear ambitions, ending the bloodshed in Syria require more dialogue and more creativity to find common cause as well as manage our differences, and a new phase of political change in Russia also will present challenges. But tougher challenges create opportunities for doing meaningful work, and that's exactly why I look forward to my next assignment in Russia. As President Obama said in the State of the Union address last year: "We do big things." Creating and cementing a normal cooperative relationship with Russia, a strong, prosperous, and democratic Russia, is a big thing, and I'm eager to keep working at it.[5]

Privately, I was more nervous about prospects for continued momentum in our bilateral relations. In public, I had to sound upbeat. And as an eternal idealist, I remained hopeful that building a closer relationship with a strong, prosperous, and democratic Russia was still possible.[6]

A few days later, we were off. In my swearing-in remarks, I called this next chapter a great adventure for my family and me. In reality I had no idea what kind of adventure actually awaited. I left Washington as Mr. Reset. I landed in Moscow as Mr. Revolutionary.

15

PUTIN NEEDS AN ENEMY— AMERICA, OBAMA, AND ME

Between May 2011, when Obama asked me to stay on his team to advance the Reset as his ambassador in Russia, and my arrival in Moscow in January 2012, politics inside Russia changed dramatically. Two events were most consequential — Putin's announcement that he planned to run for election as president for a third term, and the eruption of popular demonstrations against the Russian government. Together, these developments effectively ended the Reset. Both were factors over which the U.S. government had zero influence or control. They also would shape profoundly my time as U.S. ambassador.

On September 24, 2011, Putin announced his intention to run in the March 2012 election for a third term as Russia's president. If Putin ran, his victory was assured. Medvedev, our main Reset partner, would soon be demoted to prime minister. Publicly, we downplayed the significance of this change for U.S.-Russia relations. In the *New York Times,* quoted as "a senior Obama official," I stated that we had "very deliberately sought to avoid playing favorites." I added, "Everyone knows that Putin runs Russia . . . Remembering this obvious fact means that Putin has supported the reset with the U.S."[1] Senior Russian officials, including First Deputy Prime Minister Igor Shuvalov and Medvedev himself, came to the U.S. and delivered the same message to us privately. Many Russian experts both inside and outside government shared that assessment. Testifying before the U.S. Select Committee on Intelligence on January 31, 2012, Director of National Intelligence James Clapper predicted "more continuity than change" after the transition from Medvedev to Putin.[2] Medvedev was just the front man, with no independent political power, or so conventional wisdom held at the time, so why would Putin's return to the Kremlin lead to significant changes in policy?

Privately, I held a different view—one that departed from the official U.S. government position I was articulating to the press. (Sometimes that happens in government work!) As I recall writing to my bosses as early as January 2011, Putin's likely return as president would complicate our relations with the Russian government. Trained as a KGB agent during the Cold War, Putin had a different worldview than the younger Medvedev. On good days, Putin saw the United States as a competitor. On bad days, the United States was his enemy. He had said as much publicly, including most memorably in a confrontational speech at the Munich Security Conference in 2007. During the Reset years, Putin had allowed Medvedev to take the lead and experiment with cooperation with us. I think Putin temporarily entertained the possibility that Obama might be different. But popular uprisings throughout the Arab world, punctuated by our military intervention in Libya, confirmed Putin's original instincts about the United States, and revealed, in his mind, the naiveté of Medvedev's approach to U.S.-Russia relations. Things were going to be different under a Putin presidency.

The power shift in the Kremlin also worried Obama. After a meeting in the Oval Office a few days after Putin announced he would run again, Obama pulled me aside and asked me what I thought. I told him that we would have to try to work with Putin, but with much more limited expectations. Obama agreed. He was a believer in the long game, and we had a lot of business to get done with the Russians. We both lamented, however, that the chance for a major breakthrough in U.S.-Russian relations was now over, or at least postponed. Putin was different from Medvedev—from a different generation, more suspicious of the United States, and less interested in pursuing win-win outcomes with us. Things were going to get a lot harder now.

I left that conversation very deflated. We had pursued an ambitious agenda to develop a new kind of relationship with Russia through our Reset policy, well aware of the long odds. We still could work with Putin, but we'd have to temper our ambitions. Grandiose projects like missile-defense cooperation had to be interrupted. Dreams of fostering political modernization had to end. Maintaining the status quo was all that could be hoped for.

I also worried that Putin's return to the Kremlin would make my job as ambassador a lot more difficult. Like Obama, I knew Medvedev and his team well.

Despite my pre-government reputation as a hardliner, I believed that I had convinced them of my genuine intentions in seeking to build a more cooperative relationship with Russia, based on common interests. By contrast, at the time of my appointment I had met Putin only twice as a U.S. government official. Before joining the government, I had written articles criticizing Putin's autocratic ways, some of which I know he had seen. He was not going to like me. For a short time, I pondered withdrawing my nomination as ambassador; let them find someone else more agreeable to the new Russian president. But conversations with my White House colleagues convinced me that such a move would signal weakness or retreat by Obama. My job was to go to Moscow and represent President Obama, his policies, and the interests of the American people, not to befriend Vladimir Putin. I also reasoned that my encounters with Putin would be infrequent; on most issues, I'd be dealing with Putin's foreign policy advisor, Yuri Ushakov, whom I knew well. We Montanans are optimists. I was hoping for the best.

And then the Russian people stepped in, making things even more difficult for Russian-American relations, and for me personally.

Putin expected most Russian citizens to welcome the news of his return to the Kremlin. Some, maybe most, applauded his decision, but many were indifferent. During his years as prime minister, Putin's popular support fell, gradually but consistently.[3] His announcement of a third-term run generated no uptick in the polls. Putin made no compelling argument for why he needed to return to the Kremlin. During his first eight years as president, he had been able to argue that he was the Kremlin leader responsible for economic recovery after a decade of depression, even if oil and gas prices, not his policies, had done most of the work. During that time, Putin also cultivated an image as the restorer of stability after a decade of chaos and lawlessness, even if the empirical data to support this claim were weak.[4] He and his public relations team also successfully developed a narrative that he had lifted Russia off its knees and restored some of the Soviet Union's international stature. He controlled the airwaves, so he controlled the story. But in 2011 economic growth had tapered. Following the 2008 financial crisis, educated, urban voters cared more about economic opportunities than a greater Russian role abroad (although this mood would change radically after Russia's intervention in Ukraine in February 2014). They also wanted more than just economic growth, including more political freedoms. In 2011, Putin was offering them nothing new. Even the most popular leader gets a little tiresome after more than a decade in power.

The decision by Putin and Medvedev to switch jobs angered many young, urban middle-class voters.[5] The agreement was an insult to them, denying them any real choice on the presidential ballot. Medvedev had let them down; he had promised forward-looking policies but was now stepping aside timidly. His promises of political and economic modernization seemed hollow the day he acquiesced to Putin's return to the Kremlin.

Elites with whom I met during this period expressed little enthusiasm about Putin's return, in part because they felt Medvedev was committed to a more open market economy, but also because Putin's return meant empowering further his cronies at the helm of several major state-owned companies. Russian businesspeople who came to see me at the White House during the lead-up to Putin's decision to run were especially critical of his return. They feared new challenges to their property rights, more corruption, and less impetus for market reforms.

When Putin finally made his announcement, Medvedev's senior economic advisor and close confidant, Arkady Dvorkovich, tweeted from the floor of the United Russia party congress, "There is no reason for celebration . . . now it is time to switch to the sports channel."[6] I could not believe that Arkady would dare say such a thing in public! Obviously, he and other close associates of the president had remained hopeful until the very last days that Putin might let Medvedev run for a second term. Their degree of disappointment must have been acute for him to post that tweet, which I assumed (erroneously) would mean the end of his government career after Putin returned to the Kremlin. Even Vladislav Surkov, my cochair of the U.S.-Russia working group on civil society and a former close aide to Putin, did not believe Putin should run again, as he explained to me earlier that year in Washington. Surkov was subsequently demoted for expressing his opinion to more people than just me, or so we were told.[7]

Putin even endured a few public challenges to his candidacy. Most dramatically, in November 2011 fans booed Putin at a wrestling match after he jumped into the ring in front of twenty thousand people to congratulate the winner. I remember watching the clip in my office at the White House, totally shocked. I had never seen people yell epithets at Putin before, let alone at a venue that should have been a friendly audience for him. Russia's liberal intelligentsia do not attend wrestling matches! A poll by the Levada-Center conducted in November 2011 showed that only 34 percent of likely Russian voters planned to cast their ballot for Putin in the spring.[8]

A month later, in December 2011, Putin's party, United Russia, performed much more poorly than expected in the parliamentary elections. United Russia enjoyed unlimited coverage on national television stations, abundant financial resources, the backing of regional governments, and a bump from falsification, yet won only 49.3 percent of the vote, a significant drop from the 64.3 percent it had garnered four years earlier. Given all its advantages, failing to win 50 percent of the popular vote represented a major setback for the ruling party.

The number of votes that United Russia accrued through falsification was probably no greater than during previous elections. At least that was our assessment in the U.S. government. But in 2011, the proliferation of smartphones, better-organized election-monitoring organizations, and social media platforms such as VKontakte, Twitter, and Facebook, combined to expose electoral irregularities to many more people. Compelling evidence that this election had been stolen in favor of Putin's party, in turn, triggered popular demonstrations in December — first thousands, and then tens of thousands, and occasionally hundreds of thousands.[9] Civic activism had been growing in the years before the 2011 election, but voter fraud was the spark that really ignited the discontented. The last time so many Russians had taken to the streets for political reasons was 1991, the year the Soviet Union collapsed.

Demonstrators in Russia were mobilizing in the same year that hundreds of thousands were taking to the streets against dictators throughout the Middle East in the Arab Spring. Putin had watched similar kinds of demonstrations topple strongmen in Tunisia, Egypt, Yemen, and Libya, and almost overthrow leaders in Syria and Bahrain. Al Jazeera's broadcasts from Tahrir Square made an impression not just on the protesters, but also on autocratic leaders seeking to avoid Mubarak's fate. Analysts of Russian politics in the U.S. government explained that Putin was especially surprised by how quickly elites, including the military, abandoned Mubarak, and especially disturbed by the brutal killing of Gadhafi.

Putin's first reaction to these Russian demonstrators was anger. In his mind, he had made these young professionals rich, and now they had turned against him. Even his former finance minister, Alexei Kudrin, attended one of these demonstrations. That was betrayal. Putin's second reaction was fear. He and his team were surprised by the size of the protests. Never before had so many Russians demonstrated against his rule. The message from the streets quickly turned radical, starting with outrage against falsification, but morphing into demands for the end of Putin's regime.[10]

Putin now faced pressure not only to win the March 2012 presidential election, but also to defuse these popular protests. Rather than engage with his opponents and attempt to co-opt them, he chose a strategy to repress and discredit his critics, dragging America into the process. To discredit opposition leaders, Putin portrayed them as traitors: agents of the United States. Putin always had been paranoid about American efforts to undermine his government. Years before, he had developed the view that the United States intended to foment a color revolution against his regime, just as we allegedly did in Serbia in 2000, Georgia in 2003, and Ukraine in 2004. In 2011, Putin also believed that we wanted Medvedev to stay on as president. An unfortunate quip by Vice President Biden during his visit to Moscow in March 2011 had helped to nurture that hypothesis.

Even before the parliamentary vote, Putin began to develop the argument about American manipulation of Russia's internal politics. As he explained in November 2011 — a month before the massive protests and two months before my arrival as ambassador — the United States was interfering in Russia's internal affairs as it had during the Cold War era: "We know that representatives of some countries meet with those whom they pay money — so-called grants — and give them instructions and guidance for the 'work' they need to do to influence the election campaign in our country."[11] American interference in Russian elections seemed very obvious to him: "What is there to say? We are a big nuclear power and remain so. This raises certain concerns with our partners. They try to shake us up so that we don't forget who is boss on our planet."[12]

A month later, the explosion of popular demonstrations against the government only confirmed Putin's suspicions about the United States' alleged nefarious activities against him inside Russia. Putin was particularly upset when Clinton criticized the parliamentary vote, claiming that she "set the tone for several of our actors inside our country, she gave the signal. They heard that signal and with the support of the State Department of the U.S. they began active work."[13] (Five years later, Putin seized his moment for revenge when he intervened in the 2016 U.S. presidential election to help Trump and hurt Clinton.) After Clinton's comment, Putin and his lieutenants struck back even harder, trumpeting a hardline, anti-American message to the Russian people.

Meeting in the Oval Office on December 9 to review the implications of the parliamentary elections for Reset, Obama asked a series of questions about our democracy-promotion efforts. When he asked whose idea it was to increase funding for Golos, the Russian election-monitoring NGO, I reported, "Mine." When he asked who at the White House cleared Clinton's tough statement, I answered, "Me." He supported what we had done — what I had done — but also gave us clear instructions for how to defuse tensions over these elections. We must keep true to our principles, but also remember the long game. We were going to have to deal with Putin for another five years, and we had a full agenda with Moscow. I understood my marching orders: as ambassador, I would not be attending any demonstrations in Moscow. I never did.

I landed in Russia just a few weeks after the December 2011 demonstrations began. A *Moscow Times* headline rightly declared, "McFaul Arrives to Keep 'Reset' Alive."[14] That most certainly was the mission that Obama sent me to Russia to pursue. But the Kremlin-loyal press described my assignment very differently. I was not Mr. Reset, but Mr. Revolutionary: the color revolutions specialist sent by Obama to orchestrate regime change. For the rest of my time in Russia as ambassador, I battled nearly every day to dispel that myth, and never really succeeded.

While waiting for confirmation back in Washington, I watched the popular protests in Moscow with hope and anxiety. I was inspired by the actions of the demonstrators. Russian society was alive again. At the same time, I was nervous about how these new domestic dramas would affect my work as ambassador. I knew the demonstrations would make everything more difficult — but just how difficult, I had no idea.

Spending time listening and learning seemed like the prudent way to start my tenure as ambassador. I wanted to focus first on getting to know my new colleagues at the embassy, learning the managerial side of the job, and becoming a presence inside the compound. I was an outsider, parachuting in to lead a team of State Department professionals as well as dozens from other departments and agencies. Some had been engaged in this kind of work for thirty years. They were rightly skeptical of a White House political appointee and professor, with little experience in formal diplomacy. Because I remained in my job at the White House until the day before I was sworn in as ambassador, I attended only two days of ambassadorial school at the Foreign Service Institute. I watched a video on how to drive out of a terrorist attack, picked up a copy of *Protocol for the Modern Diplomat,* learned who paid for what when we hosted dinners, reviewed our evacuation plans for 6,063 employees and their dependents in the event of a crisis, and got a bit of media training, but that was about it. I would have to learn on the job.

Of course, I also had some advantages. I knew Russia. Unlike most political appointees in other countries, or many career ambassadors at other posts, or even many of my employees at the embassy, this was not my first tour in the country. I had first lived in Russia in 1983, had logged roughly five years in the USSR and Russia since then, and devoted a good chunk of my academic career to writing about Russian politics and U.S.-Russia relations. I did not need to read briefing books on Pushkin, the Bolsheviks, privatization, or Putin. On the contrary, many of my new colleagues had read my books and articles in preparation for their tours in Russia. I also knew thousands of people in Russia, including top-level officials in the government, billionaires, Duma deputies, journalists, and leading figures in the intelligentsia. Some of these people I had known for decades. In addition, I spoke Russian—a skill that I hoped would impress the old hands at the embassy. And maybe most importantly, I knew the Obama administration's Russia policy—a big leg up compared to many new ambassadors. I had helped to author it, and had chaired for three years the Interagency Policy Committee on Russia responsible for its development and implementation. And finally, I knew all of the key decision makers in our foreign policy team personally, including the chief decision maker, President Obama.

These assets, however, would impress only some of my new embassy colleagues, because only a fraction of my new team actually worked on executing, let alone developing, policy. Being Mr. Reset was not going to mean much to the electricians, barkeeps, or Marines. At the embassy, we not only had diplomats but representatives of nearly three dozen other government agencies. (What exactly was the Department of Agriculture doing here? I had to learn.) I was the U.S. representative to the Russian government and people, but also the de facto mayor of a small village located in the compound at the embassy in Moscow. I now had army generals, personal trainers, nuclear physicists, hairdressers, and construction workers working for me. I had to win their respect and confidence. Several hundred Russian employees now worked for me as well. They were going to judge me not by foreign policy outcomes, but by how quickly I could secure raises for them, or get hot Russian food in the cafeteria. This last item —burgers versus borscht—triggered a giant debate in our embassy community, complete with petitions, protest marches, and color ribbons. My new job demanded that I seek compromises both on missile defense and lunch options.

During my first days in Moscow, I also focused on getting my family set up and moving in the right direction. It's hard to relocate to a foreign country for the first time, no matter your age. It's especially daunting when you are nine, as my younger son, Luke, was when we moved midstream in the academic year. It's difficult to change schools in junior high, as my older son, Cole, did, even if it's in the same city; this was a different country. In my first week in Moscow, securing spots for my sons on new basketball teams was a much higher priority than landing a meeting with anyone at the Ministry of Foreign Affairs.

Mindful of the time needed to adjust to my new role, I plotted a low-key first hundred days. I would meet a few key Russian government officials, but would save the big courtesy calls until after the Kremlin ceremony scheduled for mid-February when I would officially present my credentials to President Medvedev. I planned to do a few personal-interest, policy-free interviews, playing up my past experiences living in Russia and my love for Russian culture and history. To show respect for my new hosts, I also sought to do as much public speaking as possible in Russian. The last time I had studied Russian formally was 1985, and the last time I had spoken the language on a daily basis was in 1995. Ambitiously, I signed up for hourlong language tutorials three times a week. Between meetings at the embassy with section heads, agency representatives, and the embassy's Russian employees, I also planned to spend time learning another foreign language: the alphabet soup of State Department speak— EUR, AECA, FAST, T, P, D, S, RSO, CON, POL, ECON . . . I had never worked for the State Department before.

As we boarded our flight to Moscow on January 13, 2012, Donna, Cole, Luke, and I were nervous, but mostly excited for a grand, new adventure. Even the plane ride was cause for excitement—we flew business class, which my sons thoroughly enjoyed. (The State Department pays for business-class travel for ambassadors to their posts for the first time and upon return from post. Almost everything in between is economy class.)

Even before landing, I altered State Department protocol and told my new second-in-command, Deputy Chief of Mission Sheila Gwaltney, that she did not have to drag the forty-person "country team" — the heads of all major agencies and sections at the embassy — out to the airport to greet me and my family. Sitting in traffic for two hours merely to shake our hands seemed like a waste of their time. Instead, we met everyone at Spaso House, our new palatial residence. The event was a bit overwhelming, especially after twelve hours on a plane. The two dozen assembled that day worked for me, as did hundreds of others who worked for them. The enormity of the management dimension of this new gig was beginning to sink in.

We wandered around our new mansion feeling like visitors in a museum. Black-and-white photos of previous Spaso House residents — including George Kennan, the author of America's containment strategy toward the Soviet Union, and Averell Harriman, the former governor of New York — lined the walls. Photos of Nixon, Kissinger, Brezhnev, Gromyko, Reagan, Bush, Gorbachev, and Clinton testified to the incredible history of our new home. I made a few adjustments. I eventually took down one photo of Stalin and presidential advisor Harry Hopkins from July 1941. Cooperation with Stalin during World War II was a part of our history, of course, but I didn't have to look at Stalin — or compel my guests to — every time we had tea in the library.

The next day, I went to the embassy for my first day of diplomatic business. Our — "my" — huge black Cadillac pulled up to the main entrance with an American flag proudly flying from its fender. A statue of our first U.S. ambassador to Russia, John Quincy Adams, greeted my car, as it would for the rest of my tenure in Moscow. A beige Chevy suburban with my new Russian bodyguards followed us, and my driver called ahead to the guards at the embassy to make sure not a second of ambassadorial time was wasted waiting for the gate to open. The kids were wowed. Diplomacy was going to be fun! Or so I thought that day.

My plans for a slow, quiet start were interrupted by the visit of Deputy Secretary of State Bill Burns. Bill was one of my closest partners in the U.S. government, and a thoughtful mentor. We had plotted most of the moves of the Reset together. As a former ambassador to Russia, he'd shared with me much wisdom about how to approach my new assignment. Despite my deep admiration for Bill, the timing of his visit was not ideal. I wanted to start slow, but Bill did nothing slow. As I knew from my first trip to Moscow with him in 2009, Bill would try to squeeze in as many meetings as possible. As ambassador, it was my job to accompany him.

I came in on Sunday to chair my first modified country-team meeting to review the schedule for "D" — Deputy Secretary Burns — who was arriving late that evening. To make sure the KGB could not listen in, we held these meetings in a windowless vault equipped with giant steel doors and lots of locks. The setup reminded me of the opening scene from the television show *Get Smart*. After introductions, we turned to review Bill's schedule for the next two days. I glanced over the first day and saw many familiar names. I started to think that this new job might not be so hard. Having worked the Russia account for three years at the White House, I had lots of close colleagues in high places here already, I reminded myself. I was not starting from scratch. I started to convince myself that I was uniquely qualified to do this job. We then reviewed Bill's schedule for the second day, which included meetings with opposition leaders and civil society activists. Given the recent demonstrations, I expressed some anxiety about the roundtable with opposition leaders, but new staff reminded me that dual-track engagement was our (my) policy! My only suggestion was to add a Communist Party leader to the group, which we did.

The following day, Deputy Foreign Minister Ryabkov presided over a hasty, informal diplomatic accreditation ceremony. (The formal ceremony at the Kremlin with President Medvedev would take place the following month.) The credentials ceremony allowed me to participate in the multitude of meetings with high-level Russian officials arranged for Burns that day. In his meetings — the first senior engagement in Moscow with high-level Russian government officials since Putin's decision to swap jobs with Medvedev — Bill delivered a message about our desire for continuity in relations. Despite recent differences over Syria, missile defense, and Secretary Clinton's criticism of Russia's parliamentary-election procedures, the United States remained committed to pursuing a large and important agenda with Russia that reflected shared interests. No one disagreed. Bill was a pro with a long record of interaction with most of these Russian government officials. I did nothing but sit next to "D" and take notes. (When traveling with President Obama, Secretary Clinton referred to the job as playing the role of a "potted plant.") Most of these Russian government officials greeted me warmly, welcomed me to Moscow, and expressed a desire to work closely with me, as we had when I worked at the White House. It all felt very smooth and cordial.

The next day, Burns participated in the roundtables with civil society leaders and political opposition leaders. To save him time, the two meetings were held back-to-back at two townhouses on the embassy compound, rather than hosted by Russian organizations. Both sessions lasted only an hour, giving everyone around five minutes to speak. Right after they ended, Burns departed for the airport.

I don't recall anything special about these events. By this point in the Reset, it was standard operating procedure for senior U.S. officials visiting Moscow to meet with civil society and political leaders. The tone, however, was different, much more optimistic than usual. After years of activism with few results, it must have given these politicians a real rush to stand on the podium before tens of thousands of demonstrators again. They had a swagger that had been missing for years. Boris Nemtsov in particular seemed more upbeat than ever, and he was always upbeat. As I listened to one young environmental activist, Yevgeniya Chirikova, tell Burns why they were going to "win," whatever that meant, I thought to myself that something new might be happening here in Moscow. We mostly just listened. I don't recall Bill saying anything of great importance. I fulfilled my potted-plant function, a role for which I was well suited that day given my jet lag.

These two uneventful sixty-minute sessions, however, would have profound consequences for U.S.-Russian relations and for me personally. For my entire time as ambassador, news coverage from this meeting was used in attempts to portray me as an enemy of the Russian state.

As our guests entered and exited the embassy, television camera crews swarmed them. These weren't normal reporters; the camera crews were from a state-controlled network called NTV, and they had a special assignment — collect evidence that the United States was seeking to overthrow the Russian government. It's even misleading to call these harassment crews "journalists." Some of their microphones did not have NTV logos on them. They had no credentials. Several of them worked for the neo-nationalist, pro-Kremlin youth group called Nashi. The Russian government paid them all. As Nemtsov would later comment, "They were not from NTV, they were *chekisty, 'nashisty'* or specially trained so-called journalists, who work as propagandists of Putin's paranoid ideas."[15] Chirikova also described them as *nashisty,* a fusing of the word *nashi* ("ours") and "fascists." She recognized several of them, as this was not the first time that they had hounded her.[16] I too would come to recognize their faces as a result of repeated encounters.

Kremlin propaganda outlets later reported that these Russian civil society and political opposition leaders had come to the U.S. Embassy to receive money and instructions from me, the newly arrived usurper. Because I was a specialist in color revolutions, President Obama had sent me to Moscow to orchestrate a revolution against the Russian regime, or so they alleged. NTV's "special assignment" group produced numerous television clips and documentaries showing footage of Russian opposition leaders leaving the U.S. Embassy to promote that message. For instance, a month later, NTV aired *Help from Abroad,* a "documentary" tracing how the United States, including me personally, allegedly funded the opposition. The following month, NTV launched a new series, *The Anatomy of Protest,* which explicitly claimed that the United States funded the Russian protest movements. Putin's strategy was clear — discredit the opposition by depicting them as puppets of the West and rally his electoral base against these bourgeois intellectuals. He had an election to win in two months. The 2012 campaign was his toughest ever.

The *NTV-shniki,* as these pseudo-journalists were called in Russian, filmed our guests outside the embassy to ensure the American seal featured prominently in the frame. A deep, menacing voice narrated a story about the visitors' mission inside the embassy. The video went viral. In just a few days, more than seven hundred thousand people had watched the clip.[17] Some of these viewers were likely supporters of the protesters, as many admired the opposition leaders for pushing back on their harassers. Some on my embassy team, therefore, thought that the film made the Kremlin look bad, and we should not worry too much about long-lasting consequences. I quietly disagreed. The clip would continue to be used against me for years, well after I had left Moscow.

Not all media outlets portrayed my arrival in Moscow in negative terms. For instance, an article in *Kommersant,* one of Russia's truly independent newspapers at the time, reminded readers that I was not a traditional diplomat: my unconventional approach to government work had produced the Reset, and it was possible that I'd pursue some untraditional but beneficial things in my new role.[18] Other commentators praised my academic credentials, my knowledge of Russia and Russian, and my success at resetting U.S.-Russian relations.[19] Outlets of all political orientations underscored my close connection to President Obama; I was, in the words of one journalist, "Obama's Man."[20] Some still hoped that my arrival in Moscow would help to advance closer relations between our two countries. I most certainly did.

Russian state-controlled media outlets, however, furthered a very different message about my mission to Russia: revolution. Judging by the detailed analysis of my biography and academic writings that Mikhail Leontiev presented on his television show on my second working day in Moscow as ambassador, research for the development of this narrative about me started well before my arrival. I had met Misha, as I called him, twenty years earlier when he worked as a journalist for independent, liberal-leaning papers such as *Nezavisimaya Gazeta* and *Segodnya.* But like several others from that era, he flipped. Privately, he still enjoyed his trips to America with his daughter, as he told me proudly when we bumped into each other on the promenade at the Sochi Olympics one day. But professionally he had evolved into the Kremlin's chief hatchet man. He was a talented, clever polemicist who produced a very popular television segment on Channel One called *Odnako* ("However" in Russian) that usually appeared during the evening news broadcast. Stylistically, *Odnako* was like *60 Minutes,* but without fact-checkers. Leontiev gave the impression that he was uncovering some hidden truth for his viewers, revealing how things behind closed doors really worked. On January 17, 2012, he devoted his entire show to me.[21] He told his viewers that I used to work for the National Democratic Institute (true), an organization with close ties to special intelligence services (untrue). During my last mission to Russia in 1990–91, I had come to promote revolution against the Soviet regime (untrue). My new assignment was to do the same against the current Russian regime (untrue). He suggested that the "internet-Führer," opposition leader Alexei Navalny, was a good friend of mine (again, untrue). Despite my many years of living in Russia, and my lengthy portfolio of writings on Russia, Leontiev explained to his viewers that I was not an expert on Russia or U.S.-Russia relations, but rather a specialist on revolutions. He reminded his viewers that the last noncareer diplomat sent to Moscow, Bob Strauss, had also come to the country to destabilize the regime. (Strauss arrived in Moscow two weeks before the August 1991 coup began.) I was sent by Obama to overthrow the Putin regime. He also added erroneously that in one of my writings I had proudly compared Putin to Milošević, a dictator whom the U.S. had also toppled. He made no mention of other books and articles of mine with such boring titles as "Reengaging Russia: A New Agenda for U.S.-Russia Relations" or *Power and Purpose: U.S. Policy Toward Russia After the Cold War.* He ended his show by citing another one of my works, *Russia's Unfinished Revolution,* and then asking provocatively, "Did Mr. McFaul come to Russia to work on his specialization; that is, to finish the revolution?"

I was amazed by Leontiev's hit piece. As my embassy team explained, Leontiev would not have aired a segment of that nature about the new U.S. ambassador to Russia without instruction from senior Kremlin officials. That the piece even aired suggested that leaders in the Kremlin were assigning a much higher probability to regime change than we were.

We at the embassy were not the only ones taken aback by this new Kremlin line. Several of my old Russian acquaintances, including even some loyal to Putin and his government, told me that they too could not believe the venomous, paranoid tone of Leontiev's commentary. Some journalists even wrote about the significance of this message from the Kremlin. "If someone needs proof that the reset epoch between Russia and the U.S. is over," Konstantin Eggert wrote in *Kommersant,* "he/she should watch *Odnako*." He added:

> I can't remember such an attack on the head of a diplomatic mission, especially on the U.S. embassy, even during Soviet times. Is it a joke — the ambassador of superpower number one on his third working day is being accused of the things for which others would be declared persona non grata and deported? This attack, without any doubt, was coordinated from above, which adds extra tension to this situation.[22]

Eggert was right. The Reset era was over, and my fate was sealed. From that day forward, I would be forever known as a fomenter of revolution.[23]

As reports from my press team about attacks on me piled up, my first reaction was outrage. Most of the claims were false, what we now call "fake news."

I was not funding opposition organizations. The CIA was not running a covert operation to pay people to show up on the streets of Moscow. The Obama administration did not believe in promoting "regime change." Ask our Republican critics, I wanted to shout! How can they propagate such bold lies? I was genuinely surprised by how blatantly these media reports twisted the truth.

I also felt betrayed personally by being portrayed as an enemy of Russia. I loved Russia. I was not a Russophobe or a Cold Warrior. Well before my arrival in Moscow as ambassador, I had lived several years in the country. I was the architect of the Reset. I was the White House advisor who had pushed for cooperation with the Kremlin when others were skeptical. Didn't they remember Obama's July 2009 speech (which I had helped to write) in which he boldly declared that a "strong" and "prosperous" Russia was in the U.S. national interest? No U.S. president had ever said that before. How could they turn on us — on me — so sharply and so quickly?

Of course, I understood that Putin needed an enemy to rally his electoral base before the March 2012 presidential election. The less educated, less urban, and less wealthy you were, the more likely you were to support Putin. That segment of the electorate could be scared into fearing us. Upon arrival in Russia, I immediately became part of this campaign. I was the perfect poster boy for America. I looked the part. Some even criticized my blond hair and easy smile as subversive.

Leontiev's piece, along with the dozens of similar articles and television clips that followed, also suggested that our meeting with the opposition when Burns was in town might not have been as pivotal as I had originally assumed. The roundtable surely provided video footage that fed the narrative that I was in Moscow to foster revolution, but the Russians had planned this line of attack well before my arrival. They would have run with it no matter what I did. And there was very little we could do to stop it. It was their country, not ours.

I also took comfort in rationalizing that these attacks were not about me personally, but part of a larger electoral campaign strategy. Several Russians encouraged me to understand my fate along that line. Vladislav Surkov, one of the Kremlin's most important campaign specialists, explained that my arrival in January was like manna from heaven for the Putin election effort. He estimated that the campaign's use of anti-American propaganda helped them pick up several percentage points. President Medvedev delivered a similar message to me on the day that I formally presented my credentials to him in the Kremlin. As we mingled, drinking champagne at the end of the ceremony in the ornate St. George's Hall in the Kremlin with a dozen other ambassadors, the Russian president pulled me aside and told me not to take the attacks too personally. After the election, everything would calm down.

Some of the attacks during the presidential campaign, however, were impossible to shrug off. On February 12, a video began circulating on YouTube suggesting I was a pedophile. That's hard not to take personally! New to being the target of a smear campaign, I wrote to my White House friend Ben Rhodes for guidance. As a veteran Obama campaign and now senior White House communications official, Ben would know what action, if any, the U.S. Embassy should take, given that some experts at the embassy believed the Russian government had ordered the attacks. Rhodes responded, "This is bullshit. You're friend [sic] with the president. They're insulting him as much as you. These guys are out of control." We contacted Google and they took the clip down, but it later reappeared. A web search for "McFaul is a pedophile" still produces hundreds of hits.

That same week, remarks that I'd made to a small group of board members from the U.S.-Russian Business Council at a Marriott hotel in Moscow were secretly taped and then edited to make it sound like the United States government had a plan to discredit Putin's election victory the following month. I was shocked by the audacity of this act when the clip aired, as was the USRBC president, Ed Verona, who would later be subject to similar tactics.

On the night of the presidential election on March 4, 2012, a fake Twitter account that looked identical to mine tweeted out criticisms of the electoral procedures even before voting had ended. The Russian media went crazy, as did some Russian government officials, accusing me of blatantly interfering in Russia's electoral process. This stunt was so well executed that it took us a while at the embassy to realize what was happening. Even I initially thought that one of my staff had gone rogue, sending out tweets on my behalf. We eventually figured it out — the fake account was using a capital *I* in place of a lowercase *l* in the name associated with my twitter handle @McFaul (McFaul vs. McFauI, can you see the difference?). We eventually explained the origin of the spurious tweets, but only after a few hours of hysterical news coverage.

Denying that you are a pedophile, refuting accusations that you are plotting regime change, explaining to the world that you are not criticizing Putin on his election night—it all became so tedious, defensive, and exhausting. We all hoped that Medvedev was right, and that things would die down after Putin's reelection. It was a false hope.

Suppression of dissent was not the only option for the Russian government during those tumultuous times. Co-option was an alternative strategy, an approach that we got a glimpse of during the last months of Medvedev's presidency. In the spring of 2012, Medvedev tried quietly to engage with the reformers. Ironically, the only time I encountered the entire leadership of the opposition was in February 2012, at the coatrack in the basement of Medvedev's guesthouse at his country estate. Senator Max Baucus and I were leaving a meeting with Medvedev as the group of opposition leaders was coming in for its own session with the Russian president. At the sight of me, some of these political figures scooted away, not wanting to risk being photographed with me. The gregarious Boris Nemtsov, however, embraced me warmly. He asked in a voice loud enough for everyone to hear why Medvedev's bodyguards had not allowed NTV camera crews to document our secret meeting, and joked about encountering the American revolutionary at Medvedev's residence. (I really loved Nemtsov's sense of humor.)

Beyond taking meetings with opposition leaders, Medvedev tiptoed toward democratic reform during his lame-duck months. He agreed to more transparency at the ballot boxes for the presidential election in March 2012, including allowing election observers to take photos and make videos.[24] Video cameras were installed at every precinct. He also proposed simplified procedures for registering political parties, lowered the number of signatures needed to gain access to the presidential election ballot,[25] and reduced barriers for participating in regional legislatures. He put forward a new electoral law for the State Duma, marking a return to a system that had been defunct since 2003. The new law required that 50 percent of the body's seats would be assigned through single-mandate districts, and 50 percent through proportional representation. Most Russian election experts at the time thought this reform would help independent candidates from the opposition obtain seats in parliament.[26] Medvedev also pledged to set up a public television station similar to the United States' PBS or the United Kingdom's BBC.[27] Most dramatically, he reintroduced direct elections for governors, reversing a Putin decision in 2004 to appoint them.[28] Symbolically, he appeared on Russia's most watched independent television station, Dozhd TV. During the period between the first demonstrations in December 2011 and Putin's inauguration in May 2012, the authorities also cooperated closely with demonstration organizers.

At the time, our political analysts at the embassy considered these important reforms and gestures. We at the embassy wondered how long this pluralizing trajectory would last. No one was very optimistic.

Disappointingly, the Kremlin's flirtation with compromise and accommodation was short-lived. After his reelection in March 2012 and his inauguration in May, Putin cracked down. His campaign to weaken the Russian opposition was comprehensive and successful, at least in the short run.

May 6, 2012, the day before Putin's inauguration, marked a real turning point in the struggle between the government and the opposition. That day, tens of thousands assembled to march to Bolotnaya Square in downtown Moscow, just across the river from the Kremlin. Police had installed a row of metal detectors through which demonstrators had to pass to join a scheduled rally in the square, but when a bottleneck formed at this security line, skirmishes erupted between the police and protesters; more than four hundred people were detained, including the most prominent opposition leader at that time, Alexei Navalny. Putin and the Kremlin emphatically denounced the provocateurs, and used these events as a pretext to pass new laws criminalizing unauthorized protests, even though the Bolotnaya demonstration had, in fact, been authorized. Navalny was released, but dozens of others were jailed or placed under house arrest, some for years. The new fine for participation increased to 300,000 rubles, equivalent to 9,100 U.S. dollars at the time.[29] In the following months, the Duma passed one restrictive law after another, earning the legislative body the nickname "printer gone wild."[30] Another law empowered the government to take down websites considered damaging to the "public good." What constituted the public good was left ambiguous, allowing authorities to block a handful of political websites. Blasphemy, defined generously as offending "feelings of religious believers," was also criminalized.[31] Independent media came under attack. Eventually, the same Dozhd TV on which Medvedev had appeared in his final days in office was forced off the airwaves and cable packages and onto the internet, losing in the process the vast majority of its viewers and revenues. In July 2012, Putin and his supporters passed another draconian law requiring nongovernmental organizations "engaged in political activities" to register as "foreign agents" if they received financial support from foreign foundations or governments. Over time, new laws and regulations limiting foreign funding accumulated, making it nearly impossible for hundreds of NGOs to receive international assistance. A few years later, several foreign organizations, including the National Endowment for Democracy, the International Republican Institute, the National Democratic Institute (my former employer), the Open Society Institute, the MacArthur Foundation, and the U.S. Russia Foundation for Economic Advancement and the Rule of Law (USRF) were forced to leave Russia.[32]

Putin was different from Medvedev. The change in the Kremlin had immediate, negative consequences for opposition leaders, demonstrators, independent media, and civil society. This depressing trajectory for pluralism begun in 2012 has never been reversed.

The United States remained a target for Putin after his reelection. As it had during the Cold War, the Kremlin propaganda machine portrayed the United States as an imperial, predatory state, constantly undermining international stability and violating the sovereignty of other states. Our alleged support for revolutions in the Arab world generated special attention, as did our military intervention in Libya and alleged military efforts to overthrow Assad in Syria. Of course, the truth is murky. The U.S. did attack Gadhafi in Libya and did provide support for opposition groups in Syria after those conflicts began. But we did not start any of these uprisings, and most certainly never bombed Assad or his army, or supported ISIS and other terrorists groups in Syria, as the state-controlled Russian media often claimed.

In parallel to disrupting stability and violating the sovereignty of states around the world, the U.S. also continued its campaign for regime change inside Russia, or so the Kremlin propagandists charged.[33] And I continued to feature prominently in this propaganda effort. Vyacheslav Pravdzinskiy, a graphic designer, produced a calendar called "McFaul Girls 2012" that featured cartoon drawings of opposition leaders depicted as pinup girls, with a portrayal of me on the cover in a tightly fitting skirt (I wish I was that thin!).[34] In the days before the May 6 demonstration in Bolotnaya Square, another poster featured me as the "artistic director" for the "political circus" of opposition leaders that was coming back to the arena that day. The following year, someone photoshopped my head onto the body of a volunteer campaigning for Navalny during his mayoral electoral campaign. Amazingly, many people who saw the photo actually believed it was me. They said so on Twitter.

Departing from Cold War tropes, Putin's regime added a new dimension to the ideological struggle — conservative, moral, nationalist Russia versus the liberal, immoral, internationalist West. Russian state-controlled media asserted that Putin had nurtured the rebirth of a conservative, Orthodox Christian society. By contrast, the West was presented as hedonistic, godless, homosexual. Russians had to be protected from these decadent Western ideas. The Kremlin passed a law against "homosexual propaganda," which was portrayed as protection for Russians, especially Russian children, against Western values. While decaying Western countries like the United States and Ireland legalized gay marriage, Putin firmly rejected such ideas as antitraditional, anti-Orthodox, and anti-Russian. Putin unabashedly defended this conservative turn, arguing that societies fall apart without traditional values: "What's important for me is not to criticize Western values but to protect Russians from certain quasi-values that are very hard for our people to accept . . . Russia is a country with a very profound ancient culture, and if we want to feel strong and grow with confidence, we must draw on this culture and these traditions."[35]

Those of us in the U.S. government debated internally — sometimes with Obama — about whether Putin truly believed these tall tales of American subversive activities in Russia, or whether he just deployed these arguments to mobilize domestic support. Initially, I leaned toward the latter hypothesis. Kremlin spin doctors were a cynical bunch, but they weren't stupid. Surely they knew precisely what we did and did not do inside Russia. Their intelligence organizations, after all, are some of the best in the world. By the end of my tenure as ambassador, however, my position changed: I came to believe that Putin and some of his closest advisors genuinely believed that we were seeking to subvert his regime. Years after my departure, one of Putin's former advisors, Gleb Pavlovsky, said it bluntly: Putin "genuinely thinks the U.S. is trying to overthrow him."[36]

It was not just a campaign slogan, nor was it an instrumental method for discrediting the political opposition. Putin developed his theories about American foreign policy years earlier, when he was a KGB agent in East Germany. Putin assigns particular power in the making of U.S. foreign policy to what he would call the *siloviki*, power ministries such as the Department of Defense and the CIA. In his view, leaders of those agencies have a bureaucratic interest in starting wars and running covert operations, and often succeed in steering elected civilians to their causes. Since the *siloviki*, including Putin himself, play such a powerful role in the making of policy in Russia, it's not surprising that Putin might assume the same distribution of power in the U.S. political system.

And, of course, there are empirical data to support Putin's hypothesis about American foreign policy. Over the last several decades, especially during the Cold War, U.S. presidents have deployed covert and overt power to overthrow regimes.[37] In 1989 we celebrated the fall of the Berlin Wall and the collapse of the communist regime in the German Democratic Republic (GDR), where Putin was stationed at the time. That event made a big impression on the young KGB officer. As ambassador, in my discussions with Russian interlocutors on this theme, I frequently heard about American operations against leaders in Congo, Iran, Vietnam, and the Dominican Republic, as well as more recent cases of overt force to topple regimes in Serbia, Afghanistan, Iraq, and Libya. I tried to add nuance and context, explaining the different motivations involved in these wars, and the varying degrees of legitimacy of these different efforts. President Obama, for instance, opposed the war in Iraq well before it was popular to do so. He supported military intervention against Gadhafi to prevent genocide, not to overthrow the regime. The same was true in Serbia. It was long after the NATO bombing campaign that the Serbian people, not Americans, overthrew Milošević. It seemed wrong, therefore, to compare the American military effort in Libya and Iraq, or Serbia and Afghanistan.

But most Russians, including Putin, saw a consistent pattern across these cases and were quick to attribute blame to the United States. I worked in the White House during the Arab Spring and remember well how we scrambled to *react* to events triggered by brave Arabs with whom the United States had no contact. It was nearly impossible to explain to most Russians that the United States had not been involved in starting these uprisings. Obama most certainly failed to convince Putin. When tens of thousands of Russians gathered on Sakharov Avenue to demand "Russia Without Putin" and fair elections—demands very similar to what Egyptians in Tahrir Square advocated—it was easy to see these events as evidence of an American conspiracy, if you were looking to find one.

Moreover, as I learned from listening to him in private and in public, Putin does not believe that nongovernmental leaders or groups pursue their agendas and initiatives independently. He is always looking for a hidden hand guiding social movements, be they the democratic revolutions in Eastern Europe in 1989, the color revolutions in Georgia in 2003 and Ukraine in 2004, the Arab Spring in 2011, or the Russian demonstrations in the winter of 2011 and spring of 2012.[38] There was nothing Obama could say to change his mind.

Putin's grand theories about the purposes of American power sometimes became personal. In May 2012 I experienced Putin's animosity toward me personally during his meeting with National Security Advisor Tom Donilon out at the president-elect's country estate. The meeting started with a bang when Putin asked Tom bluntly, "When . . . are you going to start bombing Syria?"[39] The mood music for U.S.-Russia relations had changed dramatically since our last meeting with President Medvedev just a few months earlier in Hawaii. It was at this meeting that, as described in this book's prologue, Putin suddenly turned directly to me and blamed me for increasing tensions between our two countries. Putin has a thin, high voice, with which he shouted the last line of his comments, all the while continuing to stare at me. I was unnerved by his expression and tone, but also bewildered. Why was one of the most powerful men on the planet so obsessed with an American diplomat? I remember thinking that he shouldn't even have known my name, let alone considered me a powerful force inside his country. His attack on me that day was strangely emotional, and out of character for a leader whom I always had considered smart, rational, and strategic. His focus on me seemed to be a symptom of his own insecurities about his standing inside Russia at the time.

I didn't know if I should respond to Putin or not. This was Tom's meeting, not mine. It was rare for anyone outside the principals to speak during a meeting of this kind, but the Russian principal had spoken about me personally. I had been told that the guy hated me. Now I knew it, or at least felt it firsthand. If he'd wanted to intimidate me, he'd succeeded. When the meeting ended, as he was walking out of the room, I tried to say something conciliatory to Putin in Russian about seeking ways to build common ties. He seemed to accept the overture, or so I hoped.

Just a few weeks after that encounter, I attended a reception at the Kremlin with the who's who of Moscow to celebrate Russia's national day. It was an odd holiday in that most Russians did not know exactly what they were celebrating or why June 12 had been designated the day of celebration. I knew. I had witnessed the first commemoration of this holiday two decades earlier; it marked the day in June 1990 that the Russian Supreme Soviet voted to declare Russian independence from the Soviet Union, an inconvenient fact for Putin. What a paradox it was for me to be attending a reception hosted by Putin on this occasion, when Putin himself had called the collapse of the USSR one the greatest tragedies of the twentieth century.[40]

At this gathering, everyone knew me. Only the brave, however, would dare chat with me in the presence of Putin and his loyalists. One old friend, Vladimir Mau, approached me and offered some advice. He had recently met with Putin one-on-one on some matter related to his academy, where he was rector, but the conversation had turned to me, as Putin knew that Mau and I were friends. Putin lashed out about me, claiming I was too public, too blunt, and too popular. In the world according to Putin, I had too high a profile for an ambassador. To avoid my getting thrown out of the country, Mau advised me to lay low, keep out of the press, and try to find something positive to say about the Russian government.

As I listened to Mau's sincere advice, my first reaction was agreement. This was Putin's country. I should play by his rules. I started composing press releases and plotting new events in my head that Putin might like. But as I walked back through the gardens to my car parked outside the redbrick walls of the Kremlin, I remembered the tragic, bloody history that many Russian leaders had made from inside these walls. Did ambassadors from earlier times take their cues from those earlier tyrants? If they did, how did they feel upon their exit from Russia (or the USSR)? No, my job was not to be Putin's friend and stay silent. My job was not to be liked by him or his entourage. My car had an American flag on it, not a Russian one. My job was to represent President Obama, his administration, and the American people in this land. My job was to execute American policies that served the interests of the American people. Nothing I had done or said since moving to Moscow was inconsistent with these assignments. I had arrived at that reception shaken by my pariah status. I left all the more determined to remain true to my mission.

The grotesque lies about us reported on television, the sinister "documentary" films, the twisted government statements about our alleged revolutionary activities did get under our skin from time to time at the embassy. Many staff meetings began with someone reporting on the latest fable he or she had witnessed the night before on television, or read in a government-controlled newspaper. We would all roll our eyes. The old hands in the room assured us that these kinds of attacks went beyond anything they had witnessed in the Soviet era. Fake news is hard to endure, and even harder to counter.

At the same time, I tried to understand Putin and the challenges he faced as Russian president. Even if we were not running the outrageous covert operations described on Russian television, Putin perhaps had reason to fear the United States, Barack Obama, and even me. We did not provide financial support for Putin's opponents, but we did execute the Obama administration's policy of advocating for human rights and democratic values.[41] In our support for the aspirations of those Russians advocating for universal values, we were a threat to Russia's increasingly autocratic regime.

The public profile I adopted also made me a threat. I was conventionally American: open, accessible, and transparent. I talked about my kids. I tweeted about my losses on the basketball court. When asked about controversial topics — whether slavery, the Vietnam War, or the invasion of Iraq — during interviews and on social media, I offered honest opinions. The Russian government never admitted to any faults, past or present. Putin had no regrets (and never lost any hockey matches!). Even the Reset for them was a U.S. initiative to correct for our past mistakes. I welcomed Russian criticism of American foreign policies and domestic practices as a way to make us better; that was uncomfortable for them. I also tried to distinguish between our government's disdain for authoritarian policies pursued by the Kremlin and my respect for the Russian people, culture, and history. The Kremlin hated that. As one person close to the Kremlin revealed to me, my obvious love for Russia and its people drove the Putin government nuts. I would have been a much easier target if they could have portrayed me as a Cold War Russophobe.

16

GETTING PHYSICAL

In parallel to the media campaign against us, the Russian authorities conducted a ground campaign of harassment against my colleagues at the embassy, myself, and, from time to time, even my family.

In my first week in Moscow as ambassador, Nashi, the Kremlin-created youth group, threatened to organize a demonstration outside our residence. As the regional security officer (RSO) circulated emails in red ink urging embassy employees to stay away from my new home, Spaso House, I wondered what I was supposed to do, as my family was at the residence at the time and I was at the embassy. One of my sons wondered if we were going to end up in a modern-day siege, and therefore proposed getting the necessary provisions from Cinnabon, just two blocks away, before we were trapped inside our mansion without access to supplies. The red-letter email turned out to be a false alarm. All we saw were some grandmothers walking their dogs in the park in front of our house. The experience, however, kept us on alert that day and for the rest of our time in Moscow, just as our hosts desired. That day also reminded us of the importance of humor to defuse tense new situations. Mostly it worked, but not always.

Outside the American Embassy, the Russian government authorities also made their presence known. Formally, the Russian police officers posted at the gates of our compound were deployed to protect us, even though we had our own Marine guard just inside the gates. In reality, the Russian officers' main assignment in this new era of confrontation was to harass and subject to surveillance everyone entering the embassy. Even our American employees often had to stand helplessly in the cold, waiting for the Russian officers to take and then inspect their passports and record their data before being allowed to proceed. They even detained my wife from time to time. As one embassy report on harassment in the spring of 2012 documented, "January 27: Donna Norton, the Ambassador's spouse. Harassed and held by police at South Gate in sub-zero weather." Surely Russian intelligence was good enough to know who my wife was and that she was from Southern California! Standing outside in subzero temperatures while Russian police questioned her was no fun, especially when my driver whizzed right past her through the gates one time with me sitting in the back of my warm black Cadillac.

For Russian employees at the embassy and Russian guests, the process was even worse. It was like a border crossing. Russian visitors simply stopped coming. It just wasn't worth the hassle, especially as their passport information was being recorded for future use against them. The same kind of harassment happened outside Spaso House. One time, they even held up a Russian senator, Mikhail Margelov, for over twenty minutes because he didn't have his passport with him. Who knew that you needed your passport to have tea with the American ambassador at his residence? When we complained, the Russian authorities insisted they were just trying to protect us . . . from a Russian senator!

This level of police harassment at the American Embassy and Spaso House was new. No one could remember a time even during the Soviet era when our hosts were so aggressive. More than once, we delivered formal letters of complaint to the Ministry of Foreign Affairs about the behavior, noting that we did not station police outside the Russian Embassy in Washington. Nothing changed.

When we could have used these Russian guards, however, they suddenly seemed unable to act. One day early in my tenure, on February 10, 2012, a group of several hundred demonstrators strangely attired in white plastic lay down on the street in front of the main embassy gate, blocking all traffic from leaving or entering the compound. (I'm assuming the plastic jumpsuits were to protect the demonstrators' clothing.) My car pulled up to the embassy just after the group arrived; it was a strange feeling being denied access to the American Embassy, which is sovereign territory of the United States of America. I also worried about the embassy children, including two of my own, who were on their way back home from school at the time. Were they going to have to sit on the bus outside the gates of the compound, where their homes were located? We asked the police at our gate to clear the street, but they pleaded that they had no authority to break up a public demonstration. Of course, the irony was that Russian law made it illegal to convene a public demonstration without a permit, but the Russian authorities didn't seem too anxious to enforce their laws against these particular demonstrators. Our security cameras later revealed that Nashi leader Tikhon Chumakov had organized the demonstration. Chumakov and his Kremlin-backed comrades were above the law. One of my very frustrated embassy colleagues suggested that the demonstration violated the Vienna Convention on Diplomatic Relations. If so, the convention wasn't doing us much good that afternoon. Eventually the protesters dispersed, but the exposed feeling we all endured that day lingered long thereafter. What if they had tried to enter our compound? The low brick wall surrounding our embassy office buildings and townhouses, after all, was easily scalable. What would we do then? Our security team started making contingency plans. None of our options provided much reassurance.

Not long afterward, I encountered Chumakov face-to-face during a visit to the CEO of Rusnano, Anatoly Chubais, at his company's headquarters. It was a routine courtesy call. Rusnano is a giant state-owned Russian company that invests in high-tech companies all over the world, including in the United States.[1] Why Nashi operatives would want to harass me outside this office building was unclear. As I opened my car door, they sprang — Chumakov and two or three others with video cameras. My meeting with Chubais was not a public event, so it was clear they had obtained access to my calendar — perhaps electronically, perhaps from a Russian informant working at the embassy. They bombarded me with questions about supporting the opposition. Against the judgment of my bodyguards, I decided to answer, in Russian. I reconfirmed that the United States provided no financial support to the Russian opposition. After a short exchange, I finally recognized Chumakov, recalling that he had previously been assigned to follow and harass the former British ambassador, Anthony Brenton.[2] Nashi encounters with Ambassador Brenton were aggressive and sometimes violent; he had even been attacked in the driveway of his residence. Chumakov threatened me with similar treatment, promising that his group would chase me out of the country just as they had with Brenton. *How pleasant,* I thought. *Welcome to Moscow!*

My run-in with these Nashi agents at Rusnano ended uneventfully. None of the tape from the "interview" ever aired, because I hadn't said anything useful to them. And my bodyguards thankfully avoided physical contact with these "youth leaders" (yes, that's polite diplospeak), even though they were quite aggressive with me. But the event reminded me that I was under constant surveillance. How did Chumakov know that I was coming to Rusnano that day? Who was helping him obtain such information? Should I expect such a greeting party everywhere I visited?

We eventually learned to expect them. The Nashi posse did not meet me at the entrance of every meeting I attended. They never showed up, for instance, outside the gates of the Kremlin or in the parking lot at the Ministry of Foreign Affairs. But they showed up frequently enough that we planned for encounters.

They also lingered outside Spaso House on a regular basis. They filmed Russian visitors coming to see me, but also Americans, strangely enough. In March 2012 we convened a town hall meeting for American expats living in Russia, and the NTV-Nashi squad showed up. That same month, one of these crews made its way onto the premises of the Spaso House compound without authorization. Our security guards immediately apprehended them and escorted them off the premises, but the spectacle was sobering. If a couple of hundred Nashi activists showed up, would we be able to keep them out? Our security team put new procedures in place, assigning more officers to work every event at Spaso House, and started using metal detectors and wands for all larger gatherings.

My drivers and bodyguards were also subject to harassment. On one routine visit to the Ministry of Foreign Affairs, the police interrogated, fined, and suspended the driver's license of one of my drivers while he waited for me in the parking lot. His crime: illegal lights on our Cadillac. As I came out of my meeting, I was stranded. Slava, my driver, could no longer drive and was extremely distraught, worried that he might lose his job. I demanded that my MFA colleagues solve the problem. A senior official from the Department of North America did come out to the parking lot to see if he could help, but we both soon understood that the normal police assigned to that area of town were not the ones causing us trouble. This was a special operation. Later interactions with senior MFA colleagues showed they were clearly embarrassed by the incident, but there was nothing they could do. In court, we fought the charges against Slava. It was our car, after all, not his. And we lost. He spent the next year pumping gas at the embassy rather than driving the U.S. ambassador. It was frustrating and demoralizing to us all.

One of my bodyguards suffered a similar fate. As per embassy practice, he was assigned the task of following us in a chaser car. One day he was stopped, fined, and lost his license for driving a Chevy Suburban with illegal lights and an illegal horn, neither of which he ever used. He was relegated to riding shotgun thereafter. I was meeting with a senior official at a major Russian government media company the same evening my bodyguard lost his license. When I told him what had happened, his advice was to adjust, get used to it. There was nothing that could be done. The Reset was over; I was now the enemy. This was their country, not ours. Their rules — or lack thereof — not ours.

So we adjusted. We increasingly relied on American security officers to drive vehicles, as they had diplomatic immunity. We used different cars. But no one liked it, as these adjustments complicated our security efforts. Ambassador Stevens had been assassinated in Benghazi, Libya, earlier that year, so my Diplomatic Security officers were especially attentive to and vigilant about their work. The head of our Diplomatic Security unit tried to reason with his Russian counterparts; it was against their own interests, he argued, to leave me vulnerable. What if, God forbid, something happened to me or my family as a result? The Russians shrugged. They were just following orders from higher-ups.

As per guidance from our security team, I also had to assume that every phone call I made, every email I sent (on the unclassified system), every website I visited, every conversation I had, and even every movement I made inside Spaso House was being monitored by the Russian government. All the large apartment buildings next to Spaso House in downtown Moscow had FOR RENT signs hanging in them, yet no one ever moved in. In our first security briefing upon arrival, Donna and I were told that we should use one of our secure rooms at the embassy if we ever needed to have a serious argument. (Thankfully, we never needed to use that service!) The technological advances in cyber surveillance over the last decade, as well as voice and video monitoring, are mind-boggling. We had to operate in Russia as if we were being monitored all the time. I had adjusted to a life with minimal privacy as a White House official. Living in Russia, I had no privacy at all.

I also had to deal with death threats. I'm sure most of them came from kooks. It's easy for someone to get drunk at home and fire off a belligerent tweet. But kooks sometimes do kooky things, so my security team rightfully treated every threat as a serious one. Threats from Chechens occasionally popped up. Especially after Benghazi, these death threats unnerved everyone at the embassy. Some seemed more credible than others, and would compel us to change my travel routes, as well as the number of cars and bodyguards accompanying me. We sometimes reported these threats to the Russian government authorities, who took them seriously and cooperated with our security officers. Many other threats just lingered.

Harassment was not limited to my immediate security team and me. Anyone who worked at the embassy could become a target. They slashed the tires of one of my junior staffers. They broke into the homes of embassy employees, oftentimes just rearranging the furniture or turning on all the lights to let people know that they were vulnerable. During my second year on the job, the State Department's Office of Inspector General did a comprehensive review of all activities at the embassy. On security, their final unclassified report noted, "Across Mission Russia, employees face intensified pressure by the Russian security services at a level not seen since the days of the Cold War."[3]

Russian officials also regularly interrogated our Russian employees, pressuring them to report on us. My Russian bodyguards were brought in for questioning. We assumed that some of our Russian employees were informants for Russian intelligence.

The FSB, Russia's state security agency, also aggressively recruited informants among our American staff, offering large sums of money for sensitive information, just like one sees in the movies. And "honey traps"—the deployment of beautiful young women and men to lure American employees into doing things that could make them vulnerable to blackmail—occasionally work. One of the hardest parts of the job as ambassador was signing papers to curtail someone's assignment in Moscow because they had become a counterintelligence risk.

These harassment techniques were not new, but the number of incidents spiked noticeably in the winter of 2012. A memo prepared by my regional security officer and his team counted "nearly 500 additional instances of harassment against U.S. Mission personnel" between January 17 and March 30, 2012. Even during the Soviet era, no one on our staff could remember a period of harassment so intense.

• • •

The FSB routinely follows subjects without being detected. They are pros at this. But sometimes their intention is to make their presence felt. In the fall of 2012, for reasons that remain mysterious to me, it became clear that we were being followed. As I wrote to the head of our security team on October 7, "My guards informed me that I was followed today while attending my son's soccer game. And they then kept with us as we went to McDonald's." My head of security replied that if we saw them, it was because they wanted us to see them. A few weeks later, FSB agents, or so we assumed, sat in the pew behind us in church, which truly unnerved my wife. They followed us on the streets, and closely tailed our Cadillac. On one occasion, one of my drivers overreacted to being followed. With my family in the car, he began driving faster and more erratically, weaving through Russia's crazy traffic until I finally intervened and urged him to relax. After all, our situation was not like in the movies. We could never lose them for good. They knew where we lived!

The car chase episode scared me, as it illustrated that the Russian intelligence officers were succeeding in getting under our collective skin. It also was getting dangerous. Car chases often end in car crashes.

The worst form of harassment, however, was when my children were followed. The drive from Spaso House to the Anglo-American School that my sons attended was forty minutes in the morning and an hour or more, depending on traffic, on the return leg. My two sons were driven back and forth in an embassy vehicle, with a Russian driver and no security detail. State Department regulations for Russia prevented us from assigning security guards to my wife or children. As a new political appointee, I was not going to challenge these procedures. One day, in the spring of 2013, my security team reported that a car was following my kids' vehicle to school, an activity we verified through a proper investigation. It wasn't hard to confirm that my kids' car had been followed. Whoever was responsible wanted us to know. I was upset, but also helpless. I could understand why they might need to follow me. But my children? They obviously wanted to intimidate me, scare me, and make Russia an uncomfortable place for me to work. And they succeeded.

Then it happened again. On May 6, 2013, one of my senior staff members reported to me that another car had been observed following my sons to school. I concurred with the plan to issue another note of complaint to the Ministry of

Foreign Affairs, but also added, "I do not want us to assume that we know this car is the FSB. We should also keep open the possibility that this is a security threat to my children." There were many people in Russia who didn't like me or the United States. Anyone could have been driving that car.

We did push back with the Russian government regarding these family-related incidents. Through various government-to-government channels, we presented evidence of harassment to our Russian hosts. Even President Obama got involved a couple of times, asking both President Medvedev and later President Putin to stop harassing "his guy" in Russia. We could never really tell if these pushbacks worked. The harassment, especially the overt surveillance, would come and go. We could never detect a clear pattern related to my activities as ambassador or developments in U.S.-Russia relations. And maybe that was the plan.

At times, my wife, like my older son regarding the threat of a siege, employed humor to deal with these psychological-warfare tactics. For instance, she expressed relief that it was Russia's state security service and not some crazy nationalists on our tail. Surely the Russian state was not going to harm us, she reasoned, so let's just think of them as an extra layer of security! With the NTV-Nashi teams, I would chat politely, and even offer them cookies when they were hanging around outside my house for hours. I also started filming them as they filmed me. At times it felt like a game; sometimes they were even friendly in return.

One time, however, I made a big mistake. On March 29, 2012, I went to see Lev Ponomarev, a former member of parliament turned human rights activist. We pulled up to his office to find that a giant crowd had assembled between the curb and the entrance — Nashi kids, *NTV-shniki,* and even Cossacks, for some reason, adorned in their big hats and decorative knives, and various other protesters holding signs denouncing America. Seeing this motley assembly, I again wondered who had tipped them off about my meeting. As I exited the car and faced the multiple cameras, I decided, against the advice of my bodyguards, to stop and talk. I had spoken with demonstrators cordially during previous encounters, and nothing had happened. I believed that showing no fear was the right course of action. Not this day.

For several minutes in Russian, I successfully pushed back on their behavior.

I explained that it was the job of ambassadors all over the world to meet with representatives of all different sectors of society, including civil society. My visit to meet with Mr. Ponomarev was routine diplomatic work. When they asked if they could interview me on other topics, I agreed, but requested they go through my press office instead of stopping me on the street in the cold when I had no coat on. As they kept pressing me, however, I became agitated, and stated that Russia was a *dikaya strana* ("wild country") for tracking and harassing diplomats the way it did.[4]

Some opposition leaders who later watched the exchange on video on social media loved it. Alexei Navalny, for instance, tweeted that I should have punched the agitators, being that I had diplomatic immunity.[5] Russian officialdom had a very different reaction, and I agreed with them. I had wanted to say that the behavior of those Nashi activists was inconsistent with international norms. American political groups do not obtain the Russian ambassador's calendar and then follow him wherever he goes. And Nashi and NTV were instruments of the Russian regime, so their treatment of me was a Russian government operation. But in the heat of the moment, those words had not come to me. Of course, after the "*dikaya strana*" clip aired I apologized immediately on Twitter, tweeting, "Just watched NTV. I misspoke in bad Russian. Did not mean to say 'wild country.' Meant to say NTV actions 'wild.' I greatly respect Russia."[6] But the tape of that sound bite would loop a long time.

I reached out to some friends of mine at the White House to apologize for letting "the boss" down. Obama, I was reminded, had made his share of inappropriate remarks: He'd once referred to "Polish concentration camps." On another occasion, he compared his bad bowling to the "Special Olympics."[7] And he had spent a much bigger chunk of his life in the public eye; I was new to this arena. Still, I was disappointed in myself. Of course, the Russian government *was* behaving wildly at the time. It was doing things that normal governments simply don't do, both to their own citizens and to me. And wilder stuff was yet to come. But diplomats should not say such things in public, and I was now a diplomat. The Nashi strategy of constant harassment had generated dividends for the government that day.

On Russian television, I was portrayed as a demon. In private, the Russian government continued to engage with me actively on a number of issues related to our bilateral relations. The U.S. practiced dual-track engagement with the Russian government and society. The Russian government employed separate tracks of engagement and containment with me — that is, engage me in government channels but seek to contain my contacts with nongovernmental actors. Just as our administration did not deliberately link progress in one channel to progress in the next, the Russian government showed skill at keeping these channels separate.

The Russian government official with whom I worked most closely was Deputy Foreign Minister Sergey Ryabkov, the MFA's point person for the United States. Ryabkov was also the MFA's senior official on all things nuclear, and represented the Russian Federation in the P5+1 group negotiating with Iran, so we had a lot of business to do. Several senior officials in the U.S. government had Ryabkov as their main interlocutor in the Russian government. My business with Sergey was often damage limitation, as our bilateral relations were on a downward spiral. If the substance of our work was often tough, the nature of our interactions was always cordial and professional.

At the Foreign Ministry, I also met frequently with Deputy Foreign Minister Mikhail Bogdanov for long discussions on Russia's policy in Syria and the larger Middle East. Bogdanov had served for over a decade as Russia's ambassador to Syria, and had held postings in Yemen, Lebanon, Israel, and Egypt prior to that. He spoke fluent Arabic, had met all the major players in the Middle East, and knew Syria as well as anyone. Our two governments mostly clashed on the substance of Syria policy, but the process of engagement with Bogdanov and his deputies remained informative and friendly. In fact, I would describe my interactions with almost all officials at the Ministry of Foreign Affairs in similar, positive terms. At the deputy foreign minister level, the Russian MFA had real talent, and I always found it interesting and pleasant to engage with them.

I also enjoyed constructive relations with individuals in the Kremlin. Yuri Ushakov, Putin's foreign policy aide, always treated me well. We had some very tough meetings, especially surrounding our efforts to get Edward Snowden back to the United States and our decision to cancel a summit in Moscow planned for September 2013. Yet we always maintained contact and frequent engagement. Yuri accurately reflected the intentions and views of his boss. I think I did the same in representing the preferences of my boss.

In the Russian government, I had developed working relationships over the years with several senior deputy prime ministers and ministers, but my main interlocutor was First Deputy Prime Minister Igor Shuvalov. During our difficult but successful negotiations over Russia's WTO accession, Shuvalov played the pivotal role. His position gave him tremendous power within the Russian government over all economic issues, but he also had a very close personal relationship with Putin. Shortly after Russia joined the WTO, we enlisted Shuvalov as our main Russian government partner in promoting trade and investment between our two countries. He was always practical, efficient, and focused on outcomes.

An exhaustive account of my Russian government Rolodex would be tedious. But the point is an important one. For issues that mattered to our government-to-government relationship, I never felt constrained in doing my job. Those senior government officials who had business with the U.S. government saw value in engaging with me. They understood my relationships back in Washington, especially at the White House, and with President Obama. Many of my meetings with Russian government officials involved managing negative developments in our bilateral relationship, not making new progress. But in the margins, I believe that my personal relationships with key government officials helped prevent bad situations from becoming worse.

The Russian government, however, tried to limit my contacts with those whom they decided did not have a legitimate reason to meet with me. The Kremlin prevented regional government officials from sitting down with me, a practice that continued under my successor. The pattern was the same for every trip I took. My staff would initiate contact with local officials. We had large consulates in Vladivostok, Yekaterinburg, and St. Petersburg, and therefore solid contacts with government officials in all of these cities. Midlevel Russian staffers would signal to our consulate officers that the governor or mayor looked forward to meeting with me during my visit. Two or three days before travel, scheduling issues suddenly arose. It became comical how the same script was repeated for almost every planned trip. More than once, our friends in regional government told our local employees that it was a Kremlin call that triggered these sudden scheduling conflicts.

In addition to limiting my contacts with Russian regional officials, the Kremlin sought to limit my meetings with government ministers periodically. In the spring of 2012 we learned that Foreign Minister Lavrov had sent a message to all government ministers, discouraging them from meeting with me without MFA approval. Some ministers later informed me that they had chosen not to heed Lavrov's guidance. I suspect that others, however, did abide by the Foreign Ministry's instructions, as some of my planned courtesy calls, which my staff tried to set up in my first months as ambassador, never happened. I didn't lose much sleep over Lavrov's effort, since I had no real business with many of these ministers anyway. But I did think the tactic was petty and another sign of the Russian government's insecurity. Why did Lavrov (or Putin) care if I met with the minister of communications or the minister of agriculture? Did he really think I was trying to enlist their support in a revolutionary movement against the government?

My U.S. government counterparts discussed whether we should implement a similar policy for Ambassador Kislyak's contacts in Washington. I, of course, was against this. I wanted the Russian ambassador to the U.S. to be as informed as possible about American policy, and about America more generally, since I found the Russians' misunderstanding of our political system to be an obstacle to cooperation.

The Kremlin also deterred civil society leaders and nongovernmental leaders from meeting with me. The NTV-Nashi television teams played a frontline role in this mission. Who, after all, wanted to show up in a "documentary" as a puppet of the American imperial masters? For instance, after our roundtable with opposition leaders during Burns's visit in January 2012, Vladimir Zhirinovsky, the leader of the nationalist Liberal Democratic Party of Russia, delivered a fiery speech at the State Duma calling for the punishment of those deputies who'd attended the meeting.[8] Of course, it was ironic that Zhirinovsky was attacking us. We were meeting with his fellow elected colleagues! We didn't invite revolutionaries or marginal fanatics to that session. Everyone in the room was either an elected official or a former elected official, including one former first deputy prime minister and a prominent Communist Party leader. But that nuance was lost in the Kremlin's attacks.

Over the next two years, several human rights activists, politicians, businesspeople, and even pro-Kremlin government officials would admit to me that they did not want to come to Spaso House or even meet with me at a restaurant for fear of being filmed and later accused of collaborating with the Americans. As one Russian businessman emailed me in agreeing to lunch at Spaso House, "Michael it's my pleasure :) although it becomes more and more dangerous these days to come over to you for a friendly chat — I'm ready."[9] Many others were not ready. The NTV-Nashi squads tried to provoke guests as they entered and exited my residence, asking them how much money they had received from me. I found it most deplorable when they hunted down guests leaving the embassy's Fourth of July celebration, literally chasing some of them on the street. Most of our guests handled the pressure well. But others found it to be too much. Before one of our July Fourth receptions, I DM'd a political activist on Twitter, who sheepishly apologized for not attending, confessing that he didn't want to be filmed by the NTV crews. At another event we held for a group of human rights activists, one of our invited guests emailed to say that she hoped I would not think any less of her for not being brave enough to attend an event at the U.S. ambassador's home. That was depressing. It meant that our opponents were winning. On the evening of another reception at Spaso House, a civil society leader emailed me to say that she had made it to the small park in front of our residence, saw the motley NTV-Nashi pack lingering around the entrance, and turned away. This upset me, and I decided to go out and confront the crew outside the residence. I took photos of them and asked them to be more respectful of my guests. They were riled, and called in to headquarters for new instructions. But they never left.

On the evening we hosted a birthday dinner for human rights activist Lyudmila Alexeyeva, our security team allowed for a change in protocol so that Lyudmila's car could enter the Spaso House compound and proceed directly to the doorstep. She was turning eighty-five at the time and would have been an easy target for NTV-Nashi had they been able to intercept her as she made her way into the compound.

We did the same one evening when political leaders Boris Nemtsov and Ilya Yashin stopped by for a beer. Boris told my staff that they wanted to drive their car onto the Spaso House premises to avoid the Russian guards and the possible NTV-Nashi crews that might await them. We accommodated them. After they arrived, they thanked us for the special treatment, but also noted that two cars had followed them right up to the gate of the Spaso House compound. As they got ready to leave that evening, Boris's usual upbeat, confident demeanor shifted to agitation and edginess. He told me to go upstairs and watch from my balcony the car chase that would soon ensue. So I did, and he was right. As he and Ilya turned out of our driveway, two cars, their tires screeching, pulled up aggressively right behind them. *What bastards,* I thought. Nemtsov was a former first deputy prime minister, but now he was being chased around town like a criminal. On principle, Nemtsov refused to let the state authorities deter him from meeting with me. We had been friends for twenty years. Our friendship, in his view, should not end simply because I had become ambassador. He often chastised the Obama administration for not doing enough to criticize Putin's authoritarian ways. But he never made it personal; we always remained friends.

A few years later, on February 27, 2015, a Chechen gunman assassinated Nemtsov as he walked home with his girlfriend, just a few hundred meters from the Kremlin. Who ordered the murder remains a mystery.[10]

As public assaults on me from the state-controlled Russian media persisted, I worried that some back home might begin to doubt my effectiveness. Not everyone in the U.S. government agreed with the president's decision to send a political appointee to Russia. My public profile, my critics asserted, resulted from my desire for attention and my "undiplomatic" ways. My predecessor, Ambassador John Beyrle, so the narrative went, never had these kinds of problems.

Of course, this line of analysis downplayed the radically different context into which I'd parachuted. I quickly became a well-known public figure in part because of my engagement with society and my willingness to speak out against democratic rollbacks. But the more important driver of my new fame (or infamy) was the Russian state-controlled media. I had no power to stifle or stop its campaign against me.

I eventually confirmed that my bosses at the highest levels understood the situation and stood behind me, President Obama most of all. In March 2012 the White House asked me to attend the Nuclear Security Summit in Seoul, South Korea; Obama would be meeting with President Medvedev on the margins of this multilateral gathering and wanted me to participate. I was surprised by the invitation; it is highly unusual for an ambassador to attend a bilateral meeting, in a third-party country, between a head of state and the president of the United States. But I eagerly followed orders. It was a long way to fly for an hourlong meeting for which my only role would be to help brief President Obama beforehand. But I was eager to see friends and colleagues from the White House again, including first and foremost the president. When I walked into the briefing room, President Obama gave me one of those half-handshake, half-hug embraces, asked about my family, and made it crystal clear, in front of all of his senior staff, that he supported the work I was doing in Russia under very difficult circumstances.

As per usual, Obama saved two of the most sensitive issues to discuss with Medvedev for the end of the meeting. Normally, the rest of us would have left the room and let the two presidents and their translators talk alone. In this case, however, there was nowhere convenient for the delegations to go. And we were running late. So the presidents improvised, pulling their chairs closer together so that they could speak confidentially. Someone mistakenly let the press in before Obama and Medvedev were done talking, and the boom microphones picked up part of their private discussion, which later made news. Their exchange wrapped up quickly, and we raced back to our hotel so that Obama could attend his next meeting, with the Chinese president.

As we approached the motorcade, the president asked me to ride with him — another break with protocol given that many U.S. government officials in our entourage that day outranked me. It was an obvious show of support for me, and that was Obama's point. As we drove back to our hotel, the president, National Security Advisor Tom Donilon, and I debriefed on the entire meeting, including a tough exchange on Syria. Obama still felt like the meeting went well, as did Medvedev, who was effusive in his public statement, reporting that U.S.-Russian relations had never been better. That sure wasn't how it felt to me! Obama then reported to us that he had discussed two issues in his pull-aside with Medvedev: missile defense and me. "I told them to stop screwing around with McFaul," I remember the president reporting to us. (He actually used a more profane verb.) I initially felt embarrassed that President Obama was using his precious time with Medvedev to talk about me. My job as ambassador was not to be a subject of U.S.-Russian relations but a facilitator of progress in our bilateral relationship. But after that first emotion passed, I was pleased. Obama had my back. I didn't need to worry about anyone else in the U.S. government who didn't.

Regrettably, those boom microphones only picked up the second agenda item of their private discussion, on missile defense, and not the moment when Obama was pushing Medvedev hard about me. Had those two items been reversed in his talking points, Obama would have come off looking like a tough guy; rather, he took a lot of heat for his perceived eagerness to cooperate with the Russians on missile defense after the elections.

Secretary Clinton also stood firmly behind me in my role as U.S. ambassador. I'd always felt that she agreed with my analysis of Russian domestic politics and its influence on U.S.-Russian relations. She also understood clearly the regime's motivations for attacking me, and wanted the government to know that she had my back. One evening, she called me out of the blue at home on a nonsecure line. She posed a few policy questions, but mostly asked about how my family was adjusting to life in Moscow. She closed our conversation by affirming emphatically her support for me. I was pleased to hear these kind words, but also surprised by the passion of her remarks. A few months later, I saw Secretary Clinton in Washington and asked her about this unusual call on an open line. She told me that the call had been a deliberate move to let all of Russian officialdom know about her views on my performance. As she wrote in her memoir, "I made a point of calling him on an open line one night, and speaking very clearly so all the eavesdropping Russian spies could hear, I told him what a good job he was doing."[11] Secretary Kerry, Deputy Secretary Burns, and National Security Advisors Donilon and Rice were equally firm in their support. At the highest levels, I never experienced any loss of confidence.

The Kremlin had pivoted on us, portraying the United States as Russia's enemy. The Kremlin was the one rolling back the Reset, not the U.S. administration. Its attacks on me were part of this larger campaign. It felt personal at times, but it wasn't only personal. I eventually came to understand that these negative trends in our relationship were much bigger than myself. I was not a cause of the problems, but my troubles as ambassador were a symptom of larger forces over which I had little, if any, control.

17

PUSHBACK

Putin's sharp turn against the United States had begun before I arrived as ambassador to Russia, but intensified during my first few months in Moscow. There was an election to be won in March 2012, and casting himself as the defender of Russia against imperial, liberal, regime-change-promoting America generated votes, especially in the rural heartland. We had hoped that this anti-American campaign would simmer down after Putin won the presidency. Russian officials had told us as much. It did not.

In the embassy and in Washington, it took us a while to realize that this new anti-American campaign was not just an instrument in Putin's reelection bid, but a strategic shift in the Kremlin's orientation. Unlike other spikes in anti-Americanism from Moscow, the 2012 campaign was not launched in reaction to a specific U.S. foreign policy action: President Obama had not expanded NATO, bombed a Slavic nation, or occupied a country in the Middle East. Rather, the Kremlin launched this new assault on America in response to the domestic challenges to Putin's regime. It was a strategy to increase Putin's popular support and weaken his liberal, Western-oriented opponents. The campaign also reflected Putin's theory of international relations, which differed from Medvedev's approach. (Although, to survive as Putin's prime minister, Medvedev would eventually adopt and cheerlead for Putin's perspective, especially after the Russian invasion of Ukraine in 2014.) In Putin's analysis, the United States was an enemy, not a partner; he saw the U.S. as a promoter of regime change around the world, including in Russia. Putin blamed the United States for everything bad in the world and in Russia.

I believed that we had to respond. But how? Even after we agreed on the diagnosis, not everyone in the U.S. government embraced the same prescription. Many did not want to push back, especially in the early months of Putin's presidency. They hoped things would return to normal after the presidential vote. Some on our team clung to the hope of continued cooperation with Putin on a host of issues on our bilateral agenda. Therefore, we should not overreact to Kremlin mythmaking. I held a different view. I did not believe that remaining silent in response to Putin's barrage of lies was going to increase our chances of getting a new arms control treaty or securing Russian cooperation on a political settlement in Syria. Saying nothing looked weak, and Putin does not respect weakness. At the time, some Russian elites already perceived Obama as too soft, too accommodating. We did not want to reinforce that image with silence.

By the time of my arrival as ambassador in Russia, the Reset was over. I lamented this fact, but also recognized it as a fact. The needed pivot in our policy was painful for me: I was the author of the Reset, after all, so I should have been the person most vested in maintaining that policy course. But I am also an academic, who values analysis and data, and the evidence that Putin had abandoned the Reset was overwhelming. We could not continue to pursue Reset on our own.

Back home, our *siloviki* — the power ministries — were quickest to embrace the need to push back. Secretary Clinton also understood early that Putin was going to make cooperation difficult, although some of her subordinates at the State Department worried that too much pushback would undermine the Russian cooperation we needed on our arms control agenda or regarding Iran and Syria. Eventually, my side of this argument gained ground with most (not all) of our senior foreign policy team back home in Washington, and most importantly, with President Obama. But it took a while.

As we gradually took a harder line in response to Putin's new tone and actions, I maintained that the change in strategy did not mean the Reset had failed. On the contrary: we'd accomplished a lot under the Reset. I used an analogy from my fifth-grade basketball coaching days: if we ran a play that produced twenty points in the first half, but couldn't run that same play in the second half because the defense had changed, the play in the first half was still successful. And we got to keep the twenty points. We could remain proud of our past achievements, but also adapt to the new circumstances in Russia. It takes two to tango, and we no longer had a dance partner in the Kremlin.

After winning some agreement on the need to push back, we had another intragovernmental squabble about who should be the messenger. Some of my State Department colleagues, both at the embassy and back home, believed that it was not our job at Embassy Moscow to criticize our hosts publicly. Our mission, the argument went, was to build relationships with our Russian government colleagues. We had to interact with them and gather information from them, so wouldn't it be best not to criticize them? I disagreed. Of course, I always desired and cultivated close, personal ties with Russian officials. By the time I left my post as ambassador, I had logged roughly seven years in the USSR and Russia: pretty powerful evidence that I did not hate Russia or Russians. During that time, I nurtured close relationships with several senior Russian government officials, including some who despised our policy and generally disliked the United States. But despite those bonds, I thought it was too easy for us to leave the difficult talking points to our colleagues back in Washington. If our new strategy was to push back on anti-Americanism and antidemocratic Russian policies, then we at the embassy should be executing the new policy as well. I did not respect diplomats who would deliver some difficult démarche, and then whisper under their breath that they personally did not support the message from Washington. I played that game a few times, casting those of us in Moscow as the "understanding" diplomats, in contrast to the "misguided" bureaucrats back in D.C. But generally I found it to be bad practice, because I did not think such maneuvers won us any respect or goodwill, but rather displayed weakness and disloyalty. Former secretary of state George Shultz had once warned me about the dangers of diplomatic clientelism, and I took his message seriously. Our job was to explain but not represent the Russian point of view back in Washington, and to explain and implement the Obama administration's policy in Moscow.

I also wanted to protect President Obama. Back home, Republicans were attacking the president for being too soft on Russia. As the author of the Reset, the senior Russia advisor at the White House for the first three years of the Obama administration, and Obama's representative in Russia, I believed I had to be part of the pushback. I also hoped that my reputation as a defender of human rights might help us with our critics back home.

I incurred costs for my stance. Most disturbing to me, Russian government officials sought to sow divisions within our government by portraying me as a "rebel" or a "rogue," representing my position as independent of Obama and the U.S. administration. Lavrov was particularly sly in playing this game. Lavrov knew that I had worked for five years with Obama before coming to Moscow (three years at the White House and two years with the Obama campaign), had developed and executed the Reset, and maintained close ties with everyone in Washington who had anything to do with U.S.-Russia policy. But it was convenient for him and his government to pretend that others in Washington did not endorse the statements that I made or that our embassy put out. At times, I wondered whether some lower-level officials in our own government were backing up this portrayal of me to their Russian counterparts. Some of my sources at the White House and State Department would warn me about this ankle biting from some midlevel officials in our government. But if the Russian intelligence community was as good as we thought they were — that is, listening in on my phone calls and reading my unclassified emails — then they also knew that I had the complete backing of top government officials in Washington for what the U.S. Embassy in Russia said locally.

We knew that we were going to lose the battle for the hearts and minds of many Russians, at least in the short term. The Russian government devoted huge resources to shaping citizen attitudes, vastly eclipsing our paltry budgets for public diplomacy. They owned or controlled all major television stations, which reached tens of millions every night, while we just had my Twitter and Facebook accounts. To articulate their message, they had Putin at their disposal every day. We had an American jazz quartet for a weeklong visit. They had hundreds of talking heads, bloggers, social media trolls, and television documentary teams dedicated to propagating anti-American ideas. We had a dozen people in our public diplomacy shop. Moreover, we at the embassy did not control some instruments in the U.S. arsenal that might have been useful to our cause, such as Voice of America or Radio Free Europe. Unbelievable as it was to most Russians and many Americans, I had zero editorial power or influence with either of these news outlets working in Russia. That firewall had been erected long ago. Nonetheless, we used our limited resources to do what we could to refute the Russian regime's narrative about our administration and our country, and to tell a positive story about America.

Our most important pushback message was to remind Russians about the win-win outcomes we already had achieved during the Reset era. I was not convinced we were getting that story out, so I told my entire embassy team — State Department diplomats as well as representatives from all the other agencies at our embassy — that they were all deputized as "public affairs officers" charged with identifying and formulating positive public messages about their work. Our public diplomacy team would then look for ways to circulate these good-news stories, both through traditional media outlets and social media platforms.

Initially the response was tepid. My team produced many memos documenting success stories, but their audience was generally their home offices back in Washington, whether at State, Commerce, Defense, or the FBI. I wanted to change the focus of their reporting efforts to include the Russian people. Every day, hundreds of Americans and Russians were cooperating on projects that benefited both of our societies, be they doctors collaborating on tuberculosis prevention, astronauts living and working together on the International Space Station, or professors teaching side by side at Skoltech. Public-private partnerships also produced positive outcomes for both of our societies, from Text4baby in Russia — a social media network designed to get information to new mothers in remote areas of Russia — to the creation of the Renova Fort Ross Foundation to help renovate a Russian fort built in the early nineteenth century in northern California. Even as the Russian regime slammed the U.S. as a sinister force in the world, Russian and American paratroopers were conducting counterterrorist exercises together. I wanted the Russian people to know about all this cooperation.

Another of my top priorities as ambassador was promoting trade and investment, so we made a special effort to advertise these win-win outcomes. From the very beginning of the Reset, we aimed to develop a multidimensional relationship with Russia, and the economic dimension was a critical underdeveloped component. During the Reset, our trade with Russia had grown by 170 percent from 2009 to 2012, but still totaled only $40 billion,[1] while U.S. investment in Russia was only $13.4 billion.[2] Russian investment in the United States was growing, but still amounted to only $6.3 billion in 2012.[3] We aimed to increase these numbers, as the gains would not only profit American companies and the employees engaged in the related economic activity, but also would create more connective tissue between our two countries. I believed that deeper economic ties between Russia and the United States would make it more difficult for either government to take stupid foreign policy actions. I also assumed that deeper ties between our business communities would generate greater understanding between our societies. Stories about trade and economic cooperation also generated a positive image of the United States. Russians may not have liked Obama's airstrikes in Libya, but they loved the iPhone.

Our administration had achieved concrete outcomes and increased trade and investment: Russian WTO membership, a liberalized visa regime for businesspeople, and new working groups within the Bilateral Presidential Commission (BPC), which aimed specifically at bolstering our economic relations. As ambassador, I looked for ways to further promote this agenda and advertise these economic success stories. The United States rightly has laws and regulations defining the conditions under which U.S. officials can promote the interests of individual economic entities, but within these legal constraints, I did all that I could to lobby on behalf of American companies. Sometimes this meant helping ExxonMobil's major Russian partner, Rosneft, expedite visa applications for travel to Houston. Other times it meant lobbying directly on behalf of an American company for a major contract, such as helping to sell American jet engines made by Engine Alliance to the Russian airline Transaero.[4] At first, these activities seemed strange to me. *Marx was right,* I thought. *The state is an instrument of capitalism!* But this trade and investment would support hundreds of jobs back in the United States. Unlike many other aspects of ambassadorial work, these economic outcomes felt very tangible.

I found Boeing's operation in Russia particularly interesting. Boeing was the largest American exporter to Russia before I entered the government, but exports skyrocketed during the Reset. My old friend and head of Boeing's operation in Russia, Sergey Kravchenko, argued that the improved atmosphere in U.S.-Russian relations helped Boeing acquire new contracts from the Russian government. I'm not sure our diplomacy helped Boeing—I know the difference between causation and correlation — but I admired Sergey's determination to frame Boeing's work and our diplomatic efforts as synergistic. I also admired Boeing's strategy in Russia, as they employed hundreds of Russian engineers in both Russia and the United States, and created a joint venture with a Russian titanium company to produce parts for their airplanes.[5] There is no clearer example of a win-win for our two countries than Boeing's operations in Russia and the United States.

I tried to put in similar efforts to help other American exporters to and investors in Russia, from eBay and poultry farmers to California wine exporters and the Ford Motor Company. Henry Kissinger and I shoveled dirt together to break ground for a General Motors plant in St. Petersburg. I posed with a giant bull flown in from Montana, whose assignment was to "engage" with Russian cows and expand the cattle population at a new Russian ranch. I smiled and munched California almonds — the best in the world, we claimed — with cameras clicking. While Putin and his media were bashing us for fomenting revolution, we were helping to promote deals that benefited Americans and Russians alike.

I even touted the benefits of Russian investment in the United States. While I was ambassador, the Department of Commerce was running a program called SelectUSA to promote foreign investment in America, including from Russia. Some of our old-timers at the embassy, the Cold War veterans, scoffed at the program, grousing that we didn't want corrupt Russians polluting our free enterprise, especially if some of them — for instance, in the high-tech business — might bring spies along with them. But I believed in business development, and I went out of my way to get to know some of the big Russian companies investing in the United States, such as the TMK pipe-making company, the Severstal steel company, the Lukoil oil company, the high-tech investment fund Rusnano, and Mikhail Prokhorov, the Russian billionaire who owned the Brooklyn Nets.

If the Nets were the most novel win-win investment for Americans and Russians, Putin had a different pet project: the joint venture between Rosneft and ExxonMobil. In one meeting with President Obama, he called this cooperative project the most important achievement in U.S.-Russian relations in decades, predicting that the megaproject would grow into a joint venture worth $500 billion.[6] Rosneft presented complicated ethical challenges for our government. The company obtained its most valuable asset in a "purchase" — which everyone knew more closely resembled appropriation — of property belonging to the oil company Yukos after Yukos was forced into bankruptcy in 2007. When ExxonMobil and Rosneft initialed their multibillion-dollar deal, the former CEO of Yukos, Mikhail Khodorkovsky, was still in prison on charges we knew to be politically motivated. Within the government, we considered our public posturing options, but eventually decided not to criticize this deal. As ambassador, my job was to praise the joint venture. If I was seeking to undermine Putin's regime, as the Kremlin accused, praising multibillion-dollar investments championed by the Russian president seemed an odd way to go about it.

Putin's propaganda machine filled Russian imaginations with sinister ideas about what we were doing in their country and inside our embassy walls. I wanted to open the embassy to Russian society so that they could see who we really were and what we were doing. I often quoted the former U.S. ambassador to the Soviet Union, George Kennan, who said, "Diplomacy, after all, is not a conspiracy. The best diplomacy is the one that involves the fewest, not the most, secrets."[7] The more we could tell the Russians about our intentions and actions, the better. We began to publicize all aspects of our work, whether it was driving over to the Ministry of Foreign Affairs, attending holiday receptions, touring American companies operating in Russia, or giving talks to local American and Russian organizations. A lot of diplomatic work was mundane, so very unrevolutionary.

Our greatest messaging challenge was to explain to the Russian people that we did not support regime change. Obama did not send me to Russia to foment a revolution, and I did not hand out bags of cash to opposition leaders. I rarely even met with them. I had met oppositional leader Alexei Navalny once while I was working at the White House and he was a visiting scholar at Yale. While I was ambassador, we never met once. Our only encounter during my tenure as ambassador was a fifteen-second chat at an awards dinner celebrating the twentieth anniversary of the *Moscow Times*. Once I tweeted to him, "BTW @navalny, we should meet someday. Odd that everyone thinks we hang out every night when in fact we have never since I came to Moscow."[8] He responded, "xaxaxaxa, Terrific!"[9] But in truth, he did not want to be seen with me, and I respected his reasoning.

From time to time, I did meet with nongovernmental leaders, but so did President Obama when he visited Moscow in 2009 and St. Petersburg in 2013, as did every other senior Obama administration official visiting Russia, including Vice President Biden, Secretary Clinton, Secretary Kerry, and Deputy Secretary of State Bill Burns. That was our policy, and I thought it would be odd if I suddenly stopped following it just because I became ambassador, especially as the previous ambassador, John Beyrle, had also invited opposition figures to his events. In describing my encounters with opposition leaders, I always stressed that a meeting did not equal an endorsement. And I stressed the same when describing Ambassador Kislyak's meeting with "the opposition" in America. (Kislyak actually attended a Trump campaign speech and the Republican Party's national convention; I would never have dared to appear at similar events in Russia.)

To further prove that point, I also met with anti-American political leaders, including the nationalist leader Vladimir Zhirinovsky. He was even invited to our July Fourth receptions, but no one ever accused me of aiding him. I frequently invited deputies from United Russia, the pro-government party in charge of both houses of parliament, to dinners, receptions, and concerts at Spaso House, but was never accused of funding Putin's party.

While I served in the Obama administration, the United States government did not give direct financial support to the Russian political opposition. However, we did provide funding to Russian civil society organizations. Explaining this policy was without question the greatest public diplomacy challenge I had to tackle as an ambassador. Most of this financial assistance went to civil society organizations working on nonpolitical issues. For instance, one of my favorite Russian NGOs was Perspektiva, which "aims to promote independence and an improved quality of life for persons with disabilities in the Russian community."[10] Every holiday season, we hosted an event for disabled children and their families, one of the best events of the year. How could anyone view USAID's support for Perspektiva as subversive?

Defining what is political and what is not, however, was not straightforward. Civil society organizations must engage in the political process. If they did not, they wouldn't really be working on behalf of the citizens they claimed to represent. In that sense, one could argue that support for any NGO was a political act.

To help navigate these difficult calls, I relied on classic political science categories, especially the typology that Stanford professor Larry Diamond had developed twenty years ago, which distinguishes between "political society" and "civil society."[11] We developed an argument that it was legitimate to support civil society when it was nonpartisan, but not legitimate to aid political society, meaning political parties, which were partisan by definition. In dozens of television and radio interviews, tweets, and Facebook posts, I explained why the United States government was not providing support to the Russian opposition or any political group, and published an article in *Moskovskiy Komsomolets* explaining what U.S. democracy assistance is and is not.[12]

Despite our best efforts at categorization, the lines between civil society and political society, or "nonpartisan" and "partisan," sometimes blurred. For instance, the U.S. government funded some NGOs whose leaders openly criticized the Russian government. Lyudmila Alexeyeva, the courageous octogenarian who headed the Moscow Helsinki Group, never shied away from chastising Putin's antidemocratic policies, and we funded her organization, allowing the Russian government to accuse us of supporting critics of their president. The lines also were fuzzy when it came to political party training. Two American NGOs, the National Democratic Institute and the International Republican Institute, placed party development at the core of their work in Russia, and both organizations received USAID grants. NDI and IRI never funded parties directly and tried to be nonpartisan, but they worked more closely with Russia's liberal parties than they did with communists, nationalists, or Putin's party, United Russia. I defended their activities because I believe that political parties are necessary for democratic development. The research is pretty clear: no parties, no democracy. If we supported democratization in Russia — or more precisely redemocratization in Russia — we had to advocate for party development as well. Privately, I pushed both NDI and IRI to work with as many parties as possible, including Putin's party, and to their credit, both tried to develop working relationships with United Russia, but with limited success. Many at the embassy, as well as many back in Washington, believed NDI and IRI were too political for Putin's Russia. Before I left my post as ambassador, Putin chased both organizations out of the country.

The U.S. government was not the only American source of financial assistance to Russian civil society organizations. The National Endowment for Democracy also provided direct assistance to Russian nongovernmental organizations. NED received funding directly from the U.S. Congress, and therefore operated independently from the executive branch of the U.S. government. We tried to explain that NED did not report to us at the embassy. I don't think we convinced many. Some private American foundations, such as the MacArthur Foundation, the C. S. Mott Foundation, and Open Society Foundations, also supported Russian civil society, but faced the same accusations about working as our agents.

American support for regime change in other countries in earlier eras undermined our claims of not supporting the same in Russia in the Obama era. Our explanations about the differences between nonpartisan civil society and partisan political society were convoluted. And most importantly, Putin and his media offered a compelling, straightforward story of our revolution-promoting activities that was broadcast to millions every day. These broadcasts and articles were filled with falsehoods and distortions, but the sheer volume and repetition of these messages drowned out our tweets, blog posts, op-eds, and press releases. Propaganda works. Russian propaganda in the Putin era works extremely well.

While trying to explain that we did not advocate regime change, we did criticize the Russian government when the Kremlin took actions that weakened democratic practices or violated human rights. This practice was always an element of the Reset, but became significantly tougher and more controversial to execute in 2012, as compared to 2009.

Throughout his presidency, Obama discussed human rights and democracy issues with his Russian counterparts, and over the years we issued public statements about human rights violations, election fraud, and the like. Despite what our critics claimed, we were not silent on issues of democracy and human rights, and most certainly not silent enough for Putin.

As ambassador, I continued our policy of issuing critical statements about human rights abuses. We criticized arrests of peaceful protesters and antidemocratic laws. We devoted special attention to the new law that forced Russian NGOs to register as "foreign agents" if they received international funding. We also lamented the passage of the so-called Dima Yakovlev Law, banning adoptions by American parents. On that one, we were in good company; even Foreign Minister Lavrov spoke out against the draft law.[13]

The government's response to our critiques changed dramatically after Putin returned to the Kremlin. During the cooperative days of the Reset, President Medvedev did not seem to take offense when Obama raised the issue of human rights violations or discussed the relationship between democratic institutions and economic growth.[14] In 2012, however, Putin reacted very differently. Putin was always more suspicious of democracy than Medvedev was, and always perceived our advocacy for democracy as a threat to him personally. He feared those tens of thousands of demonstrators on the streets of Moscow and other large cities in 2011 and 2012, and believed our statements encouraged them. Nothing we could say or do would undermine his theory about our revolutionary intentions.

When I first arrived in Moscow, I said little about the popular protests against the government or the Kremlin's reaction to these demonstrations, and I never got within ten miles of any demonstration, political speech, or party congress during my entire time as ambassador. Instead, I watched the live broadcasts of their meetings on Dozhd (TV Rain), "the optimistic channel." In those early days, it was indeed an optimistic time.

As Putin executed increasingly autocratic policies, however, we had to respond. When protesters were detained illegally (in our view), we spoke out.[15] We also denounced what we believed to be politically motivated arrests and convictions, such as those of the Pussy Riot performers in the summer of 2012 or of Alexei Navalny and his colleague Pyotr Ofitserov in the summer of 2013.[16]

Deciding when and how to criticize the Russian government was always difficult. We overused the words "concerned" and "very concerned." In one discussion, I joked that we should say we were "very, very concerned" because we were only "very concerned" in the last statement. I sometimes felt like our government was trying to say just enough to appease our critics back home, and not truly aspiring to affect Kremlin behavior inside Russia. Still, I leaned toward making these statements, and as quickly as possible. In making my case, I was guided by three principles. First, it was the Obama administration's policy. Our commitment to standing up for universal values was codified in the 2010 National Security Strategy, which stated explicitly: "We are supporting the development of institutions within fragile democracies, integrating human rights as a part of our dialogue with repressive governments, and supporting the spread of technologies that facilitate the freedom to access information."[17] Second, I believed that all countries had the right to criticize other countries' behavior in both foreign and domestic affairs; I actively encouraged Russian officials and nongovernmental leaders to scrutinize U.S. policies. Why not? Such criticism made our democracy stronger. Third, I knew that those fighting for democracy and human rights inside Russia would notice our silence. Even if our words did not change the actions of the regime, they could help inspire those being oppressed. At least, that's what I heard hundreds of times while serving as ambassador.

For some Russian activists, the U.S. never said enough. Former chess champion turned opposition leader Garry Kasparov was one of our most vocal critics, berating us for trying to work with Putin and for not doing enough to help Russian civil society.[18] Kasparov published a book called *Winter Is Coming,* which blamed the Obama administration for Putin's consolidation of power. Garry got under the skin of a lot of my colleagues in the embassy and back in Washington, since he always seemed to be overestimating America's ability to do in Russia what he and his allies could not: defend and deepen democracy. In response, some wanted to stop inviting him to embassy events. I disagreed. That was his job: he was supposed to prod, poke, and challenge Putin, Obama, and me. For those who believed in the possibility of Russian democracy, these were frustrating times. Someone had to be blamed. But it sometimes seemed like the two people in the world who most grossly overestimated American influence in Russia were Putin and Kasparov.

We were in the middle of an information contest, if not an all-out war, and our Russian opponents played by a different set of rules, completely separate from the norms that bound us. Sometimes, they would just circulate blatant lies. I'm an academic, and in my professional world, data matter. Not so in the world of propaganda.

Countering blatant lies, fake news, and disinformation is not easy. Some believed it was better to ignore outrageous statements and not dignify idiotic claims with a response. I disagreed. Reflecting perhaps a White House mentality, I pushed for rapid reaction to reports that were blatantly false. We put out all kinds of statements on the facts, on everything from missile defense to the livestock-feed additive ractopamine.[19] I pushed my USAID colleagues to publish as much factual information about their activities in Russia as possible. I always believed that transparency served our interests.

To combat misinformation, we made explaining "What is America?" a central message of our public pushback strategy. It drove me nuts to see how my country was portrayed on Russian television, or described by some on Twitter. Most Americans do not murder their adopted children, or start their mornings by saluting Hitler, but you might develop that impression watching the state-controlled television channels. And most Americans, including President Obama, do not spend their days thinking up ways to weaken Russia and oust Putin. To offset these distorted images, I wanted to offer snapshots of American life — our history, our traditions, and our concerns. Facts were our friends. So I wrote blogs on the American foster care system, the role of NGOs in the American political system, the evolution of LGBTQ rights in America, and my favorite vacation spots in the United States. I tried to tell these stories with sensitivity to the parallel narrative in Russia. For instance, when describing foster care in the United States, I did not cite statistics about the number of Russian children who die every year in orphanages. Instead, I focused on explaining the nuts and bolts of our system, highlighting that we no longer maintain orphanages, which often surprised Russians. In telling the story about the struggle for LGBTQ rights in the U.S., I reminded my readers that just a few decades ago we still had many discriminatory laws on the books. We were not suggesting that America was superior to Russia, but wanted Russians to know the real America. As part of our messaging campaign, we also tried to associate the embassy with popular American culture. When Will Smith came to town for the opening of *Men in Black 3,* I was there. When the NBA sent a delegation of players to Moscow, we hosted them, Russian and American players, at Spaso House. Anything to do with popular technology invented in the United States, we promoted vigorously as part of the American way.

We sometimes struggled to find the right way to explain our policies regarding human rights back home. In the United States, demonstrators sometimes do get arrested. As Kremlin aide Vladislav Surkov pointed out to me in a contentious discussion on the topic, the District of Columbia requires permits for demonstrations; he even handed me printed regulations from the D.C. government's website. Russia was no different, he argued. When I criticized the arrests of peaceful protesters in March 2012, the Minister of Foreign Affairs Twitter feed shot back, "The police on Pushkin [Square] were several times more humane than what we saw in the break up of the Occupy Wall Street protests or the tent camps in Europe." The state-controlled Russian media especially loved to replay the footage of a police officer from the University of California at Davis who pepper-sprayed student protesters. Accurate news reports of American police brutality also presented real challenges to our public relations efforts. My fallback was to admit that we were imperfect, but we had institutions — democratic institutions — to help us get better. Sometimes that worked; sometimes it didn't.

I hoped that finding common ground between our cultures would be less controversial, so we embraced all things Russian that we admired and respected. Our critics constantly labeled us Russophobes, but I wanted to prove that some of us, including Obama, were Russophiles. How could a man with a daughter named Sasha not love Russia! In fact, Obama did have a real appreciation for Russia's cultural achievements and its difficult history, especially World War II, when Russians endured incredible sacrifices to defend not just the Soviet Union but the entire world. As ambassador, I wanted to do the same, even as we were confronting very difficult issues in our bilateral relations. So when an opportunity arose to meet and honor Russian cosmonauts, I was there.[20] When I could attend an academic conference dealing with Soviet or Russian history, I attended. I also tried to attend as many Russian cultural events as possible.

Almost every time I showed up at a ballet, a concert, or a theater, Russians would recognize me, come take a photo, and beam with pride that I was experiencing their culture. I especially liked to pop into Igor Butman's jazz club. At Igor's place one night, we saw a fantastic performance by Trombone Shorty, a young African American player who blended New Orleans jazz with funky beats. As the band snaked through the club at the end of their performance, my family joined the chain, and so did some Russians. Our differences that night seemed small.

Even my jazz outings were not uncontroversial, however, because Butman was also close to Putin. He performed at Putin's public events and personal parties, and they played hockey together. Butman was even a United Russia party member.[21] Russian opposition leaders despised him. So when Igor would on occasion recognize me in the audience at his club and the crowd would applaud, some — including one person on my staff — wondered whether my appearance there made me look pro-Putin. I never saw it that way. My job was not to endorse Putin's team or his critics; I had to engage everyone. It shows just how polarized life in Putin's Russia had become: even listening to jazz had become political.

I also tried to attend performances of American artists in Russia that highlighted our respect for Russian culture. In April 2012, I attended the Chicago Symphony Orchestra's electrifying opening night at the Great Hall of the Moscow Conservatory. This world-renowned group of musicians was in town for the first time in over twenty years.[22] Because we Americans had such a weak reputation in Russia as a country of culture, the mere presence of Americans performing classical music helped our image—especially when they played Shostakovich! I sat next to the composer's widow, looking for any hint of approval. I'm still not sure what she thought of the performance, but the audience roared with pleasure at the end, standing and clapping for a long time. On that evening, our message of respect for Russian culture was loud and clear, as was Russians' respect for American culture.

At Spaso House, we devoted most of our concerts to highlighting American artists, but I deliberately added Russian performers to our programming to highlight our respect for their achievements. Conductor Vladimir Spivakov brought his chamber orchestra to our house to help celebrate the Christmas holidays, and Bolshoi Opera singer Svetlana Kasyan lent us her voice to recognize International Women's Day on March 8—the first celebration of that date at Spaso, according to my staff. One of my American Twitter followers ridiculed me for celebrating this "communist" holiday, but I was happy to do so, as I understood how much it meant to Russians and to my wife, who works on women's rights issues. And at my wife's suggestion, we added a special twist to the event, profiling women in U.S. history and inviting mostly women, and asking the invitees to bring someone who inspired them or a child they wanted to inspire. Russians loved the idea and many guests brought their mothers or daughters.

We especially went out of our way to highlight our appreciation for the sacrifices that the Russian people made to defeat fascism in World War II. Any time I showed respect for Russian heroism in the Great Patriotic War, people responded well. And of course, retelling the World War II story allowed us to remind Russians of a time when the United States and the Soviet Union had been allies. We observed numerous World War II anniversaries in a variety of ways, from working with Moscow's Allies and Lend-Lease Museum to do a display at Spaso House on Veterans Day to dining with the Elbe Group, retired generals dedicated to nurturing the spirit of Russian-American cooperation during the war.

Without question, the most emotional event of my never-forget-history campaign was the seventieth anniversary of the Battle of Stalingrad. Touring Volgograd (the current name of Stalingrad) before the military parade and official ceremonies, I was horrified to hear the factual details of the industrial slaughter that took place in this decisive battle. The average Soviet private lived only twenty-four hours after arrival at Stalingrad; officers survived only days longer.[23] After listening to this devastating, depressing barrage of statistics, I met some heroic survivors. Their memories were fading about many events in their lives, but not about the Battle of Stalingrad. When they learned that I was the U.S. ambassador to Russia, they wanted to grab my hand and tell their stories, in full detail. I bore witness, and shed a tear. I later wrote on my blog that Stalingrad was the turning point in a war in which the fate of our civilization was at stake. So it was not hyperbole to write that I had met some of the people who had saved the western world. That evening, we attended a cultural smorgasbord of dancing, theater, opera, and patriotic sing-alongs hosted by Putin. During that tense time in U.S.-Russia relations, my usual reaction to Putin's presence was either slight fear or quiet disgust. That day, however, I felt comfortable celebrating with him. At a classic Soviet-style banquet later that evening, some of the Russian veterans, perhaps fueled by vodka, began gushing about their love for America. Several veterans came up to me, clutching my elbow, and implored me to convince Obama of the importance of close ties between the United States and Russia, just like there had been when we were fighting together in World War II. I eventually got up the courage to deliver a toast in Russian, which prompted thunderous applause. I left that evening more optimistic about the future of relations between our two peoples, even if our governments could not get along.

Another message of our pushback on anti-American propaganda was to go out of our way to praise positive domestic developments in Russia. After the presidential election process in March 2012 appeared to be freer and fairer than the parliamentary election the previous year, we noted the difference. I also tried to praise instances of economic progress, be they individual company achievements or good macroeconomic numbers. When the Russian parliament passed a law banning smoking in restaurants, I was elated, as were my sons. We made sure the Russian public knew. And I made a point of speaking positively about Skolkovo, Russia's closest equivalent to Silicon Valley. I even tried to cheerlead for Russian successes in sports. My sons and I adopted the CSKA basketball team — the Red Army's team (which, by the way, included several American players) — as our favorite, attending as many games as we could, even joining in their cheer, "CSKA, CSKA, CSKA!"

By the end of my time in Moscow, we had developed a sophisticated portfolio of arguments, facts, and messages to push back on the Russian propaganda campaign against our country and our administration. We tried to keep our communications positive, emphasizing the mutual benefits of cooperation, the goodness of America, and those aspects of Russian history and culture worthy of respect. Our efforts were constrained by our meager methods to communicate our messages directly to the Russian people. Putin controlled all television stations and deployed hundreds to attack us in the media; his team had the home-field advantage. And most importantly, Putin had Putin, and the popular leader of the country was a powerful main messenger.

It wasn't a fight we could win on our own, but I still believed in fighting. We had to push back. We had to remain true to our values and support others inside Russia who shared those values. All Russians did not think alike. All Russians did not believe the propaganda attacks against us. And all Russians did not want the United States as an enemy. Our messages did have an audience. We had to develop the content of our message, and then seek creative new ways to get the word out. For me, that meant learning a new verb: "tweeting."

18

TWITTER AND THE TWO-STEP

Before I took up my post as ambassador, Secretary of State Clinton had given me a set of final instructions. As she wrote in her memoir, "I told him that he had to find creative ways to get around government obstacles and communicate directly with the Russian people. 'Mike, remember these three things,' I said, 'be strong, engage beyond elites, and don't be afraid to use every technology you can to reach more people.'"[1] She was my new boss. I took her advice. That phrase, "be strong," really stayed with me. Did she worry that Mr. Reset—a White House guy—would look too quickly for compromise? Or was she anticipating that our relations were going to get a lot tougher? If the latter, she was right.

In Moscow, I spent all my workdays and many of my evenings practicing conventional diplomacy, delivering démarches to deputy foreign ministers, accompanying visiting U.S. officials to meetings with their Russian counterparts, attending formal receptions, and participating in dinners hosted by other diplomats. In this work, there is no substitute for old-fashioned, off-the-record direct contact with government officials. But along with these traditional methods, we did experiment with some novel ways to get our story about U.S. foreign policy, and America more generally, out to Russian society. Most unconventionally for the times, I embraced social media—Twitter, Facebook, and a personal blog on LiveJournal—as essential new tools of our public diplomacy. I became one of the first American ambassadors to actively tweet.

Before leaving Washington, Alec Ross, Secretary Clinton's senior advisor for innovation, gave me a few tips. At the time, I had only a few hundred friends on Facebook, and had never even seen a tweet. I had lived almost my entire adult life in Silicon Valley, but remained a social media Neanderthal. When I got to Moscow, I reported my instructions from Secretary Clinton about social media to my senior staff. Most responded with nervous glances. Almost no one at the embassy was on Twitter, and none at all had ever thought of using such a tool for diplomatic work. I did have a few immediate allies, including my press secretary, Joe Kruzich, and most of his team, both Russians and Americans. Most, however, did not want to be part of this revolution in digital diplomacy. I wondered if diplomats had the same reaction to radio and television when these technologies first became available.

Roughly 143 million people live in Russia, stretched out across the largest country in the world. Even if I traveled to a different city every week, I still would have touched only a tiny fraction of the population using conventional public diplomacy methods. Twitter, LiveJournal, and Facebook provided me with a way to reach Russians all over the country immediately and often. With internet penetration at 64 percent of the population in 2012, that wasn't insignificant.[2]

Unlike conventional media, social media platforms offered the chance for two-way communication. Virtually every night, I startled Russians by replying to their tweets or jumping into their Facebook chats. Russian tweeters and Facebookers often did not believe it was really the U.S. ambassador when my words appeared suddenly on their screens. At first, many believed that my staff wrote my posts. Over time, my frequent typos and grammatical errors in Russian affirmed personal validity. Eventually, those who followed me on Twitter and Facebook developed a feel for my voice, humor, and informal ways. Some, including certain American followers, hated my social media presence; diplomats are not supposed to crack jokes. But most on social media loved it. I was also struck by the diversity of followers I had, from deputy prime ministers to high school students. Big cities had higher concentrations of people online, but I would often be surprised to get a reply on Twitter from a small town in Siberia or a comment on LiveJournal from the Caucasus.

With time, my staff and I began organizing Twitter chats, specific times when Russian readers knew I would be online and ready to answer any question. I tackled everything from my views on missile defense to my take on the best restaurants in Moscow, in 140 characters or less. In a single, one-hour chat session, we could reach hundreds of thousands of Russians.

Social media platforms also provided my staff and me with tremendous information about Russian society and culture. Before becoming ambassador, I thought I knew Russian society as well as any American. Yet once I opened a Twitter account and started following some of Russia's most important personalities there, I discovered a whole new dimension — a virtual dimension — of Russian society. I had never even heard of some of the biggest Russian microbloggers until I started reading them on Twitter. And back then, one could not fully understand Kremlin critics such as Alexei Navalny, Ksenia Sobchak, "Darth Putin," or the satirical duo known as "Kermlin" — or pro-Kremlin bloggers such as Konstantin Rykov and Anton Korobkov-Zemlyansky — without following them on Twitter. Reading the feeds of Russian political and social elites made me a better-informed diplomat.

At first, engaging on Twitter made me nervous, especially as a freshman ambassador. Digital diplomacy was brand-new in 2012, and so there were no codified protocols within the State Department for how to use this new tool; we had to learn along the way. The first decision was whether to use Russian or English. In the beginning, we set up two accounts in my name, one in Russian and one in English, but that proved too cumbersome to maintain. So I bounced back and forth between languages on my one account, trying to do as much in Russian as possible. Tweeting in Russian, however, proved difficult, and sometimes dangerous. I generally engaged on Twitter and Facebook only in the evening, as I wanted to be a real, live ambassador during the day. But tweeting at night meant that I was usually doing so at home, with no Russian staff to help correct my grammatical mistakes. And the Russian language offers tremendous opportunities for making grammatical errors! As I learned bitterly, it is much easier to cover up your mistakes when speaking. When you write things down, your confusion about which *i* to use (Russia has two letters for what we think of as one vowel) or which case ending is needed becomes much more evident.

Russian tweeters also deploy tons of slang. Bloggers creatively invent verbs, use words from other languages, and abbreviate. I could not always follow the flow. Early in my Twitter career, I would cut and paste phrases from Russians' tweets as a way to save time and get the grammar right. One night, however, this shortcut produced a disaster. I copied in a slang name for the city of Yekaterinburg, "yoburg."[3] I just assumed that this invented word was a cool way to refer to Russia's third-largest city. I had no idea that the word I had cut and pasted from someone else's tweet actually meant "f***burg." Of course, that tweet went viral, just a few days before I visited the city for the first time as ambassador. As I learned later, it was OK for residents of Yekaterinburg to use this word, but it was definitely not OK for others, and especially not for me. After that mistake, I stopped cutting and pasting and put in place a new system that had Russian staff members on call during the evening in case I wanted to have them edit my Russian before tweeting.

The "yoburg" gaffe, however, had an upside. Of course, I quickly apologized for my error. Thousands, including many residents of Yekaterinburg I met the following week, accepted my apology as sincere. I was a smiling American, trying to engage with common people in their own colloquial language. As one Twitter follower reported to me, my response stood in stark contrast to what Russian officials would have done — they never admit to mistakes. And as another American friend in Moscow consoled, if you are going to live on the frontier, you have to be ready to take some arrows.

Drawing lines between appropriate and inappropriate content presented another Twitter challenge. Our social media experts all recommended combining U.S. policy statements with personal vignettes. That was the Twitter culture. Just reposting State Department press releases was a quick path to obscurity. At first, I resisted this advice. As a foreign policy expert, I had decades of experience on television and radio, but I had never been a public figure. I didn't feel comfortable talking about my personal life. I also did not feel comfortable bantering about trivia in public, and believed that the title of ambassador was a dignified one that should not be cheapened by tweets about the spices in my plov. But I gradually waded in. For months, all news about my family was off-limits. With time, however, I acquired a better feel for the medium and grew bolder. I began incorporating several themes from my private life: my upbringing in Montana, my professorial life at Stanford, fatherhood, basketball, and music. Glimpses into our family life always generated the most positive attention. A photo from the summer of 2012 of my mom, dad, brothers, and sister in Red Square collected more likes on Facebook than any other post, until we posted a

Christmas family photo later that year with my wife and two sons. Posts about my fourth-grade-basketball coaching career defied many of my followers' assumptions about diplomats in general. I was challenging stereotypes; I liked it.

Eventually, even the most mundane facts of my life became news. For instance, when I broke my finger playing basketball at the embassy, I tweeted about going home for surgery. News of my injury even trickled into the mainstream media. When I bumped into Prime Minister Medvedev a few weeks later at a Skolkovo event and began to explain my cast, he said, "I know, I know," as he had already read about my mishap on the internet.

When I was in the government, ambassadors did not have to get tweets or blogs cleared with the mothership (as the professional diplomats called it), the State Department in Washington. On occasion, though, embassies did tweet out information that later caused problems for our colleagues back home. I don't remember a time when my tweets were inconsistent with policy (though I'm sure some working in Washington during my time as ambassador have a different impression). But we certainly made statements — at times controversial statements — before Washington had seen or approved of them.

For instance, in the spring of 2012 we criticized the arrests of government protesters without coordinating our message with Washington. In August of that year, when a judge sentenced Nadezhda Tolokonnikova and Maria Alyokhina, singers with the punk rock band Pussy Riot, to two years in a penal colony for hooliganism, we reacted immediately on Twitter. I personally questioned the wisdom of the women's actions — performing an anti-Putin song in the Cathedral of Christ the Savior, one of the Russian Orthodox Church's holiest places. But the sentence doled out to these free expressionists, both mothers of young children, was grossly excessive.

My most controversial tweet was one posted in the summer of 2013 about Alexei Navalny, one of Russia's most popular opposition leaders. At the time, he was accused of embezzling money from Kirovles, a timber company in Kirov, while serving as an unpaid advisor to the governor of that region. To our team of political analysts at the embassy, the charges were obviously political, designed to silence this Kremlin critic. On July 18, 2013, a judge in Kirov was scheduled to rule on the case. Everyone knew that Navalny was going to be convicted of embezzlement. The only remaining question was what the sentence would be. One year? two years? Suspended sentence?

I woke up that morning not intending to watch the reading of the verdict, but then happened to click on the live feed on my home computer, and couldn't stop watching. In Russia, in reading the verdict, the judge basically recaps the entire trial. The process can go on for hours, and this case was no different. As the judge speed-read without emotion, Navalny tweeted in parallel, until his phone battery ran out. At one point, he tweeted innocently, "hello everyone."[4] I tweeted back, "privet, smotriu," meaning "hello, I'm watching." Immediately, my two words went viral. Navalny supporters loved it; his detractors accused me of meddling in Russia's internal affairs.

When the judge finally got around to the sentencing, everyone was shocked: five years in jail and a five-hundred-thousand-ruble fine. As I sat in front of my computer screen, watching Navalny, his wife, and his supporters react to this unexpected result, my first thought was to wonder how Joseph Davies, our ambassador to the Soviet Union in 1937, had reacted to Stalin's show trials. And how was what I witnessed on my screen any different? My second thought was how many hours it would take our government to issue a statement on this injustice. Washington was still sleeping when the verdict was announced, meaning we were several hours away from getting the interagency team to agree on an official response. So I decided to take to Twitter. No long statement. Just those 140 characters. I wrote, "We are deeply disappointed in the conviction of @Navalny and the apparent political motivations in this trial." This tweet also went viral, while also being quoted in television, radio, and print news stories, both in Russia and the United States. Other embassies later told us that our decision to condemn the verdict loudly and quickly put pressure on their governments to do the same.

Because of the time difference, all of this public diplomacy occurred before our bosses in Washington were in their offices. I was nervous about how some people in Washington might react, but I did not want to be on the wrong side of history on this one, and I did not want President Obama to be on the wrong side either. I felt like a witness to a momentous event: the return of political prisoners in Russia. Even if I had to endure criticism from my hosts in Russia or my bosses back home, it was the right thing to do.

Millions of Russians had a similar reaction to Navalny's jailing. The Levada-Center's public opinion poll showed that 57 percent of Muscovites and 44 percent of all Russians believed that the charges against Navalny were politically motivated. That day, tens of thousands risked arrest to join an unsanctioned demonstration in downtown Moscow.[5] Amazingly, in the evening of that same day, the prosecution petitioned the court to substitute arrest with house arrest, and the petition satisfied the court. Even the Russian government seemed divided on how best to deal with Navalny. The demonstrators helped tip the balance in favor of those in the regime who did not want to lock him up. Later that year, Navalny was allowed to run in the Moscow mayoral election and came in second with 27 percent of the popular vote, despite not having access to television or radio.[6] Obviously, he could not have participated in this election from jail. Popular reaction to politicized rule of law produced a positive result.

In the aftermath, some Russians accused me of organizing the demonstrations in support of Navalny that evening. A photo circulated widely on the internet showed me sitting in front of a giant window at the National Hotel next to Manezh Square where the demonstration took place, allegedly watching my work unfold. Of course, it had been photoshopped, but the truth never slows savvy propagandists. Privately, some Russians thanked me for speaking out, asserting that my reaction helped embolden the protesters, who in turn helped keep Navalny out of jail. I was pleased to hear that — every now and then, digital diplomacy could feel meaningful — but I doubt there was any connection. I thought it was the right thing to do, even if it had no effect. This was a political show trial, and the United States had to call it for what it was. On that day, I was proud to be the representative of the United States in Russia on Twitter.

During my time in Moscow, the Russian government also had started to use Twitter. President Medvedev himself was active on social media. He'd opened a Twitter account well before me, while visiting the company's headquarters in San Francisco in the summer of 2010. His aide, Arkady Dvorkovich, was also an avid user. He and I sometimes traded tweets on policy issues, a new turn in U.S.-Russia relations that George Kennan, a champion of transparent diplomacy, would have appreciated. During this same period, Deputy Prime Minister Dmitry Rogozin also became active on Twitter and Facebook, putting out blunt anti-American messages. Had I used the same style and language to criticize

Russia in my social media feeds, I would have been kicked out of Russia. Eventually, even the stodgy Russian Ministry of Foreign Affairs joined the virtual conversation. Early on, the MFA mostly just tweeted out the basic facts on this meeting or that phone call. Whenever possible, I retweeted MFA messages to underscore our common causes. Occasionally, however, the virtual voice of the MFA turned nasty. Its special representative for human rights, Konstantin Dolgov, used Twitter to criticize what he claimed were American human rights abuses. Russia's commissioner for children's rights, Pavel Astakhov, also used Twitter to distort and mislead about the abuse of adopted children in the United States.

My most regrettable engagement with Russian officialdom on Twitter occurred one evening in May 2012. The Foreign Affairs Ministry had sent out a rapid-fire series of tweets that berated me personally for a lecture I gave that day at the Higher School of Economics. Ironically, the central argument of my lecture was how the Reset was producing results for both the United States and Russia. I delivered an hourlong presentation in Russian, without reading from a text. My American accent may have been heavy, and my grammatical mistakes abundant, but it was evident from the enthusiastic, packed auditorium that the students appreciated the effort. They even laughed at my jokes. But then, during the question-and-answer period, I let my guard down, feeling relaxed in a familiar setting, a university lecture hall. Someone asked me about American and Russian cooperation in Central Asia, since I had claimed in my talk that our collaborative work in developing the Northern Distribution Network — supply lines to Afghanistan through Russia and Central Asian countries — represented an important, but poorly understood, achievement of the Reset. To illustrate my point, I reminded my audience that we started the Reset era competing with Russia in Central Asia. Just weeks after Obama entered office, Kyrgyz president Kurmanbek Bakiyev had traveled to Moscow, met with Putin, and received a pledge of $2 billion in economic assistance, and then promptly announced his decision to close our air base. We worked hard to keep the base open, and eventually found a way forward with both the Kyrgyz and the Russians. In my lecture, I described the Russian financial offer that nearly closed the base, and our counteroffer. But instead of using a proper diplomatic phrase, "economic assistance package," I used the word "bribe" to describe both the American and Russian efforts. My word choice, constrained by the fact that I was speaking

Russian, was poor. I knew well the Russian word for bribe; "economic assistance package" did not roll off the tongue so easily.

As I learned later that day, a journalist had found her way into my lecture. For all the content I'd covered in my hourlong address praising Russian-American cooperation, she'd homed in on the word "bribe." Journalists! That evening, the Russian Ministry of Foreign Affairs Twitter account lit up, firing off a dozen tweets within an hour, accusing me of deliberately trying to damage U.S.-Russian relations.[7] One tweet screamed: "Ambassadors' job, as we understand it, is to improve bilateral ties, not to spread blatant falsehoods through the mediasphere."[8] Another added: "Michael McFaul's analysis is a deliberate distortion of a number of aspects of the Russian-U.S. dialogue."[9]

After fretting for a while, I decided to fire back. I tweeted out the slides from my talk so that Twitter readers could decide for themselves if distortion was my aim. The truth is that my talk was not an accurate portrayal of U.S.-Russian relations. At the time, our bilateral relationship was in much worse condition than I'd suggested in that talk. Without question, I was being too diplomatic. Swedish foreign minister Carl Bildt, one of the world's pioneers in Twitter diplomacy, was reading my Twitter exchange with the Russian foreign ministry that evening. At one point, he jumped into the fray to tweet: "I see that Russia MFA has launched a twitter-war against US Ambassador @McFaul. That's the new world — followers instead of nukes. Better."[10] At the time, it did feel like a war to me, especially as I could see the ministry's towering Stalinist building from my office window at Spaso House.

The next month, I gave the same lecture at the New Economic School, hosted by the school's rector, Sergey Guriev, one of Russia's most prominent economists, now living in quasi-exile in Paris. In that talk, I repeated the same good-news story about U.S.-Russian cooperation in Central Asia, but went out of my way to use the phrase "economic assistance package." The student audience broke out in thunderous laughter. What the MFA scorned, the best and the brightest in Russia loved.

In fact, most of my Russian followers were supportive of my efforts. On March 2, 2012, one tweeted, "Michael, u're an outstanding ambassador Russia have never seen [sic]. Thanks for all your incredible effort&hard work. Reset works." On February 12, 2013, another wrote, "To my mind you're the greatest

U.S. ambassador we ever had." I enjoyed such public messages as much as the Kremlin hated them.

I embraced digital diplomacy with enthusiasm, but also tried to use more traditional media — Russian television, radio, and print — to get our messages out, especially regarding the false claims about our government and me as agents of regime change. This was not an easy task in Putin's Russia. The government strictly limited my appearances on Russian state-controlled television networks. After all, they were pushing the exact opposite line. From time to time, I thought about approaching Dmitry Peskov, Putin's press secretary and close confidant, to plead my case. But what could I say? "Pretty please, let me on television more. I promise to say only nice things about Putin"? Just like many others in Russian society, we had to work within the constraints of life in Putin's Russia, however unpleasant and unsatisfying.

From time to time, we got opportunities on the Russian state channels. In my first week on the job, I appeared on Vladimir Posner's show, a sixty-minute, one-on-one interview on Channel One, Russia's most watched television channel. Posner, a skilled and sophisticated interviewer, pushed me hard on allegations of my meetings with the opposition and U.S. support for Russian opposition leader Alexei Navalny. I was grateful for these questions; Posner was giving me a chance to refute these ridiculous claims. After taping the show, we worried that some segments would not survive the censors, but nothing was cut. It was still my first week as ambassador. Medvedev was still president.

It took my press attaché more than ten months to get another appearance on Channel One, this time on Ivan Urgant's *Vecherniy Urgant* (*Vecherniy* means "Evening"), Russia's closest equivalent to *The Tonight Show.* The excuse for inviting me was Obama's reelection. Urgant traded in cynical comedy; he was really funny, but not very political. I decided to do the show in Russian, though I was given an earpiece that I could use for translation if I didn't understand his jokes. I was nervous. Tens of millions watched Urgant's show. By November 2012, relations between the United States and Russia were much more tense than when I did Posner's show. Urgant was nervous, too. For the taping of the show, his boss, Konstantin Ernst, sat in the audience, just to make sure we didn't go too far. As I walked out onstage, his band played a Led Zeppelin song, and then

Urgant allowed me to banter about my love of Zep's music, an immediate bridge builder with middle-aged Russians. I just barely kept up with his wit, but did find places to land a few jokes of my own. He made fun of my heavy American accent, which made "Lenin" sound like "Lennon," but all in good fun. And the audience applauded wildly when I named some Russian musicians that I liked, such as Akvarium and Vysotsky. Judging from the Twitter traffic that followed my appearance on Urgant's show, many Russians came away with the impression that I was not an evil, secretive revolutionary, but a friendly, upbeat Zep fan.

And that, from the Kremlin's perspective, was exactly the problem. It didn't serve their interests to have me on national television, speaking in Russian about things that joined our societies. I had to be cast as someone foreign, as a threat, as an enemy, and chatting about my love for Led Zeppelin and Akvarium undermined that image. I never appeared on a major national television program in Russia again.

I made several appearances on the independent Russian radio station Ekho Moskvy, because I respected the journalists there. Ekho's charismatic anchor, Alexei Venediktov, usually interviewed me; he was fair but always tough. Doing an hourlong interview in English is hard; doing it in Russian is downright torture. We sometimes dove into the details of difficult issues like missile-defense technologies and American legal issues; these occasionally left me searching for words in Russian and wondering if I would have been better off doing these interviews through a translator. Ekho listeners, however, cut me a lot of slack. The week before I left Russia, Ekho conducted a poll about my performance as ambassador. Over 70 percent of its listeners gave me positive marks. Of course, Ekho's listeners in no way represent Russians as whole, but I was still proud to have won over Russia's middle-class liberal intelligentsia.

I also did several interviews with Russia's print media. *Kommersant, Vedomosti, Nezavisimaya Gazeta,* and *Vlast* were good to me in that they let me articulate our policy positions at some length. At times, I also published op-ed pieces, including one in *Moskovskiy Komsomolets,* the rough Moscow equivalent of the *New York Post,* with the goal of explaining our assistance programs to a wider audience. We seized every chance we got to explain our policy and our country.

A friend of mine with close ties to the Kremlin reported to me once that Putin did not approve of my public profile. I was too well known, always drawing media attention, the Russian president allegedly complained.

If Putin did not like seeing me in the media, then maybe that was reason enough to keep engaging all platforms to spread our messages.

I grew to like Twitter, but my most pleasurable tool for engaging with Russians and communicating our messages about policy and America was old-school public diplomacy at Spaso House. The residence for the U.S. ambassador in Moscow is spectacular. Our entire house in Palo Alto could probably fit into the chandelier room at Spaso. Taking a page from the Obamas at the White House, my wife and I decided that we would use the residence as much as possible, and also try to expand the kinds of events we hosted, and the kinds of people we invited. By the end of our time there, the Spaso House staff gave us a certificate for what they believed was the "world record" for the residence: twenty-two thousand guests in two years.

We convened traditional events, like concerts, dinners for visiting American dignitaries, receptions for trade delegations, and, of course, the biggest party of the year for all American embassies everywhere, July Fourth. At all these events, we made a special effort to expand the guest list beyond the usual suspects: more women, more young people, more employees from the embassy (including Russian employees), and more spouses. I believed in maximizing the number of people we fit into a concert even if that meant minimizing the fanciness of the hors d'oeuvres and drinks. On occasion, we even served popcorn. What's more American (or less expensive) than popcorn?

I *was* a true revolutionary in Russia in one sense: dancing at Spaso House concerts. It started when a country band from Montana called Wylie and the Wild West were scheduled to perform at our residence just a few weeks after our arrival. I suggested that we clear some chairs to create a small space for a dance floor. My father was a country-western musician in Montana and I remember him coming home after a gig from time to time and complaining that it had been a slow night because people weren't dancing. I didn't want my musician visitors from Montana to go home and complain about some sleepy diplomatic function they were compelled to play at the ambassador's residence. Some on the Spaso House staff looked confused when I explained my reasons for removing some chairs. "Mr. Ambassador, people don't dance at Spaso House concerts," I was told. But when Donna and I got up to dance the two-step, others jumped right in. As the *Moscow Times* reported, "The ambassador and his wife,

Donna Norton, drew applause for their enthusiasm and stamina, out-twisting and two-stepping all but a few of the several hundred guests . . . The festive mood was infectious, and at least one guest even found it restorative," reporting that "it's amazing what good energy can do to the psychological condition."[11] U.S.-Russian relations looked tense and confrontational on Russian television at that time, but inside Spaso, the mood of Russians and Americans dancing together could only be described as joyous.

Cultural events, especially concerts, were always great bonding moments for Russians and Americans. Because of Russia's deep cultural traditions, our Russian guests appreciated more than most Americans the virtuosity of the quintet assembled from the touring Chicago Symphony Orchestra that played at Spaso. Likewise, hosting a jazz legend like Herbie Hancock made it easy to be a proud U.S. ambassador in a room filled with appreciative Russian guests. And sometimes our American visitors filled the ballroom to overflowing with irresistible sound, talent, and exuberance. One evening, Irvin Mayfield and his New Orleans Jazz Orchestra electrified our five hundred guests with their enthusiasm and energy. After Irvin's moving final encore, a tribute to the victims of the floods in New Orleans, people stood with tears in their eyes and smiles on their faces. Everyone left our residence that evening feeling just a little bit better about humanity, including the shared humanity between Russians and Americans.

Sometimes I used Spaso House events to convey more overt political messages, especially about democracy and human rights. At our first Fourth of July reception, for instance, I went out of my way to explain the meaning of the Declaration of Independence to our two thousand guests, emphasizing why people rebel if not offered basic rights by their rulers. I thought it was a very subtle criticism of Putin's regime. Afterward, a Russian guest came up to me and growled that I had insulted his government with my celebration of the American system of government, but then burst into a smile and commended me for my bold statement. I guess the message was not so subtle. On November 6, 2012, our election party had a similar, indirect message. As we waited for the results to trickle in, I reminded our packed house that we still did not know who would win, and that was the essence of democracy: competition and uncertainty. Several people joked to me that they already knew who was going to win Russia's next presidential election in 2018. A few months later, we decided that it wasn't overkill to do an inauguration party. As several hundred Russians gathered to watch the live translation of Obama's second inauguration before hundreds of thousands on the Mall in downtown Washington, I felt a deep sense of pride about our nation. We could still inspire. A few times, the audience even clapped in response to Obama's address. Many of our guests noted the contrast with Putin's inauguration ceremony. For Putin's big day, they cleared Moscow's streets completely to allow him to drive solemnly from the Russian White House to the Kremlin undisturbed. No cheering crowds. No parades. No address to the people, just the elite gathered at the Kremlin. Moscow was turned into a ghost town.

The guest list for any Spaso House gathering was always a political statement. I wanted to invite the widest range of people possible, including both government and nongovernment officials. For our Fourth of July reception, we debated whether to include some more extreme politicians. My approach was to err on the side of inclusiveness. We invited both Putin and his nemesis, Navalny. Neither ever came. In 2012 we invited nationalist leader Vladimir Zhirinovsky for the first time ever, or so I was told by my staff, to attend the embassy's July Fourth event. Some on my team thought this gesture was a mistake, but I couldn't come up with a good reason to keep him off the list, which also included many government officials and United Russia deputies with views very similar to Zhirinovsky's. Two decades earlier, I had worried that Zhirinovsky might become a fascist leader. That concern proved wrong; he instead played an important role for the regime as a loyal nationalist. I did not respect his political views, but he had not violated the constitution; he played by the rules. He showed up with an entourage (he wanted to videotape his appearance but we said no) and ate sugar cookies coated with red, white, and blue frosting. A photo of him eating our American cookies went viral on Twitter, since the democratic opposition always was accused of eating *Gosdep* (State Department) treats. Standing right next to him in line for hot dogs — another innovation by my wife at Spaso House — were human rights activists who detested him. My younger son, Luke, had a long, animated chat with Zhirinovsky, which worried my bodyguards but ended without incident.

My big-tent approach did not always succeed. At times, political polarization was simply too intense. For instance, one night in the spring of 2012 I hosted a dinner for our visiting assistant secretary of state for European and Eurasian affairs, Phil Gordon. Phil was in town for only two days, so I wanted to give him a flavor of the energized political debate still percolating in response to the massive demonstrations happening then. Among the table of ten or so, we had in one corner Lilia Shevtsova, a brilliant political scientist and Putin critic who worked at the time for the Carnegie Moscow Center. At an opposite corner sat Alexey Pushkov, a Putin loyalist and television host, who at the time was the newly elected chair of the International Affairs Committee of Russia's parliament. For me, Alexey was a complicated figure. On his television show, he had a strong, blunt delivery style, but in Washington, or at embassies in Moscow, Pushkov projected a much more refined, sophisticated image. He speaks perfect English, wears stylish outfits, and understands Western culture. That evening, however, the heat between Lilia and Alexey could not be dampened by the formal, polite ambience of a candlelit dinner at Spaso House. They held diametrically opposed views about Putin and his regime, and expressed them sharply. At one point, Shevtsova interrupted Pushkov, waving her watch around and asking how much longer he intended to berate us with his blather. I used all my diplomatic capacities to keep Pushkov from walking out. The event was a good lesson. Some schisms are too wide to try to bridge.

Television appearances and tweets reached a wide audience, and events at Spaso House blended traditional diplomacy with new ways of forging connections. We also used lectures, speeches, and roundtables to explain our policy and our country, and they provided opportunities for personal engagement. I tried to accept as many invitations as possible to speak at clubs, conferences, and universities — natural venues for me, as a professor. Despite the anti-American propaganda on Russian television networks, these events always were packed. Russians wanted to engage, even if some in their government did not.

On trips I took around Russia, I always sought to include a public event, often at little nooks of American culture called American Corners, usually located in libraries that had been set up around the country in the 1990s — though Putin eventually closed them after my departure. I democratized the way we invited people to these events, using Twitter and Facebook to advertise them. My security detail hated this practice. Just as they feared, we received information about a death threat to me before one of my first appearances at an American Corner in Yekaterinburg in the summer of 2012. Some embassy staff lobbied to cancel the event. I pushed back, but gave the green light for my team to beef up security. It was an odd sight to watch the audience pass through metal detectors to attend a talk by me — something that had never happened at Stanford — but that inconvenience was the necessary price to pay to hold the event. The local police were extremely cooperative, providing a presence on the street, but not entering the building. And as it turned out there was no trouble. Just the opposite: schoolteachers, small-business owners, students, and pensioners crowded the room, amazed that a U.S. ambassador was there to entertain their wide range of questions, and in Russian. Afterward, one participant remarked that I didn't look or sound like a revolutionary rabble-rouser. That was a small victory.

I also loved speaking at universities, where politically engaged students generally packed lecture halls and gave me a rousing reception. In my only appearance at Moscow State University — where I had studied in 1990–91 — students even swarmed my limousine to snap selfies with me as I arrived. This young, educated, urban demographic was not influenced by Russian state television. These settings were very familiar to me; I had spent the last two decades talking to twenty-year-olds. It got to the point where my talks became too popular in the estimation of the authorities, and I was banned from most campuses. Most frustrating to me was being blacklisted by the Moscow State Institute of International Relations, or MGIMO, Russia's premier institution for training diplomats. What would such an organization possibly have to fear from me? Eventually we got creative, and simply invited students from MGIMO and many other universities to lectures at my house. I couldn't be prevented from lecturing in my own ballroom. Some students were reluctant to come to Spaso House; an appearance there could be recorded in an unfavorable way in their file. But when we advertised these events on social media, they filled within hours. We transformed our ballroom into a giant lecture hall for the hundreds of students in attendance, who asked questions on everything from Syria to Obama's jump shot. These were some of my favorite events at Spaso House.

It's hard to measure the impact of our pushback efforts on Putin's false messaging about the United States, President Obama, and me. In January 2012, according to public opinion data, 52 percent of Russians had a positive view of the United States; only 23 percent did in January 2014, and that number dropped even more dramatically after my departure when Russia invaded Ukraine.[12] But you'd have to hold a highly inflated view of my role as ambassador to assign me responsibility for these declining numbers. Larger forces were at work. Putin's campaign against the United States, fought primarily on the airwaves of Russia's major television networks, succeeded. Our tweets and jazz concerts were no match for his media empire.

Nonetheless, as I left Moscow in February 2014, I felt like we had had some success in strengthening our ties with the Russian people. When I became ambassador, we had zero followers on Twitter, and just a few hundred following our embassy's page on Facebook. By the time I left, I had over seventy thousand followers on Twitter, thirteen thousand "friends" on Facebook, and a solid readership of my blog on LiveJournal. (Those numbers sound small now, but at the time, in these early days of social media, they were big.) I frequently made the list of the top ten cited bloggers in Russia, as rated by Medialogia.[13] Appearing on the same list with opposition leader Navalny and Prime Minister Medvedev meant that we were doing something right in connecting with our Russian audience. One year, *Foreign Policy* magazine recognized me as one of the most effective "Twitterati" among politicians, putting me on a list with the Clintons and the pope.[14] We were reaching millions of Russians directly, not filtered through Russian state media outlets.

Reach and presence, however, do not translate neatly into impact. We struggled even to find useful metrics for measuring that. We reasoned that our social media voice, growing as it was, registered as just a faint chirp in the screaming cacophony of Russian conventional and social media. To my thousands of followers on Twitter, Channel One alone had tens of millions of viewers, and television remained Russians' most popular news source by far.[15] We knew national polls about Russian attitudes toward the U.S. would not be sensitive enough to capture the individual impact of our public diplomacy work, so we collected anecdotes. Nearly every night on Twitter or Facebook, I witnessed someone — a student in Krasnodar or an engineer in Zelenograd — confess to changing her thinking about the United States due to our virtual engagement. After two years, I came away with an impression that most of my followers appreciated my openness, humor, and patriotism, which we believed moved people in a slightly more positive direction toward the United States. The outpouring of goodwill messages tweeted to me in the days before my departure — measured in the tens of thousands — left me with a strong sense that we had connected to Russian society in a unique, massive, and profound way.

Whether these efforts helped us achieve American national interests is hard to know. My intuition, supported by anecdotes, makes me think our efforts were worthwhile. One anecdote in particular stuck with me. After I announced that I was leaving the ambassador post, a prominent pro-Kremlin nationalist told me privately that he was glad to see me go: he considered my understanding of Russia, my respect for Russian culture and history, and my charisma dangerous to Russian national interests. That he feared me — that Putin feared me — made me think we were doing something right. Upon my departure, the *Moscow Times* op-ed editor, Michael Bohn, wrote, "One of the Kremlin's main problems with McFaul was that, unlike previous U.S. ambassadors, he greatly expanded direct contact with ordinary Russians."[16] Even if we didn't move the dial on support for America throughout the country, I left Russia in February 2014 thinking that we had touched millions of Russians, many of whom had never had any contact with a U.S. government official before. That in and of itself was an achievement.

19

IT TAKES TWO TO TANGO

Well before arriving in Moscow as ambassador, I had become skeptical of the United States' ability to work constructively with Putin. As prime minister, Putin expressed modest support for, if not indifference to, the Reset. He endorsed New START, allowed new sanctions to go forward against Iran, reversed course and tolerated our staying at the Manas Air Base in Kyrgyzstan, and eventually acquiesced to Russia's WTO membership. He gave Medvedev running room to cultivate relations with Obama, seeing the thaw in relations with the United States as a Russian national interest. In 2011, however, Putin began to withdraw his tepid support, first by publicly criticizing Medvedev's decision to allow the U.S.-led military intervention in Libya, then by chastising the United States for fomenting the Arab Spring, and finally and most dramatically by blaming us for Russian protests against his regime. Popular demonstrations on the streets of Cairo, Damascus, and Moscow confirmed for Putin our sinister intentions. We were regime changers, not trusted partners. He pivoted hard against us, and hard against those protesting against him.

How could we continue the Reset with a leader who thought we were seeking his overthrow? How could we continue to pursue business as usual with this anti-American autocrat? I didn't see how. Putin was making clear that he thought the Reset was over. Accordingly, in my view, we had to adjust. We could not continue the Reset alone.

• • •

In place of Reset, I advocated a new strategy of selective engagement and selective disengagement. We should sustain cooperation with Russia on issues with which we already had some momentum, such as Iran, Afghanistan, and arms control, but stop chasing the Russians to cooperate with us on every initiative, and forget about the more ambitious goal of a strategic partnership with Russia. While Putin ruled Russia, strategic partnership was impossible.

During my first months in Moscow as ambassador, I worked with my team at the embassy to articulate this new strategy. As is typical in diplomatic practice today, many of the cables issued by the U.S. Embassy in Russia during my tenure were written largely by staff. But to develop my concept of selective engagement, I held the pen for a couple of important texts, including cables drafted before National Security Advisor Tom Donilon's visit to Moscow in May 2012, and for a special note to the incoming secretary of state, John Kerry, in the spring of 2013. I also worked with Secretary Clinton and her staff—Deputy Chief of Staff Jake Sullivan and Assistant Secretary Phil Gordon, in particular—to craft a departing note from Secretary Clinton to President Obama on Russia policy.[1] And I pushed hard for this policy of selective engagement and selective disengagement in all our interagency meetings, in which I participated via secure videoconference.

With Putin, I continued to believe that we should sustain cooperation on issues of mutual interest where possible. In 2012 I was not advocating a return to full-blown containment, let alone sanctions; that would come later. I never recommended the disruption of policies or programs that served U.S. national interests as a way to punish Putin for his new authoritarian ways and anti-American rhetoric. I appreciated the importance of the P5+1 negotiations with Iran and wanted our Russia policy to support that process. We needed to maintain Russia's cooperation on the Northern Distribution Network supply route for our soldiers fighting in Afghanistan. I continued to believe that increased trade and investment would benefit the United States. I entertained the hypothesis, pushed on me by the American business community in Russia, that Russians working for Western companies in Russia would eventually adopt Western norms about competition, rule of law, and anticorruption, likely becoming advocates of more accountable government over time.

While supporting engagement with the Russian government on a select, well-defined set of issues, I also advocated deliberate disengagement on other issues. I wanted us to change our analytic framework, our body language, and our public discourse about cooperation. If Putin did not want to work with us, if Putin continued to spew vile untruths about us, so be it; we would stop courting him. I argued that a little less engagement — a little less chasing after the Russians — might help us in some of our negotiating positions. A little less attention and a little more indifference might produce better responses to our proposals. After all, on most international issues of greatest importance to us in 2012, we did not *need* Russia.

For instance, with respect to missile defense, I believed that with Putin back in the Kremlin, cooperation was impossible. Under the best of circumstances, cooperating with Russia on missile defense had been an ambitious goal. Even increasing transparency as a way to reduce misconceptions proved extremely difficult. We did not want Russia sharing information about our missile-defense capabilities with their Iranian partners. Moreover, how could we collaborate on this most sensitive military program with a leader openly spouting anti-American sentiments? In the long run, I continued to believe that missile-defense cooperation with Russia would serve the security interests of the United States and our allies — but not while Putin was in power.

I also cautioned against the pursuit of a new arms control treaty. New START served American national interests well, not only in reducing the number of deployed nuclear weapons, but in preserving intrusive inspections and thereby reducing uncertainty. For me, the treaty's inspections regime was paramount. Obama wanted to do more — deeper cuts beyond the limitations laid out in New START. I embraced Obama's aspirations, but Putin did not, and there was nothing we could do to change his mind. The Kremlin was content to implement New START over the next decade, so it was going to exact a very high price — constraints on missile-defense deployments — for any new reductions in warheads or delivery vehicles. That to me was a bad trade. I recommended that we disengage on the issue of a further round of arms control negotiations.

A third issue on which I recommended disengagement with Russia was Syria. Already in 2012, as I discuss in detail in the next chapter, I worried that we were spending too much time trying to get the Russians to cooperate with us on ending the Syrian civil war, when our interests diverged too widely. President Obama did not want to use force against the Syrian regime without UN

Security Council approval, so we continued to ask Russia to support a resolution. That was a waste of time. Unlike Medvedev on Libya, Putin was never going to allow let alone endorse international intervention of any sort in Syria. We also assumed that we needed Russia to support the Geneva I Process, a plan to bring together the Assad regime and the Syrian opposition in a roundtable negotiation that would eventually produce a political transition in Syria. I never believed that Putin would commit to such a plan, because we had a fundamental disagreement with the Russian government about the final objective of this process. The Russians wanted Assad to remain in power. The Americans were committed to Assad's removal. So why continue to engage with Moscow on this subject? I also believed that we grossly overestimated the Kremlin's leverage over Assad; even if the Russians eventually agreed with our objectives, they would not have the power to help us achieve them. Instead, I argued in favor of pursuing measures that did not require Russian endorsement, like providing military support to the Syrian opposition, without a UN Security Council resolution or establishing contacts with potential defectors in Assad's regime.

Fourth, I wanted to end formal conversations with the Russian government about civil society. With Putin in power, these engagements inevitably would devolve into defensive shouting matches, not constructive dialogues. I advocated that we dissolve the working group on civil society within the Bilateral Presidential Commission, which I had originally cochaired while at the NSC. In the Medvedev era, we encountered challenges making that group work. In the Putin era, the effort seemed futile. We still needed to raise our concerns about human rights, but it was best to do so publicly, and not in these government-to-government channels.

My one worry about pursuing this new strategy of selective engagement and selective disengagement was the Russian practice of linkage — that is, the strategy of tying unrelated issues together (for example, Russian withdrawal of support for the P5+1 negotiations in response to our unilateral actions on Syria, or the interruption of NDN supply routes in reaction to our declarations on Russian human rights abuses). In the Medvedev era, we worked hard to avoid linkage, and mostly succeeded. Putin played by a different set of rules. Eventually, linkage became the default Russian strategy in dealing with us.

• • •

I had ample opportunity to advocate for my new strategy with administration officials through individual consultations and the interagency process via videoconference. Deliberation, however, does not always produce change, especially in the fifth year of an administration. Many of my colleagues in government back home had been pursuing their individual projects, policy goals, and strategic objectives for years. They did not want to abandon them just because there was a new leader in the Kremlin. In our first Deputies Committee meeting after Putin's reelection, on March 9, 2012, the theme was still on reengagement, reinvigoration, and turning the page on a difficult period in our bilateral relations.

There were also philosophical differences among some of my colleagues back in Washington about the strategy of partial disengagement. As I learned, usually indirectly, some thought that the Kremlin's rough treatment of me in my initial weeks as ambassador had affected my judgment: I was responding emotionally to the Kremlin's heat, not thinking rationally about U.S. interests. I understood the argument. In fact, I was quite sensitive to this line of analysis, and tried to play down the level of attention we gave to the Russian government's personal attacks against me. I also thought some of my colleagues back home were slow to appreciate how quickly opportunities for cooperation with Russia had closed in the new Putin era. They hadn't grasped how quickly and definitively Putin had pivoted from the Reset. They hadn't watched the virulent anti-American clips on Russian television. They did not see the protests at our embassy, or experience the FSB's increased scrutiny and harassment of our staff.

My recommendations also cut against the grain of State Department culture. Diplomats are trained to engage, not disengage. What I was advocating challenged the institutional impulses of our diplomatic community back home, even if most State Department officials working at the embassy eventually agreed with me. Philosophically, I too favored engagement. I was not a Cold Warrior thrilled to be back in the old game, but a reluctant disengager, driven by my analysis of Putin. We had not returned to the Cold War, but we were definitely in a new, more confrontational era: a hot peace. In one exasperated email sent on November 7, 2012, to my colleagues at the White House and the State Department regarding an upcoming Deputies Committee review of our Russia policy, I wrote, "As the author of the reset, I frankly am not sure that reaffirmation of reset is sufficient. It takes 2 to tango, and the govt. here doesn't seem to be all that interested in continuity right now. Just going back to the strategy we developed

4 years ago is not enough. Expectations also need to be adjusting. And frankly, I think that less engagement, not more, is an option that should be considered."[2]

Some in the Obama administration remained hopeful about continuing the Reset in the Putin era, because senior Russian leaders kept telling us to remain so. In his first conversation with President Obama after learning of his demotion to prime minister, Medvedev insisted that Putin was on board with the general direction of Russia's standing foreign policy. In October 2011, First Deputy Prime Minister Shuvalov came to Washington to deliver this same message.[3] (Of course, both of these conversations were before the massive protests in Russia that began in December.) In his meeting with Vice President Biden, Shuvalov emphasized that deeper cooperation with Russia was now possible because Putin wielded greater authority inside the Russian government. He was trying to get our attention about the possible benefits of allying with a strong leader. Prime Minister Putin sent a letter to Vice President Biden in December in which he pledged deeper cooperation. When I arrived in Moscow as ambassador in January 2012, most Russian government officials echoed a similar theme. Medvedev made one last plea for continuity in U.S.-Russia relations when he met with Obama at the second Nuclear Security Summit in Seoul in late March 2012. They made little progress on any of the tough issues weighing down U.S.-Russian relations at the time. Medvedev was not going to make any bold moves just weeks before stepping down. The tone of the conversation, however, was remarkably upbeat. Despite our differences over Libya, Syria, missile defense, and Russian electoral procedures, Medvedev still sounded like a Reset true believer. There was no bluster in his voice, no mention of America's evil ways, which I heard constantly on Russian television at the time. Medvedev still saw the glass as half full. As he spoke, I heard a man disappointed that he could not continue in his role as Obama's partner in the Reset, but also a retiring president who wanted to preserve his legacy. The Reset may have been his greatest achievement over the last four years, given how modest his record of accomplishment was on other domestic and foreign policy issues. During the press availability at the end of their meeting, Medvedev referred to Obama as his friend. Realizing that this would be his last meeting with Obama as president of Russia, Medvedev then reflected on the Reset policy the two presidents had pursued together and assessed that they "probably enjoyed the best level of relations between the United States and Russia during those three years than ever during the previous decades."[4] The best level of relations in decades — that was a bold, remarkable statement for the leader of Russia to be making in March 2012. Medvedev closed by expressing the hope "that the same high level of our relations will remain between the United States of America and the Russian Federation when the new President steps in office." Needless to say, I didn't share Medvedev's optimism on that score.

Obama responded kindly, reporting, "The last three years of my work with President Medvedev has been extremely productive." He went on to repeat the list of accomplishments in U.S.-Russian relations over the last three years, ending with a flurry of personal praise for the outgoing Russian president: "At a time of great challenges around the world, cooperation between the United States and Russia is absolutely critical to world peace and stability. And I have to say that I could not have asked for a better partner in forging that strong relationship than Dmitry." He concluded by wishing Medvedev well in his next job, ending his parting remarks with "Good luck, my friend."[5]

As I rode back to our hotel with President Obama after the meeting, I couldn't help but think that I would never witness such a warm exchange between our presidents ever again. Tragically, I was right.

Despite my personal misgivings about the prospects for success, President Obama's desire to reduce the number of nuclear weapons in the world, as he'd underscored with Medvedev in Seoul, compelled our government to engage quickly and directly with Putin. Obama wanted another arms control agreement, or at least amendments to New START that would mandate deeper reductions. Our task was to figure out a strategy to achieve that goal with the new man in the Kremlin. It was a tall order. The Russian government seemed perfectly content to execute the existing treaty, and not accelerate the timetable regarding new agreements.

We tried to come up with creative solutions. In the spring of 2012 our interagency team wrestled with various ideas for how far we could go in providing new initiatives regarding missile-defense cooperation in return for deeper nuclear weapons cuts. We aimed to time our opening offer to National Security Advisor Tom Donilon's trip to Moscow on May 3–4. Since the Russian parliamentary election in December 2011 and the subsequent mass demonstrations the following winter and spring, no senior U.S. official had met with Putin. We anticipated seeing Putin at the G-8 summit at Camp David on May 18–19, 2012. To make that meeting a productive one, we decided to have Tom come to Moscow first and lay out a framework for U.S.-Russian relations for the next several years. Central to Donilon's presentation was going to be our new ideas on further strategic arms reductions, including some very initial ideas for missile-defense cooperation.

Tom also came to Moscow with an invitation for President Putin to visit the White House right before the G-8 summit. Offering up an Oval Office visit to Putin was not an easy decision for our government. Putin had just stolen a parliamentary election in December 2011, and then won a presidential election in March 2012 that many considered not free and fair. He was harassing his political opponents and civil society leaders. On the campaign trail, he was in full America-bashing mode. Our more politically minded staffers in the White House worried that an East Wing press conference with Putin could be awkward and embarrassing for the president, who was only months away from his own reelection bid. But ultimately Donilon and other senior leaders of the national security team decided that the benefits outweighed the risks. Formally, our invitation to Putin was for a "working visit" as opposed to the more elaborate "state visit." But our offer to meet included almost all the bells and whistles of a state visit short of a state dinner, including an offer for Putin to stay at Blair House, across the street from the White House. We were going the extra mile in our effort to engage the new Russian president.

I too advocated for the meeting, based in large measure on what I was hearing at the time from Putin's advisors. Yuri Ushakov, Putin's foreign policy advisor, had repeatedly stressed to me the importance of personal relationships for Putin. To have any chance of maintaining continuity in U.S.-Russian relations, we had to create opportunities for Obama and Putin to interact one-on-one in ways that would nurture a personal chemistry between the two leaders. Echoing Shuvalov from the previous fall, Ushakov reminded me that Putin had the authority within the Russian government to take much bolder steps than Medvedev. Putin could achieve breakthroughs. Medvedev could not. Ushakov hinted to me that Putin even could do a deal with us on missile defense, as he had the standing with Russia's military to pull it off. But he would only do so after developing some personal rapport with Obama.

We couldn't just assume that a U.S. national security advisor would be given the go-ahead to meet Putin. Even as prime minister, Putin hadn't taken many meetings with foreigners, and just a few Americans. That Putin did agree to a meeting with Donilon, just weeks after his election and before his inauguration, was a positive sign.

By the time he boarded his flight for Moscow, Donilon was armed with what I thought was a sophisticated, comprehensive presentation for how to frame our negotiations during the next phase of arms control. Tom came prepared to talk about other issues as well, and had in hand an elaborate preamble to share with Putin about the necessity of maintaining constructive cooperation between our two countries based on common interests and mutual respect. Tom took pride in his ability to deal with leaders of the great powers, especially the more autocratic ones. It was kind of a special portfolio for him, and he was good at it. He was always well prepared, and there was never any doubt that he was speaking for Obama. This was his shot at trying to establish a rapport with one of the biggest global players of them all.

Many of our visitors from Washington wanted to fill up their schedules with as many meetings as possible. Tom had the opposite approach. He understood rightly that only one meeting mattered — his appointment with Putin. He rejected most of the proposals for meetings that my embassy team suggested, agreeing to sit down only with Foreign Minister Lavrov and the secretary of the Security Council, Nikolai Patrushev. The rest of his time in Russia would be used to prepare for his meeting with Putin.

In the run-up to the Putin-Donilon meeting, we endured the usual drama about how many people could attend. Putin always preferred small delegations; the fewer people in attendance, the better. For this meeting, Russian protocol was especially strict. In addition to the translator, the Kremlin told us that Tom could have only one other person with him in the meeting (I was designated that plus-one). Putin's rules. Worse yet, when our motorcade pulled up to the gates of Putin's compound, located an hour or so outside Moscow, the guards informed us that only one vehicle would be allowed on the premises. The rest of our very senior delegation from Washington would have to wait in their cars in the compound's wooded surrounds, with no food, water, or restrooms. And then it turned out that Putin had invited two people to join his side — Ushakov and Patrushev—not one. Even on these small matters of protocol, Putin did not play fair.

We drove to the guesthouse—the same building where Obama had met Prime Minister Putin in July 2009—and waited. True to form, Putin was late—this time, two hours late. But we had warned Tom about Putin's perpetual tardiness, so we used the time to rehearse his presentation, talk through some of the more intractable issues like Syria and Libya, and sip a lot of tea. Putin eventually arrived, apologized for being late, and invited us to sit in the chairs he normally used for his meetings with heads of state. These chairs were configured in a semicircle, which always struck me as an odd format for holding a serious conversation. It meant that the two main interlocutors—in this case, Donilon and Putin—rather than facing each other, were side by side and angled somewhat away from each other. As the note taker, I especially disliked this format for meetings with Putin, since he spoke softly and the angle made it hard to hear him. For this meeting, we luckily did not have simultaneous interpretation with earpieces, but consecutive interpretation, which meant I got to hear everything that Putin said twice, first in his Russian and then in our translator's English.

Tom started at the strategic level. He worked through his laydown of the special role and responsibility our two countries shared in maintaining the global order. He spelled out our intellectual framework for maintaining cooperative relations between Russia and the United States, and then hinted at some of our new ideas on further nuclear reductions.

Putin listened politely, but not closely. When Tom made his eloquent pitch for the importance of deepening U.S.-Russian relations, including further engagement on missile defense and arms control, Putin seemed disinterested. Instead, he wanted to talk about Syria, and the more general theme of alleged American support for regime change in the Middle East. We clearly were now dealing with someone in the Kremlin with a radically different perspective on our policies in the Middle East, and a very different style of interaction than Medvedev. Putin did not dismiss Tom's ideas for cooperation, but they obviously did not excite him. Deepening the Reset was not going to rank as a top priority for this new president.

Putin was generous with his time, talking with Donilon for over two hours. He wanted to make a positive impression on Tom, even if he was openly hostile toward me (this was the same meeting, mentioned earlier, at which Putin accused me of trying to ruin U.S.-Russian relations). But only at the end of the meeting did I understand why Putin was being so cordial, and only in retrospect did I realize why he'd agreed to meet with our national security advisor in the first place. As the second hour wound down, Putin hinted to Tom that he might not be able to attend the G-8 summit or accept Obama's invitation to visit the White House in May. In an unconvincing, meandering manner, he tried to explain the difficulties of putting together his new government. He suggested there were complications in the process that likely would require him to remain in Moscow to complete the job of naming his new government. Putin's usual style is to speak directly and bluntly. This part of the conversation was not that. In fact, it took us some time during the ride to the airport after the meeting to reconstruct the conversation and make sure we had understood Putin's message correctly. It was both odd and unexpected. But we had heard him correctly. Putin was not coming to the G-8 — the international club Russia had fought for two decades to join — and he was not accepting Obama's invitation to meet at the White House.

What was going on? For the next several days after the Putin-Donilon meeting, the smartest analysts on Kremlin politics both at the embassy and back in Washington wrestled with this question. His formal excuse sounded lame. Couldn't Putin make decisions about who was going to serve in his government before the trip to Washington, or even during his trip, using a telephone? Moreover, he'd decided to send Medvedev in his place. Didn't the prime minister need to be involved in forming his government? Something else had to be shaping this decision.

After weighing various competing hypotheses, I decided that Putin's decision was not about slighting Obama. Rather, he felt that his hold on power inside Russia remained shaky. On May 6, just days before the G-8 summit, a major political demonstration was planned in Moscow and elsewhere. Was Putin worried that it might undermine his authority? I personally did not think so, but Putin was in a paranoid state of mind at that time. Or maybe he anticipated that these May 6 demonstrations could turn violent, and that people would be arrested. If he appeared at the White House or with the leaders at the G-8 summit at press conferences, Western reporters would ask Putin tough questions. Maybe he did not want to be hanging out with seven other democratically elected leaders of the free world just days after arresting protesters?

Whatever the reasons, Putin's refusal to accept Obama's invitation represented a big missed opportunity. Some in the White House were relieved to not have to host such a controversial guest at that moment, but I knew that sustained, one-on-one meetings between our two presidents were hard to arrange. I also felt misled by my Russian government counterparts who had been stressing the importance of getting our two leaders together. When offered the chance to come to the White House just weeks into his third term as president, Putin had rebuffed Obama.

Equally strange was Putin's decision not to attend the G-8 summit. Gorbachev and Yeltsin had fought hard to gain Russian admittance into this elite club. Russia really did not qualify for membership. Its economy was comparatively underdeveloped. During the Putin years, Russia became increasingly autocratic, in contrast to the seven other liberal democracies in the group. The pageantry and expense on display when Putin hosted the G-8 in St. Petersburg in 2006 suggests that he too valued acceptance in this club, at least back then. To not show up in 2012, therefore, seemed inexplicable. We wanted engagement with Russia. But we could not force Russia — or specifically, Putin — to engage with us. It takes two to tango.

Although I would not have predicted it at the time, Putin never visited the White House while Obama was president. And Obama never made it to the Kremlin while Putin was president.

After Donilon's meeting with Putin, I lost all hope regarding another nuclear arms reduction treaty, or even an amendment to New START lowering the ceiling for deployed weapons. Putin simply was not interested. My view was that we should just forget about it. We gave the Prague Agenda our best shot. We did as much as we could when conditions were favorable for advancing the vision of a nuclear-free world. But now those conditions had changed. We had to adjust accordingly.

Those in the U.S. government responsible for arms control were not so eager to give up. I understood and even empathized with their perseverance. Some had joined the Obama administration inspired by his commitment to reduce the number of nuclear weapons in the world. Others, including professionals at the State Department, were paid to do arms control. That was their vocation. We continued to poke and prod, but nothing serious ever came out of these meetings.

Instead, senior Russian government officials, as well as Russian media outlets controlled by the government, increased their attacks on our missile-defense deployment plans in Europe, claiming that our interceptors had the capability to shoot down Russian ICBMs, and therefore threatened "strategic stability," a euphemism for mutual assured destruction. The Russian Ministry of Defense even convened international conferences on missile defense to bring attention to this growing danger. I attended these meetings, and sat in the audience dumbfounded as Valery Gerasimov, chief of the general staff, presented "evidence" about the growing American threat.[6] I continued to believe that Gerasimov and others in the Russian military were too smart, and their intelligence community too sophisticated, to not understand the real capacities of U.S. interceptors planned for deployment in Europe: they had no capabilities against Russian missiles. So I reasoned that it was all posturing in advance of a future negotiation over a new treaty that would include limits on both offensive and defensive weapons. Still, their propaganda statements reflected how far our exchanges on these issues had deteriorated.

After President Obama's reelection in November 2012, we made one last push to restart negotiations over deeper nuclear weapons cuts. In his State of the Union address in February 2013, Obama stated boldly that "we'll engage Russia to seek further reductions in our nuclear arsenals."[7] That spring, the president decided that Donilon should travel to Moscow again to make our best case. Tom brought an even more elaborate proposal for new arms control negotiations, including greater details about possible ways to increase transparency about missile defense.

On this trip, however, Tom received a very different welcome. His host for this visit was General Patrushev, the secretary of Russia's Security Council and former head of the FSB. Patrushev was considered a hardliner, close to Putin. He showed little interest in arms control. We barely engaged with him during the negotiations over New START. He brought Russia's arms control experts to the meeting, and they took copious notes as Donilon outlined our new ideas. But Patrushev wanted to talk about combating terrorism, as well as lecture us on our naive efforts at promoting regime change in the Arab world. He also scolded us for trying to foment revolution inside Russia, a theme he would develop over the coming months and years. He listened politely to Tom's ideas on nuclear reductions, but was not interested in engaging seriously.

During our meeting with Patrushev at the Kremlin, Putin "dropped by" and stayed for nearly an hour. Tom delivered a letter from Obama to the Russian president, which he read right then and there. Putin listened to Tom's presentation on arms control, but did not seem to want to engage in a substantive manner. Putin was done with arms control. Maybe he was done with us as well.

20

CHASING RUSSIANS, FAILING SYRIANS

I was disappointed that Putin did not want to engage with us on another round of nuclear weapons cuts, but I didn't interpret the rebuff as a major setback for American national interests. It would be nice to get down to 1,000 deployed warheads each, but we could blow up the world just as easily with 1,000 nuclear weapons as we could with 1,550. I was not indifferent, however, about our unsuccessful efforts to engage the Russians on ways to end the Syrian civil war. The failure of that strategy of engagement with Russia in part resulted in tens of thousands of innocent civilian deaths. Every day we wasted trying to get the Russians to work with us, more Syrians died. By the end of my time in government, roughly 150,000 people had been killed in Syria, most of them civilians; hundreds of thousands more died after I left government. We did try to work with Russia on ending the fighting in Syria. In retrospect, maybe we tried too hard. While I was in government, I considered our Syria policy, including the Russian component of it, to be the most frustrating foreign policy challenge of our administration. Of course, it is impossible to say that other strategies would have been more successful. We had no good options. But our strategy clearly did not achieve the intended objectives. We wanted to end the civil war in Syria; we failed to do so.

One of our mistakes was to make Russia a central player in our Syria strategy. As the conflict turned violent, we concluded that the only way to pressure Syrian president Bashar al-Assad to negotiate with the opposition over a new political system was to lean on Moscow. We did not have close ties with Assad, his government, or his military. The Kremlin did, or so we believed, and therefore might be convinced to use that leverage to broker a deal between elements of Assad's old guard and the opposition. In retrospect, that assessment was wrong. We overestimated Moscow's influence over the Assad regime, and misread Putin's intentions. On August 18, 2011, President Obama declared, "Assad must go." Putin wanted the opposite: to keep Assad in power. That fundamental disagreement haunted our diplomatic efforts with Russia regarding Syria for my time in government and beyond.

Our second key decision with long-term negative consequences was not providing significant military equipment and other assistance to the moderate opposition, the Free Syrian Army, once peaceful protest morphed into civil war. We eventually did provide assistance, but not significant help. This policy choice did not require Putin's blessing or a UN Security Council resolution. We could have acted alone or in close coordination with our partners in the Middle East and Europe. Some senior officials in the Obama administration recommended greater military assistance to the Syrian rebels in the spring of 2012, not as a strategy for helping them defeat Assad, but as a way to force the Syrian government to the negotiating table. We needed a stalemate on the battlefield to compel the regime to negotiate. Arming the moderates also would help them maintain an advantage over the rapidly emerging extremist groups. At that time, President Obama made the decision that arming the Syrian opposition would make things worse, and not help us pressure Assad to negotiate.[1] The president also worried about our weapons falling into the hands of terrorists. How would the world react if one of our antiaircraft rockets shot down a civilian aircraft? And Obama was rightly concerned about the dreaded slippery slope, having the conflict become another Vietnam or, even more aptly, Afghanistan. In the 1980s the Reagan administration armed the mujahedeen in Afghanistan fighting the Soviet occupiers and their puppet regime, eventually achieving victory. After the Red Army went home and their surrogates in Kabul lost power, however, the Taliban filled the power vacuum, and later invited al-Qaeda to take up residence. Al-Qaeda in turn attacked us on September 11, 2001. Obama wanted to avoid a replay of that tragedy. More recently, our military intervention in Libya did not produce the desired results, making many in Washington, including Obama, skittish about getting involved in another civil war. More generally, an

Obama-commissioned study reported that the outcome of aiding insurgents is rarely democracy.[2] Our actions in Syria also might impact our negotiations with Iran regarding its nuclear weapons program. Military escalation by us could lead eventually to direct military clashes between our Syrian partners and Iranian soldiers and militias helping to defend them. Finally, it must be remembered that Obama was reading numerous reports — generated both inside and outside the U.S. government — that predicted the opposition was on the verge of winning. As Clinton states in her memoirs, "In the early day of the fighting many assumed Assad's fall was inevitable."[3] Based on this assessment, why complicate matters — why discredit the opposition — by becoming involved militarily?

But then Assad did not fall. Our judgments about his weakness were incorrect. So, as the war continued, we eventually got into the fight, albeit incrementally and marginally. The president approved assistance to the rebels, but the desired outcome of defeating Assad or more minimally bringing the war to a stalemate was not achieved while I was in the government and still has not been achieved to this day.

A third key decision we made was to seek UN Security Council approval for all Syrian actions, including external interventions. Obama believed in the United Nations and did not want to act alone. In adhering to this principle, we gave the Russians a veto over our actions, power they used to prevent any military activity, stall the delivery of humanitarian assistance, and block many political moves aimed at pressuring Assad to negotiate. Even a UNSC resolution condemning Assad for human rights violations proved impossible to pass. In disgust one day, Secretary Clinton called Russian intransigence on Syria at the United Nations "despicable."[4]

This decision to act in Syria only with UN Security Council support contradicted another Obama principle — the right to intervene militarily to stop human rights abuses. As he stated very explicitly in his Nobel acceptance speech, "I believe that force can be justified on humanitarian grounds, as it was in the Balkans, or in other places that have been scarred by war. Inaction tears at our conscience and can lead to more costly intervention later."[5] We applied this norm to Libya, in part because Russia acquiesced at the UN Security Council. In Syria, however, Russia never supported the use of force, and we were not prepared to do so without UN Security Council backing. Well before Putin returned to the Kremlin in the spring of 2012, Medvedev refused to cooperate with us at the UN Security Council on any kind of military action in Syria because of his bitter experience supporting our military actions in Libya. The collapse of the regime, as well as the execution of Gadhafi in the aftermath of our bombing campaign there, marked a major turning point in U.S.-Russia relations. By abstaining on UNSCRs 1970 and 1973, Medvedev tacitly had endorsed our military plans for Libya, but then felt betrayed by us — not without reason — when the mission crept beyond what he thought those resolutions mandated. Events in Libya crept away from us as well. As the Libyan government and state started to collapse, we did not have a comprehensive postwar reconstruction plan for Libya, in large part because we had no intention of deploying our soldiers on the ground there. Nor did the Libyans want us there. Early on, the transition to a new political order in Libya looked like it might succeed. There was no immediate need for international peacekeepers and nation builders. But as the state imploded and fighting erupted between warring paramilitary groups, our relations with Russia grew strained. Putin was especially offended by our actions in Libya.[6] He was not going to allow a replay of Libya in Syria. Despite many attempts, there would never be a UN Security Council resolution on the use of force in Syria. We were wasting our time trying to pursue one.

The first meeting between President Obama and President Putin happened not in Washington or Moscow, but in Los Cabos, Mexico, on the sidelines of a G-20 summit in June 2012. Again breaking protocol, Obama asked me to participate in this meeting as I had in Seoul. I welcomed the opportunity to witness firsthand how the meeting between the two leaders would go. The main agenda item was Syria.

We stayed at the gorgeous Esperanza Resort in San José del Cabo, with a private beach and stunning views of the ocean splashing against the rugged Baja coast. For security reasons, we had the entire resort to ourselves. The evening of our arrival, we did the pre-brief with the president in his suite. As the president walked in, he greeted me warmly, and I grew nostalgic for my days at the White House. Moscow was a long way from D.C.

As the briefing progressed, I felt that we were not spending enough time talking about the new leader in the Kremlin, President Putin. I had flown all the way from Moscow, after all, to warn the president and this group about the very disturbing antidemocratic trends happening internally in Russia, and how these domestic developments in turn were going to make Russian foreign policy more belligerent toward us. In the few months between Obama's last meeting with President Medvedev in Seoul in the spring of 2012 and this first summit with President Putin, there had been a dramatic change in the Kremlin's posture toward us. We were now the enemy again. Living in Russia, I was witnessing Putin's rapid pivot away from the West every day. It was clear to me from this briefing session, however, that some in our government were slow to pick up on just how quickly and dramatically Russia was changing under Putin. I had spent the entire flight from Moscow to Los Cabos rehearsing my main talking points, but never really got the chance to deliver my soliloquy. The next day, however, Putin delivered the message just fine on his own.

Obama's meeting with Putin was first up on the schedule the next day. We had arranged to host the meeting at the resort where we were staying. True to form, Putin was late. Some on our team became agitated. Our head protocol person for this trip, Asel Roberts — a savvy, incredibly organized, fluent Russian speaker — began to bark questions into her cell phone and push texts through to her Russian counterparts. We were told that the Russian president was "on the way." The room was small and getting hotter, both because the sun was getting higher in the sky and because our delegation was becoming impatient. At one point, Obama got up and walked outside, alone. I had nothing to do at that moment — no one to call, no one to text, no one to shout at — so I joined him. I apologized that Putin was late. He looked at me with a grin on his face and said, "Do you think I care?" It reminded me of his first phone call with Medvedev in January 2009, when our sides were playing the cat-and-mouse game of who would get on the line first. Obama ended that ritual in a heartbeat, picking up the phone and holding for the Russian president. Obama doesn't sweat the small stuff. He reminded me that he didn't get much downtime, and as a Hawaiian, welcomed the chance to sit in the sun and be near the ocean. It was a beautiful day and a fantastic view. He spent the time asking about my sons and their basketball careers in Moscow. Despite the wait, he seemed in great spirits. His mood was about to change profoundly.

Putin walked into the meeting forty-five minutes late, giving some lame excuse about the traffic. (I later asked of Obama's traveling squad if anyone had ever been that late for a meeting with the president. The answer was no.) In his talking points, Obama had several big items teed up for discussion, including missile defense, nuclear cuts, Iran, and trade and investment. But their discussion quickly gravitated to Syria and the Arab Spring more generally. In some detail, Putin explained his theory of what was happening in the region. He assigned a lot of agency to us. In Putin's view, the only way to modernize traditional cultures like those in the Middle East was gradually and from the top. You needed strongmen in power who could steer their conservative societies slowly to greater openness, more tolerance of other ethnic and religious groups, and more freedom. Putin went out of his way to put both Assad and Mubarak in this category of the modernizing autocrat. He did not approve of our pressure to push out our guy, Mubarak, and he was not going to tolerate any effort to push out of power his guy, Assad. As I listened to the Russian president, I thought that he was describing himself as well. Putin believed that he was serving, albeit with much more sophistication and success, as the good czar — the modernizing strongman — leading his subjects to a better future.

Putin emphasized that he had no close personal bond with Assad. The Syrian dictator, he reminded us, shuttled to Paris all the time but had only come to Moscow once. Assad was educated in London, not Moscow; he was more our guy than Putin's. But even if he did not like Assad personally, Putin saw him as the best hope for keeping the country together and eventually pushing Syria in a modernizing direction. Assad was also a Russian ally.

Putin then explained the consequences of letting these autocratic modernizers fall. Speaking softly and calmly, Putin used tough, disparaging language to describe the shortsighted mistakes of our policies in the Middle East. In his view, we were supporters if not instigators of regime change in the Arab world — we were encouraging the Tunisian shopkeepers, the Egyptian students, the Syrian doctors. In Putin's world, the masses never acted on their own. Rather, they were tools, instruments, or levers, to be manipulated by the state. He made reference to the CIA, his nemesis from a very early stage in his professional career at the KGB. Listening to Putin, I sometimes felt that he even believed that the CIA and the "military-industrial complex" were the real decision makers in our system. Elected officials like Obama or Bush come and go, but the power ministries — what some call the "deep state" — really run the show behind the scenes. After all, that's how it worked in his system; in Russia under Putin, even the president was from the KGB.

Obama pushed back, explaining that the U.S. had nothing to do with sparking protest in the Arab world. We were responding to circumstances created by citizens in these countries. Putin did not believe him. As I listened to the Russian president that day in Los Cabos, I wondered how he could be so poorly informed about our foreign policy decision-making process. The Russian Embassy and Russian intelligence run a massive operation in Washington. Don't they report accurately on how it really works? Surely they must have written cables about how we were caught by surprise by the Arab Spring, how we were reacting uncomfortably to events in the region, not driving them.

I also wondered if maybe Putin and his KGB comrades had developed an inferiority complex with respect to the CIA over the years because their country had collapsed and ours had not. In order to stomach losing the Cold War, they had to believe they had lost to a very formidable opponent, one with extraordinary powers and resources capable of manipulating people in the Soviet Union in the late 1980s. They believed we were now doing the same in the Arab world, and again in Russia. The CIA is powerful but not nearly as powerful as Putin thinks, but nothing Obama was going to say that day in Los Cabos nor for the next several years would convince him otherwise.

As the two presidents wrapped up their extensive sparring over Syria that day, I was convinced that we had no shot at crafting a political transition in Damascus with Putin as our partner. Our frameworks for understanding events in Syria were just too far apart. Our ultimate objectives in Syria were diametrically opposed. We wanted a negotiated transition from dictatorship that would leave the Syrian state intact and include some members of Assad's regime in the new government, but not Assad himself. They wanted to keep Assad in power. No amount of summitry, no number of meetings or bonding at the dacha with Putin was going to resolve this fundamental disagreement. It was also clear to me at the end of this long, tense meeting that we were dealing with an entirely different interlocutor in the Kremlin now, one with fixed and flawed views about the world in general and the United States in particular. It was going to make cooperation between our two countries much more difficult, not only on Syria but on everything.

The press spray with Obama and Putin at Los Cabos was a complete disaster. Neither leader said anything memorable. Instead, their body language did all of the provocative messaging, or so the media portrayed. Putin's posture in particular drove the news. As always, he slouched in his chair, legs open, and rarely looked at Obama. For me, that was nothing new; I had seen that pose many times. But this was the first time that our White House correspondents had seen *President* Putin with Obama. Obama also was less engaging and friendly than normal. I wondered at the time if the president was making a deliberate effort not to look too chummy with Putin. At the press briefing after the meeting, we tried hard to downplay the importance of their body language, or what we later called Bodygate. But headlines told a different story: "Obama-Putin Meeting Brings Chill to Mexico,"[7] "G20 Body Language Decoded: A Spring in Obama's Step but No Pat for Putin,"[8] "Poker-Faced Meeting: Putin, Obama Avoid Pushing Sore Points."[9] Putin's slouching and Obama's dour expression signaled the hard times ahead on most issues in U.S.-Russia relations, but especially Syria.

Los Cabos gave me little reason to be optimistic about a deal with Russia on Syria. But our administration was not prepared to quit. Diplomacy requires persistence. Even when the probabilities of success are very low, diplomats still try to engage their counterparts, even if just to keep the talks and processes going. We were no different. It next fell to Clinton to try to work with Lavrov to see if something more constructive could be done on Syria.

Clinton and Lavrov did their next deep dive on Syria in St. Petersburg just a few weeks after the Obama-Putin meeting in Los Cabos. Secretary Clinton had never been to this fabulous city, and Lavrov wanted to show the place off. Clinton found a few hours to take in the sights, and made it to the Hermitage for a tour of one of the most impressive art collections in the world. She was greeted like a rock star. On every block, people stopped to say "Hello, Hillary," as if the entire world was on a first-name basis with the American secretary of state. Her first night in town, her advisors, Jake Sullivan and Phil Gordon, and I joined Clinton for dinner at a newly opened, hip rooftop restaurant overlooking the Neva River. Clinton came to St. Petersburg at the peak of White Nights, allowing us to enjoy some good wine and easy banter until late in the evening. When off the clock, Clinton knew how to relax, be real, and laugh. After walking back to the Astoria Hotel, Phil, Jake, and I planned to say goodnight to Secretary Clinton and then watch the television coverage of the European Cup semifinal match between Germany and Italy. When Clinton heard of our "boy" plans, she insisted on joining us, to the delight of the hundreds of people (mostly Germans) crammed into the hotel bar that evening. But her main focus in St. Petersburg was not Rembrandt, White Nights, or football. It was Syria.

After it became clear that the Russians would never back a seriously punitive UNSC resolution against Syria, we pivoted to the idea of bringing international pressure on the Syrian regime to agree to a ceasefire and political transition process. Because we believed that only Putin could convince Assad to engage in such a negotiation, we expended a great deal of effort trying to secure Moscow's support for a political transition. We believed that we needed Putin to bring Assad to the negotiating table. Our hope was that these negotiations eventually would produce an exit strategy for Assad. To achieve the latter, we were prepared to accept an extended timetable for this transition, as well as the possibility of leaving in power many Assad loyalists. In the summer of 2012, our plan seemed plausible. Assad was on the defensive regarding international public opinion. His indiscriminate bombings of civilian protesters had generated indignation from around the world. It was still early in this tragic conflict; the world had not yet become numb and indifferent to the killing.

And at the time, Assad was losing the war. In the first half of 2012 his army lost control of border posts on the Iraqi and Turkish borders, and ceded territory to the rebels in several Damascus neighborhoods. By the end of that year, even Russian deputy foreign minister Mikhail Bogdanov worried publicly, "We must look at the facts: There is a trend for the government to progressively lose control over an increasing part of the territory," adding, "An opposition victory can't be excluded."[10] So going into the meeting with Lavrov in St. Petersburg in June of 2012, we clung to some optimism that Moscow might support the idea of a conference on transition, not out of any love for us but in rational response to the "correlation of forces"—an old Soviet term—inside Syria trending in the opposition's favor.

In parallel to our bilateral negotiations with the Russians, the United Nations and the Arab League had tasked former UN secretary-general Kofi Annan in February 2012 to work with all parties and develop a plan for political transition. A month later, Annan had crafted a six-point peace plan, which Assad allegedly had endorsed during a meeting with Annan on March 27.[11] Two of the six points were especially crucial: "commit to work with the Envoy in an inclusive Syrian-led political process to address the legitimate aspirations and concerns of the

Syrian people" and "commit to stop the fighting and achieve urgently an effective United Nations–supervised cessation of armed violence in all its forms by all parties." Annan then called upon all external actors in the civil war — the "Action Group for Syria" — to convene in Geneva on June 30, 2012, to agree on a common path forward. The timing magnified the importance of the Clinton-Lavrov meeting in St. Petersburg the day before. Everyone understood that Russian-American agreement would be essential for any progress in Geneva.

For dinner with Lavrov that evening, our American delegation traveled to yet another upscale restaurant near the Neva River. Before any formal discussions, Lavrov took Clinton and the rest of our delegation up to the roof of the building, which offered up a spectacular, panoramic view of the city. The blue sky was clean and crisp that evening, allowing us to see the results of Putin's investments in restoring the city. This was not the same crumbling, paint-peeling Leningrad where I had studied in the summer of 1983. This was St. Petersburg — bright, strong, wealthy, imperial.

As we descended from the roof, Lavrov took Clinton's elbow and steered her to a small room for a private chat. Because Lavrov speaks flawless English, no translators were needed. For the next two hours or so, they spoke alone about Syria. None of us in Clinton's delegation were happy with Lavrov's move. We wanted to hear the conversation. We wanted to offer guidance and support to the secretary as needed. I knew Lavrov to be excellent at his job, playing whatever cards he held to maximum effect. In one-on-one sessions, he utilized an effective combination of persuasion and stubbornness that could wear the other side down. We fretted that Clinton was at a disadvantage.

By this time in their relationship, however, Clinton was well immunized against Lavrov's charms and talents. Their personal relationship had started out strong. Despite the reset-button fiasco, their first meeting in Geneva in the spring of 2009 had given me cause for optimism. Lavrov was on his best behavior back then, eager to make a good first impression. Clinton had made an effort to connect with Lavrov, both analytically and personally. They seemed to click that first evening, spending a long dinner reviewing the possibilities for Russian-American cooperation on dozens of issues, but also bantering informally. By the summer of 2012, with the Reset in retreat, Clinton and Lavrov's relationship had grown strained. Lavrov's style of diplomacy frustrated the secretary of state. He would come to his meetings with her armed with a long list of items

— mostly complaints. These included what we called "irritants," minor issues such as visa requests not worthy of Clinton's time.

Syria, however, was no irritant. Heading into the Geneva discussions, there were a lot of substantive issues for Lavrov and Clinton to discuss. When they'd concluded their one-on-one and we were finally seated for dinner, Lavrov and Clinton appeared satisfied with the progress they had made. But then the Russian minister quickly returned to his familiar set of talking points, walking through a growing list of complaints about our policies. Alleged progress on Syria was not having a positive effect on any other issue in U.S.-Russia relations, it would seem. At that dinner, Lavrov even called Clinton's press spokesperson, Victoria Nuland, the "Minister of Disinformation." (And the Russian Ministry of Foreign Affairs accused *me* of being undiplomatic!)

When speaking to the press after dinner, however, Lavrov described his meeting with Clinton as "one of the most productive" ever. He added, "Syria dominated the international affairs section. I felt Hillary Clinton's position has changed . . . She said she understands our position. We have agreed with Hillary Clinton to look for agreements on Syria which would bring us closer together."[12] When I read these comments later that evening, I grew worried. Lavrov rarely expressed such exuberance. There must be more to this story. His statement that "Clinton's position has changed" was particularly disconcerting.

Later that evening and for the rest of her time in St. Petersburg, Clinton was churning. I had the impression that she did not like what she had heard from Lavrov in their one-on-one. She believed the draft language in the Geneva protocol did not go far enough regarding Assad's ouster. She felt failure looming. I wondered if maybe she was concerned that Lavrov had listened selectively to what she had said. Or was Lavrov playing games, trying to box Clinton in before the Geneva meeting the following day? We found out soon enough, and the answers were not good for our diplomacy or for the people of Syria.

In Geneva, Clinton joined Annan, Lavrov, several other foreign ministers, the UN secretary-general, and the head of the League of Arab States to discuss and then release the "Action Group for Syria Final Communiqué." This six-point plan included a set of actions to produce an immediate ceasefire in Syria, as well as a series of steps that would produce a "Syrian-led" political transition, which would commit all actors to "compete fairly and equally in elections," as well as a "multi-party democracy" that would go "beyond an initial round of elections." The Geneva communiqué also called for the creation of a "transitional governing body" that would lead the country temporarily while all parties negotiated a credible political agreement to end the war and guide a transition process. No Syrians — neither the government nor opposition — were in Geneva, yet the Action Group for Syria emphasized in its plan that "it is for the Syrian people to determine the future of the country."[13] The expectation at Geneva was that the West would pressure the opposition to negotiate with the ruling regime and the Russians would convince Assad and his inner circle to sit down with the opposition.

The communiqué released in Geneva was trumpeted as a major success, both for international diplomacy and U.S.-Russian relations.[14] Despite a tumultuous and tense start after Putin's return to the Kremlin earlier that year, we were now cooperating on a common plan to solve the world's greatest security and humanitarian crisis at the time, or so it was said.

The positive vibe lasted less than a day. It soon became apparent that Lavrov and Clinton were interpreting the communiqué very differently. In Geneva, Clinton and her partners had succeeded in adding language to the agreement stating that the transitional governing entity, comprised of representatives from Assad's government and the opposition, would be constituted "on the basis of mutual consent." For us, this phrase meant that Assad would have to go, since the opposition would never agree to allow someone they considered a war criminal to join the transitional authority. As Clinton stated bluntly in Geneva, "We all agreed to support Kofi Annan's principles and guidelines for a Syrian-led transition, including the goal of a democratic, pluralistic Syria that upholds the rule of law and respects the universal rights of all people and all communities, regardless of ethnicity, sect, or gender . . . the formation of a transitional governing body exercising full executive powers, which would be broadly inclusive and chosen by mutual consent . . . These are principles that have formed the basis for successful democratic transitions all over the world, and they offer the best chance for restoring peace and meeting the needs and aspirations of the Syrian people."[15] Assad in power and democratic transition were two incompatible outcomes. In her memoir, Clinton wrote more bluntly, the communiqué was "a blueprint for Assad's departure."[16]

Lavrov expressed an opposite view. He criticized Clinton for suggesting the communiqué called for Assad's departure, telling reporters "some Western participants in the meeting began to distort the agreements . . . in their public statements."[17] In forty-eight hours, we had gone from "the best meeting ever" between Clinton and Lavrov in St. Petersburg to sparring again in Geneva.

I was disappointed but not surprised. For several months already, we had been chasing the Russians, asking them to support toothless UNSC resolutions condemning Assad's slaughter of innocent civilians, begging them to pressure Assad to agree to a ceasefire, and pleading with them to use their leverage to bring Assad's government to the negotiating table, never acknowledging that they were pursuing opposite objectives all the while. We continued to court the Russians because we believed that we had no better options. We were not going to use force without a UN Security Council resolution. We had few if any effective ways to influence Assad directly. So we pivoted back to engaging the Russians, even if everyone understood the low probabilities of success.

Numerous meetings with Russian government officials throughout 2012 and 2013, including my meetings with Russia's most senior expert on the Middle East, Mikhail Bogdanov, reinforced my pessimism. At the time, Bogdanov was the MFA's deputy foreign minister in charge of the Middle East and Africa, and had served for decades in the Middle East, including two long stints in Damascus, tours in Yemen and Lebanon, and two stops as ambassador, first in Israel and then Egypt. Few people in the world know as much about the Middle East as Bogdanov. His deputy, Sergey Vershinin, was also an expert on all things Arab.

I felt like I hit it off with Bogdanov early in my tenure as ambassador. I found him to be accessible and engaging. Bogdanov was also someone I saw around town, at cultural events at Spaso House and concerts at the Conservatory. He was a real intellectual — educated, refined, cultured. He was also a skillful diplomat, who would never go beyond his talking points on policy, but freely wandered beyond the government line when it came to analysis.

We had a ritual. I would meet him at his offices to deliver my démarche. He served us black tea and stale cookies. We sat in big leather chairs and talked at length, always in Russian. After I conveyed my official talking points and he delivered his official response, I would ask Bogdanov to give me his assessment of the current situation inside Syria. He eagerly answered my analytic questions. He had a lot to say, as he was very well informed. The Soviets had devoted real resources to developing expertise about the Middle East. Bogdanov proved what that kind of investment could produce. He knew all the players in Syria, both in the government and the opposition. He knew all the regional actors in the Arab world. I learned a lot from our long conversations, even if this knowledge acquisition never translated directly into progress on policy.

My meetings with Bogdanov reaffirmed three central features of Russia policy toward Syria. First, as Putin made clear to Obama in Los Cabos, Russia was not ready to push for a political settlement that included Assad's ouster. Despite the spin by some European leaders, the Geneva plan had not changed Russia's position on this issue. In my view, we needed to accept this fundamental difference and stop hoping that we were going to change Putin's mind. Hope is not a strategy.

Second, I also came away from my tutorials with Bogdanov convinced that we were overestimating Russia's leverage over Assad and his inner circle. Of course, they had better channels of communication with the Syrian dictator and his inner circle than we did. But in diplomacy, the opportunity to talk should never be confused with the ability to persuade. Access did not equal influence. I also did not sense that the Russians had serious leverage over other senior officials in Assad's government or in the Syrian military, who might be persuaded to push him aside as a way to save the regime. From time to time, I presented such a scenario to Bogdanov. Back when the Egyptian transition still seemed like a success, I explained to him that we had used our close relations with generals in the Egyptian military to help nudge Mubarak out of power. Why couldn't Russia do the same with its contacts in the Syrian military or intelligence? As I probed, I wasn't sure that there were such contacts, both because Russia's ties to these people in Syria were not as close as we enjoyed with our Egyptian counterparts and because they had not identified any defectors in Assad's inner circle. Assad had structured a complex web of competitive, senior leaders in his intelligence community to encourage everyone to spy on each other. The minute a defector started to develop a plan against Assad, someone else in the Syrian intelligence world would report on him.

At times, Bogdanov and I also discussed the practical issues of how to arrange Assad's departure. If he stepped down, Assad could not stay in Syria, but who would take him? Iran? A country in Europe or Africa? Unlikely. How many people would go with him? Bogdanov estimated that the Syrian president would want several hundred in his entourage. Who would pay for and protect these Syrians living in Tehran or somewhere else for several decades? Assad, after all, was a young man. Because we could never provide clear answers to these questions, Bogdanov believed that we were not serious about getting Assad out safely. The memory of Gadhafi's inglorious murder in Libya hung heavily over these conversations.

Third, Russia and the United States had different assessments about the balance of power between Assad and the Syrian opposition. In 2012 we were predicting that Assad was going to fall soon. As the war dragged on, Russian officials expressed more confidence in predicting Assad could hang on and eventually even win the war. It was a genuine disagreement about facts and trends. We even arranged meetings between our leading experts on Syria to compare data. In the end, we agreed to disagree.

We argued that the longer the Syria conflict dragged on, the more radical and violent the opposition would become. It was a phenomenon about which I had written in my doctoral dissertation on southern African national liberation movements. To avoid violent, revolutionary change later, we needed to push for peaceful, evolutionary change now. As the Syrian conflict turned violent, and as more extremist groups such as al-Nusra and then the Islamic State became more powerful, our analysis was tragically validated. The Russians did not just theorize about Syria, they acted. When Assad became weaker, they doubled down on their support for him, believing that he eventually would prevail on the battlefield against both moderates and radicals. For them, Chechnya was the more apt historical analogy, not Egypt. Putin had backed Chechen strongman Ramzan Kadyrov for years. Eventually, through the use of brutal, scorched-earth military tactics, Kadyrov either destroyed the "terrorists" fighting against his regime, or pushed them into other republics in the North Caucasus.[18] Assad could do the same in in Syria.

In my assessment, the United States and Russia were never going to overcome these major analytic and policy differences over how to understand and respond to the civil war in Syria. But we kept trying, hoping that the Russians might come around. At least we could talk to them, an option we did not have with Assad or Iran, an action that gave us the sense that something was being done.

The diplomacy issuing from the 2012 Geneva meeting — or Geneva I as it came to be known — unraveled quickly. In her memoirs, Clinton reports that the Geneva conference "settled into a running argument between me and Lavrov."[19] That argument continued, consistently in private and sometimes spilling into the public, for the next several months. Kofi Annan resigned in August 2012, disgusted with the intransigence on all sides, but especially between the Americans and Russians.[20] As Moscow and Washington argued, Assad slaughtered tens of thousands of innocent civilians. All sides in the war dug in, as did their external supporters. We needed to change course. We needed to move away from our singular focus on Russia as the closer of a Syria deal. As I wrote to senior colleagues in the State Department and White House on November 7, 2012, in response to a draft paper on Russia strategy, "On Syria, we need (in my view) a basic rethink of our approach — this constant chasing of the Russians, now for over eighteen months, has been a complete failure. Based on what do we believe that we will ever get a different result?" I know some agreed with me, but the alternatives to ending the chase were not attractive either.

After Obama's reelection, we got a new leader at the State Department determined to restart the negotiations on Syria with the Russians. Secretary of State John Kerry brought new energy and focus to Syria. He believed in the power of personal diplomacy. On many issues — the Middle East peace process, Syria, Iran — he believed he could make a difference. He leaned toward persuasion, not coercion. If he could just sit down with the right leaders, he could convince them to do the right thing. In May 2013, he tested his theory of diplomacy in Moscow. His objective: reinvigorate international diplomacy to end the Syrian civil war. His strategy: convince Putin.

His visit got off to a rough start. Kerry's plane touched down early, sending the Russian protocol team into complete panic. That day, Russian tanks, ballistic missiles, and armored personnel carriers were rumbling around the streets of downtown Moscow, practicing for the May 9 parade in Red Square, one of Russia's biggest holidays, commemorating the end of the Great Patriotic War. Kerry was staying at the Ritz-Carlton, just a block away from Red Square, so all the roads to his hotel were blocked. The Russian protocol chief informed our head of the advance team for Kerry's visit, Ruben Harutunian, that Kerry would have to remain at the airport until the parade rehearsal ended. Ruben, a young, talented foreign service officer who had worked for me the year before in Moscow, looked at me and said, "Sir, I suggest you tell the Secretary this bad news." So I boarded Kerry's plane, explained our predicament, and suggested we start our briefing on the aircraft, since there would be nowhere secure in the airport for us to talk. Kerry agreed, and I jumped in with my briefing points. But the lanky secretary soon grew restless, having just flown half a day, and was eager to get off the plane, stretch his legs, and breathe some fresh air. So we did. The Russian protocol team were spun into a tizzy. I'm sure that they had never witnessed a U.S. secretary of state touring their tarmac.

An hour later, we were given the go-ahead to drive into Moscow. Then we waited again, this time for Putin, who was running late. At least Kerry had a nicer place to stretch his legs, Red Square, all spruced up and decorated with Soviet-era posters and banners in preparation for the May 9 celebration. Kerry went on an extended tour around the square, which we had to ourselves since it was blocked off in preparation for the holiday. And yet, even after we had exhausted all of our guide's many historical tidbits about the history of the GUM department store and the tragic fate of the architect of St. Basil's Cathedral, who allegedly had been blinded by Ivan the Terrible so that he would not be able to re-create his masterpiece, Putin was still not ready for us. So we returned to the Ritz-Carleton and waited. Three hours later, we were informed that Putin was finally ready to receive us at the Kremlin.

Kerry had a cordial but inconclusive meeting with the Russian president. As usual, it was a small group, just Kerry and me on our side; Putin, Lavrov, and Ushakov on the other side. Most of the meeting circled around Syria. Kerry was focused. He did not come to Moscow to cover dozens of issues in U.S.-Russian relations. He had one big issue on his mind: ending the Syrian war.

I don't recall Putin saying anything new on Syria in his two-hour meeting with Kerry. By this time, I had heard his take several times. But Putin did turn to Lavrov and encouraged him to negotiate seriously with Kerry, basically signaling that he didn't want to get involved in the details. Putin's message to the secretary as I understood it was, You want to try to work something out with

Lavrov? Go ahead, knock yourself out. Lavrov had seemed to have a strained relationship with Medvedev, but appeared much more at ease with Putin.

In his meeting with Kerry, Putin took another swipe at me personally. Staring straight at me, Putin informed the secretary of state that "your embassy" — that is, me and my team — was actively supporting those that sought to overthrow him. Once again, I was being accused by the president of Russia of fomenting revolution. It seemed evident that Putin was trying to sow discord between my new boss and me. Walking out of the meeting, Secretary Kerry turned to me and said he couldn't believe what he had just heard. As we sped outside the Kremlin walls, Kerry asked me if he had correctly understood Putin's remarks. I confirmed that he had. A year and a half into my term as ambassador, Putin still seemed bizarrely fixated on me. I thought these diatribes made him look paranoid.

From the Kremlin, we headed over to the MFA's guesthouse, a gorgeous Rococo–style nineteenth-century mansion a mile away from the monstrous Stalinist skyscraper that houses the ministry, for our follow-up meeting with Lavrov.

Because of the three-hour delay in the Putin meeting, this session started much later than planned. But I could tell that Kerry was not angling for an early evening. He planned to come away with something tangible on Syria that day. He started the meeting by strolling in the courtyard for an hour alone with Lavrov. It reminded me of the way that Clinton and Lavrov had started their conversation on Syria in St. Petersburg a year earlier. I hoped this one-on-one would produce better results.

After his garden constitutional with Lavrov, Kerry huddled with his senior team, including his most senior official at the time for Syria, Ambassador Robert Ford. Kerry thought that he'd heard from Lavrov some willingness to compromise. Specifically, he saw a chance to resuscitate the Geneva I process and get the Syrian regime and opposition to sit down and begin negotiations. Ford was nervous. He didn't feel like Lavrov had given us anything. In Ford's view, Lavrov was just doubling down on Geneva I, which, as we had learned bitterly the last time, did not mean to him or to Putin the removal of Assad from power.

Kerry and Lavrov then reconvened, this time at a large table with our complete delegations sitting on each side. The press came in and Kerry and Lavrov both made pleasant, optimistic introductory remarks, but also promised a press conference at the end of the meeting. Expectations for a breakthrough were building. At one point in the discussion, National Security Advisor Tom Donilon called Kerry on his cell phone. Kerry excused himself to take the call. This early in his tenure as secretary of state, Kerry knew that he could not commit to anything that Tom wouldn't support. White House management — some might say micromanagement — of foreign policy was on display. I couldn't hear the substance of their conversation, but Kerry's tone and facial expressions indicated annoyance.

It was almost midnight when Kerry and Lavrov opened their press conference, but they made some news for our jet-lagged press corps, announcing together their mutual commitment to reviving the idea of an international peace conference guided by the principles of the Geneva communiqué. It would become known as Geneva II. Kerry had wanted to make progress on Syria on this trip, and he did. Kerry underscored that "the Geneva communiqué is the important track to end the bloodshed in Syria, and it should not be a piece of paper. It should not be a forgotten communiqué of diplomacy. It should be the roadmap, the implemented manner by which the people of Syria could find their way to the new Syria, and by which the bloodshed, the killing, the massacres can end."[21] That evening, Kerry did not say, "Assad must go." The absence of that phrase made it possible for Lavrov to sound cooperative as well. He affirmed, "Russia and the USA will stimulate both the Syrian government and opposition groups to find a political solution," adding urgency to the assignment by suggesting that we could organize a new international conference by the end of the month. (That never happened.) Lavrov noted Russia's agreement to the principle of "mutual consent" for deciding which representatives from the government and opposition could attend new negotiations, which meant to me that negotiations would not begin, because the opposition would never negotiate with Assad.[22] To me, in both public and private, Lavrov echoed the exact same themes that he'd articulated to Clinton a year earlier in Geneva. But after a year of no progress at all, his remarks sounded like something new.

Kerry should have been completely exhausted by the time the press conference ended, but he rolled right into the banquet room to take part in the feast that had been planned to follow his meeting with Lavrov. He had stamina. Kerry and Lavrov ate, drank, toasted, joked, smoked cigars, and traded stories about their athletic achievements, Kerry in hockey, Lavrov in white-water rafting. Lavrov invited Kerry to come watch the Russia-USA hockey game at the Sochi Olympics if there was one, and Kerry said he would try. (He never made it.) The two men clicked. They had known each other for a long time, were from the same generation, and seemed to have real chemistry. In the Putin era, maybe Kerry could be our new point person on Russia? With Obama now engaging very little on the Russian account, we needed a new high-level channel.

The next day, as we drove out to the airport, Kerry confided in me that he was well aware of the risk he was taking in trying to revive diplomacy with the Russians on Syria. He knew that his critics back home, both in and out of government, would call him naive. He recognized the odds for success were low. But, he asked rhetorically, "what's the alternative?" He wanted to end the war in Syria, and believed that his best chance for doing so was engagement with the Russians.

As the secretary's flight departed — no sweeter words in the State Department than "wheels up" for a visiting dignitary — I discussed with some of my senior team the power of individual diplomacy. We had to hold on to the idea that diplomats matter. If not, why were we in the business? But the conditions under which individual diplomats influence history are very specific. Kerry had impressive diplomatic chops. He was an effective negotiator. He focused on the big stuff, and wanted results. He struck me as having the right mix of skills to get a major strategic arms treaty done or close on a deal like German unification. But he did not have the fortune of being secretary of state at the time when those kinds of deals were being negotiated. Instead, he was secretary of state when the assignment was to end the Syrian civil war, and prod Putin to help us. Syria was an incredibly complex, tragic problem and Putin was no Gorbachev. Would Henry Kissinger, George Shultz, or Jim Baker have done any better than Kerry in convincing Putin to work with us on Syria? I doubt it. This was Kerry's fate. And it was also mine — to be the U.S. ambassador in Moscow not in the glory days of the Cold War's end or even the optimistic first years of the Reset, but in the dark days of Putin's Russia. I admired Kerry's tenacity. I predicted failure for his mission.

• • •

While diplomats continued to hash out toothless communiqués, the Syrian civil war expanded, causing our government to become increasingly concerned about the security of Assad's chemical weapons. He claimed that he did not possess such weapons, but everyone knew that was a lie. If his regime broke down, these weapons could fall into the hands of terrorist organizations. We also worried that he might be tempted to use them. He was a desperate man, struggling to hold on to power. Quietly, we started a channel for dialogue between the Russian Security Council and our National Security Council on how to manage this sensitive and dangerous problem.

The first meetings in this series seemed like a complete waste of time. The head of the Russian Security Council, Nikolai Patrushev, had a reputation as a hardliner and rarely went out of his way to offer useful direction when participating in these discussions. His deputy, Yevgeny Lukyanov, was the real point person. Lukyanov was entertaining. We would meet from time to time to discuss Syria as well as other security matters. He loved to talk, and employed colorful language and stretched metaphors. He loved to provoke. He could go on for thirty minutes in answer to one question, meandering so far away from the original subject that he had to ask sometimes what the initial question was. He was great fun to banter with, but his style did not inspire hope among our visiting delegations about the prospect for serious negotiations.

In the spring of 2012 our government began to learn about Assad's limited use of chemical weapons against civilians. On April 25 we went public with our assessment. We also sent a briefer to Moscow giving a detailed account of this new information. Lukyanov and his Security Council team dismissed our findings completely, subtly reminding us of previous intelligence failures regarding the discovery of weapons of mass destruction in Iraq. It was frustrating. We were at a standstill yet again, trading competing data, as had happened so often on the Middle East during my time in government.

And then Assad acted more brazenly. On August 21, 2013, the Syrian dictator used chemical weapons to kill more than fifteen hundred people, including four hundred children.[23] Amazingly, the Russians continued to question the validity of reporting on this tragedy even after the entire world saw the physical evidence of this horrific act. You did not need to have a top-secret security clearance to know that Assad had used chemical weapons, violating one of the most sacred norms of the international system. We had to respond. A year earlier, in answer to a question from NBC's Chuck Todd about the hypothetical American response to Assad's use of chemical weapons, Obama had warned, "We have been very clear to the Assad regime, but also to other players on the ground, that a red line for us is when we start seeing a whole bunch of chemical weapons moving around or being utilized. That would change my calculus . . . That would change my equation. We're monitoring that situation very carefully. We have put together a range of contingency plans."[24]

Now, Assad had crossed Obama's red line. The world, and most eagerly Syrians in opposition to Assad, expected an American military response. Obama threatened one, as did Kerry and other senior U.S. government officials. Assad braced for an attack, as did those fighting him. Russian government officials warned me of the dangerous consequences of striking Syria, but the Kremlin anticipated military action from us. I did too. At the time, I also hoped—not predicted, just hoped—that a U.S. bombing campaign against strategic Syrian targets, especially their airplanes and runways, might debilitate Assad sufficiently to give the opposition some new momentum. At a minimum, destroying or at least weakening Assad's air power would save civilian lives. The disastrous unintended consequences of state breakdown in Libya loomed large in our government's deliberations about new airstrikes in Syria. No one wanted to see a quick collapse of the Assad regime, followed by chaos or, maybe even worse yet, a victory by the extremists in a country with giant stockpiles of chemical weapons. But a limited airstrike seemed to me to be a risk worth taking. The status quo, after all, was horrible.

Most of Obama's top national security team supported a military strike. According to Derek Chollet, at the time the assistant secretary of defense responsible for the Middle East, "Nearly everyone advocated for quick action."[25] Chollet added that "military planners at the Pentagon started working around the clock to prepare for a series of airstrikes, with the specific aims of deterring Syrian leader Bashar al-Assad from launching more chemical weapons attacks and degrading his military's ability to do so."[26] On August 30 Secretary Kerry delivered a forceful speech outlining the reasons for military action and warning against inaction: "As previous storms in history have gathered, when unspeakable crimes were within our power to stop them, we have been warned against the temptations of looking the other way. History is full of leaders who have warned against inaction, indifference and especially against silence when it mattered most . . . And our choice today has great consequences." Kerry explicitly invoked the reputational costs of inaction to the United States: "It matters deeply to the credibility and the future interests of the United States of America and our allies. It matters because a lot of other countries, whose policy has challenged these international norms, are watching. They are watching. They want to see whether the United States and our friends mean what we say. It is directly related to our credibility and whether countries still believe the United States when it says something. They are watching to see if Syria can get away with it."[27] It sounded to me like Kerry was preparing the American people and the world for American military action against Syria.

Behind closed doors, however, debate about the merits of a military strike continued. Obama expressed doubts, first and foremost because he worried about negative second- and third-order effects of our military intervention. What if Assad did not stop using chemical weapons even after our attacks? How would we escalate? And what if the regime did collapse? Who would take control of these chemical weapons? These were legitimate questions, difficult to answer. Obama also did not want to take this step alone. Putin would veto any United Nations Security Council resolution to authorize the use of force in Syria. Some proposed that Obama ask NATO, as President Clinton had before the bombing campaign against Serbia in 1999, but that idea fizzled after the British Parliament stunned us all by voting against the use of force. Days before boarding *Air Force One* to fly to Russia to attend the G-20 summit, Obama came up with a final idea: ask Congress to authorize the use of force. That was the game plan when Obama touched down in St. Petersburg.

After Obama descended from *Air Force One* for his first and only visit to Russia while I was ambassador, we piled into the Beast for the hourlong ride to the summit conference site at Constantine Palace. The atmosphere in the car was tense. National Security Advisor Susan Rice and her deputy, Ben Rhodes, were in the jump seats. We all were squeezed in tightly together, our knees knocking against each others' from time to time. (The Beast is much smaller inside than it looks from the outside, because the walls of the vehicle are so thick.) I launched tepidly into a general briefing about Russia, Putin, and our bilateral relationship, but the conversation quickly homed in on two issues: Syria, and when or how Obama was going to talk to Putin.

Rice did not like Obama's surprising decision to ask Congress for authorization to use force in Syria, because she didn't believe that we could win enough votes for approval. At that moment, both Rice and Rhodes supported airstrikes. When it seemed appropriate to do so, I concurred.

At some point during the drive, Rhodes got an update from the White House on the vote corralling, and the news was not good. The day before, on September 4, the Senate Foreign Relations Committee had approved a resolution to authorize the use of force against Syria, 10–7.[28] That was a hopeful sign. But getting majorities beyond that committee was looking increasingly unlikely.[29] Public opinion polls showed that a solid majority of Americans did not support a military strike.[30] Some members of Congress encouraged Obama to take military action but not pursue a vote. If Obama was going to strike, he was going to do so alone, not with the United Nations' backing, NATO support, or congressional endorsement.

In our discussion of when it would be best for Obama to talk to Putin, Obama was relaxed, even as the rest of us were not. He planned to find a time between meals to pull the Russian president aside to discuss Syria and a few other pressing issues. Susan confirmed that she would be meeting with her counterpart, Yuri Ushakov, on the second day; she thought it would be preferable for Obama to speak to Putin first. He nodded in agreement, not looking too concerned about teeing up Susan's meeting. Susan was also going to use her time at the G-20 summit to enlist support for a joint statement on Syria, offering us another data point on how alone we would be if we attacked Assad's forces. (The final result was disappointing; only eleven countries signed the statement.)[31]

When we pulled up to the palace for the official meet and greet with the host, Obama and Putin's body language was once again the subject of press scrutiny, just as it had been a year earlier in Los Cabos. Press assessments on the handshake were fast and furious — "awkward" and "laden with much unsaid."[32] It was Bodygate II.

The first day of the summit passed with no meaningful interaction between Obama and Putin. They held their individual bilateral meetings (thanks to Edward Snowden, Obama had a very tough one with Brazil's then President Dilma Rousseff, who berated him for spying on her), but not with each other, and then watched some over-the-top fireworks display that evening, which did not impress Obama or other jet-lagged heads of state. I wondered why the Russian protocol team thought that these world leaders would want to watch fireworks. But no serious discussions on Syria between our two presidents occurred. We were nervous. In an interview with NBC's Chuck Todd the next morning, I commented, "We've had a lot of non-progress on a lot of fronts."[33]

On the second day, Obama finally talked to Putin before lunch. They discussed in detail the idea of the U.S. and Russia working together to remove all chemical weapons from Syria. Putin made clear that he could persuade Assad to cooperate. But in return, we would have to refrain from bombing Syria. That was the deal on offer.

For months, Putin and his team had dismissed our intelligence and scoffed at our worries about chemical weapons. The specter of another American airstrike against a Russian ally in the Middle East, however, focused their minds. Assad also became more willing to discuss this idea because of the threat of military action against him. "Coercive diplomacy" is not an oxymoron.[34] Sometimes it's the only method to achieve outcomes.

After we got a quick readout from the president of his conversation with Putin, Rice met with Ushakov, focusing on Putin's proposal. Susan promised Yuri a swift answer.

Later that day, on another long drive, this time to the president's meeting with civil society leaders, we discussed with Obama Putin's offer. It was a big decision. The earlier, quiet consultations between our governments had laid the groundwork for such a deal. But now Putin was engaged, and had made a concrete proposal. Not without reason, Susan worried that the Russians might simply be stalling for time. After Obama withdrew his request for authorization of the use of force pending before Congress, and momentum for military action waned, would Putin double-cross us by feigning some excuse about not being able to convince Assad to cooperate? I thought that this bad scenario was a legitimate concern, in part because I always believed that we overestimated Russia's influence over the Assad regime. However, I explained to President Obama that this Russian offer of cooperation had one big difference — it came from Putin personally. Putin felt at ease distorting facts and manipulating myths, but also took pride in keeping his word, especially in interactions with other heads of state. Ushakov had highlighted this trait of Putin's to me repeatedly. I advised Obama that on this deal, Putin might be more likely to follow through. But to keep the pressure on, we had to maintain the specter of airstrikes as long as we could.

At some point in the discussion, Obama stopped listening and began staring out the window at the Russian countryside. He was thinking. Putin had offered him a way to avoid losing a congressional vote on military action, and a risky military intervention even if Congress did support him. Removing Assad's chemical weapons stockpile could make both Syria and the region safer, including our ally, Israel. As we pulled up to the next event, I sensed that Obama was considering this proposal the better of very bad options.

Even though tensions between our governments were mounting, Obama still continued to practice dual-track engagement, and squeezed in a meeting with civil society leaders on his way to the airport. He was the only G-20 leader to do so. Just like his first meeting with Russian nongovernmental leaders in Moscow in July 2009, Obama drew energy and hope from these brave activists fighting for LGBTQ rights, media freedom, and the environment in an increasingly autocratic Russia. Obama was one of them; he was on their side, even if there was little he could to do to help them.

On their way home, Obama asked Rice to call Kerry and update him on Putin's proposal. They would chew on it over the weekend and convene on Monday to decide how to proceed. In a press conference on Monday in London, Kerry got ahead of his skis. When a reporter asked him if there was anything Assad could do to avoid attack, Kerry suggested "that a military strike on Syria could be averted if the Assad regime handed over its chemical weapons."[35] That was a mistake. No decisions had been made regarding such a quid pro quo. Kerry knew immediately he had spoken prematurely and tried to walk it back, adding the clause, "he [Assad] isn't about to do it, and it can't be done." But his misstep forced the issue. A decision on Putin's proposal had to be made quickly.

Obama was not worried about how failing to follow through on his red line would affect his credibility. As he explained later, "Dropping bombs on someone to prove that you're willing to drop bombs on someone is just about the worst reason to use force."[36] As already discussed, he was worried about what to do the day after strikes. Acting alone, without congressional approval, meant that our administration would be blamed for all that went wrong after the first bombs were dropped. The joint project with Russia gave Obama a better option, and he grabbed it. As he elaborated, "This framework provides the opportunity for the elimination of Syrian chemical weapons in a transparent, expeditious, and verifiable manner, which could end the threat these weapons pose not only to the Syrian people but to the region and the world." The agreement, Obama maintained, was an "important, concrete" step that would prevent the proliferation of these weapons around the world.[37]

And it worked, or so it seemed at the time. Assad acquiesced, and we in turn worked together with the Russians and other partners to remove Syria's chemical weapons, and then destroy them.[38] A year later, according to the Organization for the Prohibition of Chemical Weapons (OPCW), "almost 98 percent of Syria's declared stockpile of 1,308 metric tons of sulfur mustard agent and precursor chemicals had been destroyed."[39] We proved to ourselves and to the world that the United States and Russia could still work together on the hardest of security issues when our interests were aligned. That fall, Assad also signed the Chemical Weapons Convention, which at the time also seemed like a major achievement.

Obama paid a reputational price, in both the United States and the world. He had drawn a red line, but then did not act when Assad crossed it. One of his former secretaries of defense, Robert Gates, called this a "serious mistake,"[40] and another, Leon Panetta, said, "It was important for us to stand by our word and go in and do what a commander in chief should do . . . [Obama] sent a mixed message, not only to Assad, not only to the Syrians, but [also] to the world."[41] At the time, I too believed that we were making a mistake in not acting, not because I worried about the reputational effects, but because I wanted us to do more — do something — to slow Assad's slaughter of innocent civilians. Targeted strikes against his planes, helicopters, and airports would have diminished his ability to stage attacks, at least temporarily, and maybe could have helped pressure him to become more serious about negotiating a political transition, or so I believed at the time. With limited airstrikes, we could still have reached an agreement with Assad to remove his chemical weapons. We could have threatened to keep bombing until he agreed to surrender those weapons.

The chemical-weapons-removal deal did not stop the war. Assad used other ways to continue killing innocent civilians. Two years later, our partner in brokering the chemical weapons deal actually entered the war dramatically and forcefully to help keep Assad in power.[42] Four years later, in April 2017, Assad used chemical weapons again, killing scores of innocent civilians.[43] Obviously and tragically, Assad did not fulfill his 2013 commitment to eliminate all of his chemical weapons. Putin must have known.

As this book goes to press, the warring factions inside Syria have not agreed to a permanent peace settlement and seem far away from agreeing to a pact to guide a political transition. Assad remains in power, stronger today than he was while I was in government. In the meantime, over a half million people have died, and over twelve million refugees have scattered throughout Syria, the Middle East, and Europe.[44] In parallel, the terrorist group ISIS expanded its territorial gains in Iraq and Syria, eventually compelling the United States to intervene. Since August 8, 2014, when Operation Inherent Resolve began against ISIS, we have launched some twenty airstrikes per day in Iraq and Syria, at an average daily cost of $13 million.[45] With our allies on the ground, we succeeded in retaking ISIS-occupied land in Syria, but to whom do our allies and we hand over this freed territory? Back to Assad?

I still believe that we should have pushed harder for Assad's ouster in 2011, armed the moderate opposition in a serious way just as soon as the political standoff turned violent, bombed Assad's airports and airplanes after he used chemical weapons in August 2013, and enforced a no-fly zone early on in the conflict, before the Russian air force entered the war. But I am not sure if any of these alternative strategies would have produced better results. Maybe, as President Obama has argued in retrospect, these policies would have produced even worse results. "This idea that we could provide some light arms or even more sophisticated arms to what was essentially an opposition made up of former doctors, farmers, pharmacists and so forth, and that they were going to be able to battle not only a well-armed state but also a well-armed state backed by Russia, backed by Iran, a battle-hardened Hezbollah, that was never in the cards."[46] In retrospect, Obama also has been dismissive of the positive consequences for peace of American military action in response to Assad's use of chemical weapons: "We could not, through a missile strike, eliminate the chemical weapons themselves . . . and what I would then face was the prospect of Assad having survived the strike and claiming he had successfully defied the United States, that the United States had acted unlawfully in the absence of a UN mandate, and that that would have potentially strengthened his hand rather than weakened it."[47] And Syria's sophisticated antiaircraft weapon systems raised the possibility of American casualties from a sustained bombing campaign.[48] There were no good or easy options. At the same time, I remain haunted by Obama's own observation in his Nobel Peace Prize acceptance speech in 2009: "Inaction tears at our conscience and can lead to more costly intervention later."[49] That's exactly what happened in Syria.

21

DUELING ON HUMAN RIGHTS

After Putin returned to the Kremlin, our engagement on human rights issues became a lot tougher. Medvedev saw himself as a cautious modernizer, even regarding political matters. Putin did not. Nor did Putin feel the need to prove to us his democratic credentials. Instead, he saw our two systems of government as morally equivalent, and therefore maintained that the United States was in no position to lecture him. He controlled his press, so Obama must control ours. We violated human rights in the defense of our national interest; so did he. His elections guaranteed that one policy course remained consistent; so did ours. We promoted our values abroad in the name of our national interests; so did the Kremlin. For Putin, there were no white hats and black hats. We were all the same: practicing double standards, preaching about values to camouflage the pursuit of our own national interests, and deploying propaganda to weaken foes. In 2012, however, Putin stopped playing defense and went on the offensive regarding our competition over values. He's been pushing back ever since.

In 2012 Russia hosted the annual Asia-Pacific Economic Cooperation summit in Vladivostok. We'd hoped the event would present an opportunity for Obama to meet with Putin, but the APEC conference was scheduled for September, just a few days after the Democratic National Convention. We encouraged the Russians to move the meeting date back until after the November presidential election, promising to make an Obama stop in Vladivostok his first trip abroad after reelection, but they refused, citing weather (Vladivostok is much more pleasant in Sepetember than November!)—another opportunity lost for getting our two heads of state together. In Obama's place, Secretary Clinton would head our delegation. It would be her last visit to Russia as secretary of state. It was not a triumphal bow, but a clear reminder of how far relations had devolved.

Vladivostok is a very long way from Moscow—a nine-hour journey east by plane. A visit to Vladivostok by the secretary of state represented a major challenge for the U.S. Embassy. Thankfully, the United States had a consulate in the city, which helped us line up the cars, bottled water, snacks, and an espresso maker to help make the secretary's delegation comfortable during its stay. Those whose job it is to worry about such things also worried about the accommodations. Like the rest of the participants at the summit, Secretary Clinton would be staying in a dormitory, albeit in a professor's suite, on which the Russians were still finishing construction just days before our arrival. The rest of us had to duke it out over who would get a single, and who had to double up in the student rooms. (I was lucky and got my own room, but just barely!)

But the biggest drama in the lead-up to Vladivostok was not over coffeemakers or accommodations, but face time with the Russian president. We requested a meeting between Clinton and Putin. The Kremlin said no. Clinton was not a head of state, they reminded us, and therefore fell to the bottom of a long queue behind the leaders of nations attending the summit. Everyone in the U.S. government assumed that I could solve this problem. These were the kinds of assignments I dreaded. I found begging for meetings demeaning: a sign of weakness for the United States. If Putin did not want to sit down with one of the most powerful people in the world, then that was his mistake. And how many countries at APEC had more business with Russia than the United States? Putin's foreign policy advisor, Yuri Ushakov, kept insisting that the schedule was packed, until the day before Clinton's arrival, when he ended my cajoling by offering us a pull-aside with Putin on the edges of the cocktail party before the opening dinner—standing, not sitting, and with no staff.

I met Clinton's plane at the city's new airport, and the two of us rode in her car into town so I could begin briefing her on the trip. After reporting the good news about the stand-up bilat with the Russian president, my main message to the secretary was to expect little from Putin, Lavrov, or the Russian government on our agenda for the rest of her time in government. We had to stop pretending that we were going to turn this relationship around. Putin was paranoid. Putin saw us as the enemy. Months after the large demonstrations against him had ended, Putin still worried irrationally about domestic instability, and continued to blame us for stirring social unrest against him. We would continue to engage him on issues of mutual interest, but we needed to scale back our expectations of what was possible. Clinton agreed completely. Analytically, we were on the same page.

As it happened, Clinton got more face time with Putin than expected at the APEC summit. During their "stand-up," they even sat for a while, in a corner on two couches with their translators. That evening at dinner, Clinton was seated next to Putin, in accordance with APEC protocol that the previous, current, and future summit hosts sit together, and we had hosted the last APEC summit in Hawaii. Sitting at a table right behind them, I was able to observe their exchange of pleasantries. And it did seem pleasant. Putin could be charming when he wanted to be. Later that evening, the secretary of state recounted for us a story Putin had told her about his family's experience during the siege of Leningrad. His father had been wounded; his brother died. His mother barely survived. Putin told the dramatic story of how his father found his mother on the street one day in a pile of corpses, pulled her out, and nursed her back to life. Putin's story made a powerful impression on Clinton, and maybe generated some rapport between the two of them. Putin and Clinton made no progress on any substantive issues, but the positive personal interaction further convinced me that we should get Obama to Moscow for a serious summit with the Russian president the following year.

Regarding substance, Clinton's most important meeting in Vladivostok was with Lavrov. As our delegations settled in, the secretary of state pulled out her cards; Lavrov assembled his tiny pieces of paper. Lavrov soon began to doodle, as was his habit, sometimes producing elaborate drawings by the end of a long meeting. They began working through their lists. Midstream in the conversation, Lavrov announced, matter-of-factly, that the Russian government expected the United States to close our U.S. Agency for International Development (USAID) mission in Russia by the end of the year. After Clinton queried a second time, Lavrov confirmed this bombshell. Clinton smartly stalled for more time and information, asking if Lavrov would put his demand in writing. He obliged. The following week, I received a detailed letter from the Ministry of Foreign Affairs outlining the Russian ultimatum.

Later that day, we huddled to discuss Lavrov's demand. For those in Clinton's inner circle who did not closely follow this issue, the Russian request did not seem unreasonable initially. USAID's usual mission in the world was economic development in poor countries. Russia was developed. USAID had been working in Russia for twenty years. Maybe it was time to declare victory and move on? At one point in the discussion, Clinton herself said that she would be happy to reallocate the USAID funds earmarked for Russia to more needy countries in Africa.

But in 2012 USAID activities in Russia were different from those of most other missions around the world. In the 1990s, USAID in Russia focused primarily on fostering economic reform and economic development, but shifted its shrinking resources over time to a greater focus on democracy-assistance programs. By the time I arrived in Moscow as ambassador, roughly half of its programs in Russia focused on civil society, rule of law, or other programs aimed at fostering democracy and human rights. And that's why Lavrov wanted them to end.

Upon returning to Moscow, we continued to deliberate on our potential response. Defiance was one option. We could just continue business as usual and dare the Russian authorities to close USAID down. The idea appealed to me emotionally, but I did not sense much of an appetite back in Washington for provoking more confrontation with Russia over the issue, especially since we were not on solid legal ground, as the bilateral agreement that had allowed USAID to open in Russia two decades ago had lapsed. We eventually concluded that we had to submit to Lavrov's instruction, but would slow roll our execution.

Announcing this news to the seventy-three employees — thirteen Americans and sixty Russians — at USAID was difficult.[1] Some of the Russian employees had worked for USAID for two decades. They were do-gooders who believed that USAID was helping to make Russia a better place. That their own government didn't appreciate their efforts stung. As one woman lamented to me, USAID was performing tasks that "normal" governments are supposed to do for their citizens. They should be giving USAID employees medals, not pink slips. Tears were shed. Vodka was drunk. The Russian employees at USAID were nervous about their futures. The American USAID employees would be reassigned to work in Washington or in other countries, but our Russian employees had no such exit options. The anti-Americanism whipped up by Putin and his state media made getting a new job incredibly difficult for them. They would be labeled traitors and spies.

We needed to stand by our people, as well as look for new, creative ways to continue the work. Most immediately, I asked our USAID team to spend as much money as quickly as possible before closing their doors. Give the good people big grants. I also promised that our human resources team would do everything we could to help everyone find a new job. We also made it easier for USAID employees to immigrate to the United States.

We then had several emotional meetings at Spaso House with USAID grantees and partners. Some reported that they simply could not survive without USAID funding. Others were defiant. The new Russian law on regulating "the Activities of Non-profit Organisations Performing the Functions of a Foreign Agent," the so-called foreign agent law, already had increased the risks of receiving assistance from the U.S. government, so USAID's closure had simply accelerated the efforts of such organizations to find new domestic sources of support. Some civil society organizations, however, never found new sources of funding. They don't exist today — exactly the outcome Putin desired.

We then turned our efforts to finding new ways to support civil society groups. We gave big block grants to American NGOs headquartered in Washington, who could then continue their work with Russian partners, though usually outside Russia. We created a peer-to-peer dialogue program, run out of the embassy, to provide grants to projects involving Russian-American collaboration.[2] And eventually we launched a new foundation based in Prague, in partnership with the C. S. Mott Foundation, the Oak Foundation, and the Swedish and Czech governments, whose mission was to support Russian civil society from outside the country.[3] Some in the U.S. government opposed these innovations, arguing that Putin eventually would go after these operations as well. I agreed that he would probably do just that. But we had to show a readiness to defy him. Maybe the Kremlin would be playing whack-a-mole with us for the next twenty years — we set up a new program, they shut it down — but we had to try, if only to show solidarity with those civil society activists still trying to work in Russia. New technologies made it harder for governments to assert control over their citizens and more possible to support civil society from abroad. Therefore, we needed to invest more in the transmission of ideas from without via the internet. We pursued these new ideas with conviction, while also understanding clearly that working to support human rights and democracy in Russia was going to be a lot harder now with an insecure Putin back in the Kremlin.

As our efforts to support Russian civil society became more difficult, we still hoped that deepening economic ties might emerge as an agenda of common interest in the new Obama-Putin era. We had completed a big piece of business toward this end when we helped Russia join the World Trade Organization, ending decades of negotiations. At the time, the completion of this negotiation seemed just as monumental as the New START treaty. But there was one hitch. For our companies to benefit from Russia's WTO accession, we had to end the application of the Jackson-Vanik amendment to the Trade Act of 1974, the very policy change that I had advocated in a high school debate back in 1979. This legislation denied permanent normalized trade relations (PNTR) to countries that did not allow the free emigration of Jews and had nonmarket economies.[4] Historians have debated whether Jackson-Vanik worked or not. I think it did work. But after the collapse of the Soviet Union, the issue was moot; Jews could emigrate freely, and they did so in record numbers. Logically, Gorbachev, Yeltsin, and Putin asked their American counterparts to get rid of Jackson-Vanik. But repeal required congressional legislation, and leaders in the U.S. Congress did not want to give something for nothing. Over the next twenty years, various senators and representatives added all sorts of new conditions to the repeal of Jackson-Vanik as applied to Russia, everything from increased chicken exports to ending the war in Chechnya to greater democratization. The issue was stuck.

Once Russia joined the WTO, Jackson-Vanik put the United States in a particularly awkward position. If we were still applying trade restrictions when Russia joined, we would be in noncompliance with our WTO obligations. Russia, in turn, could discriminate against our companies.

After Russia obtained WTO membership, therefore, our legislative affairs teams at the White House, the State Department, and the office of the U.S. Trade Representative (USTR) shifted into high gear on Jackson-Vanik. It was a hard fight, as no one on the Hill was prepared to do Russia any favors, even if American companies would be the biggest losers of doing nothing. And then our legislative experts discovered a lifeline — the Magnitsky Act. Senator Ben

Cardin and some allies had introduced this bill to punish those Russian officials involved in the wrongful death of Russian lawyer Sergei Magnitsky. After being arrested for trying to defend the property rights of American businessman Bill Browder, Magnitsky had died in captivity. The Russian government claimed Magnitsky's wrongful death was an accident; Browder claimed that the Russian government had murdered his friend. Bill reorganized his whole life to avenge the death of Magnitsky, bringing the same burning drive and public relations savvy that had made him one of the most successful foreign investors in Russia to his campaign to make the Russians pay for what they had done to his lawyer.[5]

While working at the White House, I met with Browder from time to time. We had known each other for decades; in my travels to Moscow I often checked in with him, as he was very well informed about Russian political and economic matters through management of his hedge fund, Hermitage Capital. In one of these meetings he discussed his idea for new legislation that would ban the individuals responsible for Magnitsky's death from travel to the United States, and freeze their assets in America. He gave me the names of several dozen people he wanted to see on what later became known as the "Magnitsky list." Bill had done his homework.

I agreed that something had to be done. Medvedev eventually pushed through a new law that reduced the categories of white-collar crimes for which those accused could be held in pretrial detention.[6] That was progress, but not enough. Publicly, Russian officials brushed off Magnitsky's death as an accident. Privately, these same officials reminded me that Americans also die in pretrial detention. But they continued to insist that Magnitsky was a criminal and, in a truly grotesque gesture, convicted him after his death. They then tried to arrest Browder, now living in the United Kingdom, through the use of an Interpol red notice.[7] Moved by Browder's passion, I began working on a U.S. government response. I quickly learned that we could put people on our visa-ban list for human rights violations, without enacting new legislation. In fact, President Obama strengthened our authority to do so, with an executive order he signed in August 2011.[8]

We had the intelligence community, State Department, and Treasury scrutinize the list that Browder had given me to see if we had enough evidence to make these individuals subject to our visa ban. Of the sixty or so people on Bill's list, we identified an initial dozen to ban from travel to the United States. After considerable deliberation, we simply did it, but without making our action public.

When I told Bill as much as I could about our actions, he thanked me but added that it was not enough. We had not announced their names, nor frozen their assets. On the former, Bill criticized our administration for not making public the list of people banned from travel. I sympathized with his point. On the latter issue, Bill and I had a philosophical disagreement. I did not believe that the U.S. government should be able to seize individuals' private property without due process. They should have the right to defend themselves in a court of law. Bill disagreed. He vowed to push on with his campaign in Congress. I wished him luck.

I also continued to look for ways to express outrage about Magnitsky's wrongful death. When I traveled to Vladimir, Russia, in May 2010 to attend a meeting of our Civil Society Working Group, we deliberately made prison reform our top agenda item. After a heated set of exchanges on the rights of defendants and prisoners, we toured the local Vladimirsky Central prison, to observe conditions. You could smell the scent of fresh paint as we walked around the facility. During that same trip, I met Magnitsky's mother in a teary exchange over tea at the Sakharov Center. For over an hour, she shared with me her recollections of her son. "He was a good boy," she wanted me to know, as she broke into small smiles triggered by fond memories in between longer stretches of sorrow and sadness. It was the hardest meeting I ever did as a U.S. government official. Back home, on November 11, 2010, I spoke at the premiere of a film, *Justice for Sergei,* that Bill Browder had arranged to screen on Capitol Hill. Senator Cardin and Boris Nemtsov were also speakers. Even if we were not doing all that Browder wanted, we were not ignoring this tragedy.

One day in the spring of 2011, I received an email from the Office of Management and Budget, through the Legislative Referral Memorandum (LRM) process, asking for comment on the Magnitsky bill, at the time being introduced as stand-alone legislation by Senator Cardin and other cosponsors.[9] I felt like we had to respond to this draft legislation, and through this OMB mechanism, tell the U.S. Congress that we already had the authority to take action against some of these human rights offenders. A Magnitsky list already existed; they just didn't know it. I wrote our response, and then got our colleagues at State to sign off. This confidential message was delivered to those staffers on the Hill working on this legislation, but then leaked in a matter of hours. Soon thereafter, Kathy Lally accurately reported in the *Washington Post* that the "U.S. Puts Russian Officials on Visa Blacklist."[10] A few days later, on July 29, 2011, Sergey Prikhodko, Medvedev's foreign policy advisor, placed a call in to my boss, Tom Donilon, to complain.

I went in to see Tom that morning to do a pre-brief before his call with Prikhodko. He was upset. Who was the idiot who allowed these Russian officials to be sanctioned? he asked. Looking at my feet, I took responsibility. When Tom asked why he was not consulted on the decision, I responded that I didn't think it rose to his level. I tried to assuage Tom by predicting that Prikhodko was not going to overreact to such a minor set of sanctions, given all the other issues on our bilateral plate. But I was wrong. On the call, Prikhodko, a man of few words with a normally calm disposition, angrily warned of retribution. Tom defended the action, reminding his Russian counterpart that Magnitsky's death was not only an *internal* Russian matter, since Americans invested in Browder's funds. And he also vowed to not let this issue dominate our larger bilateral agenda.

When I later talked to Browder about this breakthrough — the public acknowledgment of U.S. sanctions against those responsible for Magnitsky's death — he still was not satisfied. He wanted more — more people on the list, asset freezes, and a law. I thought that I had done the right thing, even if it meant upsetting both my boss and the Kremlin. For Browder and other critics of our administration, our actions were not enough.

We then handed Browder, Cardin, and others a giant opportunity: the pending bill on repealing Jackson-Vanik. It was the perfect vehicle for Bill and his supporters on Capitol Hill to get his Magnitsky Act voted into law. To most members of Congress, granting Russia permanent normal trade relations felt like a concession, even if the real beneficiaries were American companies. Adding the Magnitsky legislation as an amendment or rider to the trade bill made it more palatable, allowing legislators to vote for business and for human rights at the same time.

In February 2012 as the legislative debate started to percolate, Senator Max Baucus visited Moscow. At the time, he was chair of the Senate Committee on Finance, which had to initiate action on this bill, and I was now ambassador. I was thrilled to host my former senator from Montana, whom I had first met while visiting Washington as a sixteen-year-old Boys Nation delegate in the summer of 1980. Repeal of Jackson-Vanik was so important to the Russians that President Medvedev agreed to meet Baucus and his delegation. Baucus came away from his trip convinced that the time was ripe to finally get this legislation passed, but he also saw the utility of adding the Magnitsky amendment, both because it was the right thing to do and because it would help gain support from Republicans for the bill. He asked me what I thought. I responded honestly that I was no expert on congressional politics, but I anticipated a serious response from the Russian government. When he asked me if the U.S.-Russian relationship could handle that, I assured him that I thought it could.

Several months later, Putin told Obama the same thing. During their meeting in Los Cabos, Obama warned Putin that he was going to accept the Magnitsky language as an amendment to the Jackson-Vanik/PNTR bill. Obama asked Putin what that might mean for U.S.-Russia relations. Putin replied that they would come up with an equivalent list of Americans to sanction, but that would be the end of it. That sounded fine to me.

On December 14, 2012, President Obama signed into law the "Russia and Moldova Jackson-Vanik Repeal and Sergei Magnitsky Rule of Law Accountability Act of 2012." I was pleased. A policy change that I had advocated in 1979 had finally happened. For decades, Russian leaders had pleaded for repeal, earlier U.S. administrations had promised to lift Jackson-Vanik, and countless task forces on U.S.-Russia relations had advocated the same. A major item on my to-do list from January 2009 was now done. I also saw the Magnitsky Act as an appropriate response to our concerns about human rights violations in Russia. A visa to the United States is a privilege, not a right. Human rights violators should be denied that privilege.

Putin responded, but not as promised. His government did place several American officials on its travel ban list. For most people on the list — former chief of staff to Vice President Dick Cheney, Dick Addington, or former U.S. Army Major General Geoffrey Miller — the Russian act was symbolic. They had no desire to travel to Moscow and most certainly didn't have any assets to freeze in Russian bank accounts.[11] One might say the same about our Magnitsky list. But then, Putin escalated in a most ugly way. With Kremlin prodding, the Duma introduced the "Dima Yakovlev" law, prohibiting Americans — and only Americans — from adopting Russian children. When I heard about this action, I could not believe it. We anticipated a symmetric reaction to Magnitsky. This counter was asymmetric.

Whereas most orphaned or abandoned children in the United States are placed in foster homes, Russia maintains orphanages. Large populations of Russian children — in 2012 the number was 650,000 — spend their entire youth in these institutions because traditionally Russian families rarely adopt.[12] Consequently, over the last two decades, Russia had become a prime destination for American families looking to adopt. Nearly fifty thousand Russian children had started new lives in the United States before this law came into effect.[13] A very small number of these adoptions did not work out; an even smaller number ended in tragedy. In passing the ban, Russia focused on cases of purported abuse and neglect. The law's namesake, Dima Yakovlev, had died horribly in 2008 while left unattended in a car for several hours. To make matters worse, his American adoptive father was acquitted of involuntary manslaughter. But the overwhelming majority of Russian adoptees had found loving homes in the United States.[14] To cut off these opportunities to homeless Russian children, many of them with severe disabilities, seemed incredibly cruel.

Even Lavrov agreed, arguing passionately that the ban was "the wrong thing to do." He added, "International adoption as an institution has a right to exist . . . What really needs to be done is to make sure adoptions are carried out in a civilized manner."[15] Lavrov's statement encouraged me. Maybe there was a chance that this legislation would not move forward. Incredibly, the bill also triggered public demonstrations in Russia against the pending legislation. That too fueled hope.

The criticism by Lavrov and other prominent Russians had no effect, however, because this initiative was coming from the very top. It flew through the Russian parliament and broke the hearts of hundreds, maybe thousands, of Americans. I know because I spoke to many of them. They could not understand how Putin could be so heartless. They also could not understand why Obama, their senators, or I could not help them. "We are the most powerful country on earth," one prospective father shouted at me over the phone. Why couldn't we get his son out of Russia? Senators asked me the same. I felt helpless and angry.

Especially despondent were those Americans who already had met their children, but not yet completed the adoption procedures. Russian rules required that future parents complete several bonding sessions in Russia before finalizing an adoption. When this new law came into force, dozens of American parents were midstream in this bonding process. These people were not going to abandon their children without a fight. I vowed to fight with them.

The new law contained ambiguity about what was supposed to happen to those children already in the final stages of adoption. Russian adoption agencies and regional courts, which ultimately had to approve every adoption, were confused. The initial response was to freeze all cases. My embassy team located nearly all American families in Russia at the time waiting for court approvals of their adoptions. We then traded information about what arguments were working with local authorities. It became clear to us that some regional court officials shared our view about the law. So we worked with them. We pushed the Ministry of Foreign Affairs and other federal authorities to give us more information about how they intended to enforce the law, and then used whatever tidbits we obtained to bolster the arguments of families fighting for custody of their children in cities scattered throughout giant Russia. Three families were holed up in a small apartment in the Jewish Autonomous Republic located in the far reaches of Siberia, near the Chinese border. It was January, which meant short, cold days there in a city with few modern amenities, but these families were determined not to leave town without their children. Periodically, I called them, and several others around the country, to offer empty updates and words of encouragement. I sometimes shed a tear with them. Russia could be a cruel place.

Weeks into this drama, there was a breakthrough. One brave judge determined that the law allowed the completion of the adoption if the family had held their court hearing before January 1, 2013. We told other families about this precedent, and encouraged them to schedule their hearings as soon as possible. I informed Ryabkov at the Ministry of Foreign Affairs and pressed him to urge his counterparts to enforce their law the same way throughout the country. It worked. Eventually we got nearly seventy children out of legal limbo and on their way to new, more hopeful lives in America.

Before they could leave the country, all of these families had to bring their new children to the embassy to get travel documents. Each day during this tumultuous period I tried to make it over to the consular section to meet these new parents, who had endured so much. There were often emotional embraces — hugs, tears, handing out of little flags to our new Americans. Many of the children I met had disabilities, sometimes multiple and severe disabilities. In a country where adoptions of even healthy children were rare, children with Down syndrome were going to spend the rest of their lives in institutions. I had the chance to meet a few of the very lucky ones, who had just won the golden ticket to join loving families. Americans with giant hearts, often motivated by their religious beliefs, were saving these precious kids with special needs. Helping these families unite was some of the most rewarding work I did as ambassador.

Years later, some parents still send me photos of their growing children in iconic American scenes — visiting the Golden Gate Bridge, or enjoying a Halloween party. I also continue to hear from those aspiring parents who were less fortunate. Every now and then, one still pings me on Twitter or writes me an email, asking me not to forget "their" children, hostages in a geopolitical dispute not of their making. I still feel helpless. And I am still angry at Putin for all the pain he caused.

Setbacks and failures punctuated U.S.-Russia relations in the first half of 2013. We were trying to minimize the damage from several Russian unilateral decisions such as the closing of USAID and the adoption ban. We also were failing to make progress on arms control, Syria, or trade and investment development. Anti-American propaganda continued to pollute Russian airwaves. During this period of constant setbacks, I began to question my earlier advocacy for a presidential summit in Moscow in the fall of 2013. I still believed that Obama and Putin needed face time. If we were to have any chance of slowing down, let alone turning around, the negative trajectory in U.S.-Russia relations, we needed a full-blown summit with all the bells and whistles. Selfishly, I also wanted to host Obama. I wanted him to come to Spaso House, to meet some of my Russian friends, old and new. My Russian and American basketball buddies dreamed of having him come shoot some hoops with us at our Sunday-evening game at the embassy gym. And most importantly to me, I wanted to introduce President Obama to my new "teamers" at the embassy. My embassy colleagues were working really hard under extremely difficult, tense conditions. They deserved a POTUS visit.

So we pressed on with our Russian counterparts, looking to scrape together enough deliverables to justify a two-day visit to Moscow. We tried to leverage several visits by senior officials to build momentum toward the summit in September. In May, Attorney General Eric Holder traveled to Moscow to meet with his law enforcement counterparts, including Minister of Internal Affairs Vladimir Kolokoltsev, and the head of Russia's National Investigative Committee, Alexander Bastrykin, to try to foster cooperation between their organizations. I was not a fan of either the Ministry of Internal Affairs (MVD) or the Investigative Committee, but we were trying to engage on issues of mutual interest. FBI director Robert Mueller came to discuss deeper cooperation between the FBI and the Federal Security Service — the FSB — in combating terrorism. These two agencies have a tense history, since the FBI seeks to stop Russian espionage in the United States and the FSB carries out the same assignment against American spies in Russia. But the FBI and FSB also have a similar task within their countries — counterterrorism, and at the time we believed we could be doing more together. Just before Mueller's visit, two Chechen immigrants in the United States originally from Russia, the Tsarnaev brothers, executed a heinous terrorist attack against runners in the Boston Marathon, killing four (three runners and later a police officer during the manhunt) and wounding dozens more.[16] A dispute raged, sometimes between our two governments and sometimes in the public, over the level of intelligence sharing between the FBI and FSB leading up to the attack.[17] We claimed that the Russians did not give us sufficent information. The Russians claimed that they gave us useful information, which we ignored. After the attack, however, the FSB supported the FBI's investigation in Russia. My colleagues in the Bureau's office at the embassy said that they had never enjoyed such access to witnesses for an investigation ever. Mueller came to town to thank FSB head Alexander Bortnikov for that help, as well as look for ways to cooperate further. I viewed this set of issues as potential deliverables for our September summit. We didn't have much else.

CIA Director John Brennan also visited Moscow. I was happy to greet my old White House colleague in Moscow, both of us now with new jobs. John met the head of the FSB, Bortnikov, at the FSB's main guesthouse in downtown Moscow, just across the street from "Lubyanka," for decades the infamous headquarters of the KGB before that Soviet organization split up into different intelligence entities. (Outside the main meeting room was a small museum of KGB history and paraphernalia. It was bizarre to watch the head of the CIA glance over these Soviet spy artifacts.) The discussion focused mostly on terrorism. In many domains, the CIA and the FSB remained enemies, but regarding terrorists, specifically al-Qaeda elements operating in the Middle East and the North Caucasus, we shared some common objectives. As the Sochi Olympics approached, we had even more incentive to share intelligence, coordinate, and cooperate. We did not like the fact that Russia and the Olympic Committee had chosen to hold the games only a dozen miles from a war zone. For decades, the Russian army and police had been engaged in a major counterinsurgency campaign in the North Caucasus. The Kremlin won the war in Chechnya, but the rest of the region continued to endure hundreds of deaths from terrorist attacks every year.[18] Not surprisingly, those in our government responsible for the security of Americans at major sporting events didn't like the location. But at this point, we had to make the best of it. We were not boycotting. American athletes and their fans were going to come to Sochi, no matter the threat level. Cooperation with Russia, therefore, was key and John was advancing that cause.

That evening we drove out to a vast, secret compound where Russia trains international spies, to meet with Mikhail Fradkov, the head of the Foreign Intelligence Service (SVR), the rough parallel to our CIA. Security was tight. Again, the very fact of this meeting felt surreal — the CIA director was doing vodka shots with the SVR director. This conversation focused more on swapping assessments of various world hot spots — Syria, Iran, Afghanistan, North Korea. It was an analytic exchange, almost like a Stanford seminar. Brennan knew his brief, and I think made a positive impression on his Russian counterparts. At some point, one of the questions from our Russian hosts had embedded in it a subtle criticism of Obama. Brennan shot back, not giving an inch. I was again impressed. It felt good to have such a competent, confident, and principled CIA director on our team.

Later in the evening the conversation drifted to me. I made a weak attempt at humor by noting the "generous" protective detail their comrades at the FSB seemed to provide me everywhere I went. Fradkov said that they would be delighted to increase it, since I always kept an interesting calendar. Then he said, as I remember, to John, "The ambassador is a *boyets*" — a fighter. "That's what we like about him." I am not exactly sure what he meant by that comment. I decided to take it as a compliment.

This traffic of senior American officials was designed to build a positive agenda leading up to the planned presidential summit in Moscow in September. We still believed in engagement, even between agencies like the CIA and FBI that typically did not engage much with their counterparts. We were still looking for common interests or mutual projects that could help us reverse our negative course. As an analyst, I remained skeptical. But as a policymaker, I understood that my job was to keep trying.

After Holder, Mueller, and Brennan, I had one more visitor from the U.S. intelligence community whose visit to Moscow would have a greater impact on Russian-American relations than all these previous guests combined — Edward Snowden. On June 23, 2013, Snowden landed at Moscow's Sheremetyevo Airport, in transit on an Aeroflot flight from Hong Kong to Havana.[19] He never completed his itinerary.

I was only read into the Snowden file after he arrived in Moscow. His visit to Russia obviously produced giant, negative consequences for the U.S. intelligence community, triggered a major debate about the American Constitution back home, and generated tons of negative press for the Obama administration around the world, straining some important bilateral relations with close allies. For me personally, Snowden's appearance in "my country" ruined my summer and damaged U.S.-Russia relations even further. Many late evenings that summer, as I was driving back to the embassy for yet another secure video call about Snowden, I wondered why he could not have ended up in Beijing or Pyongyang.

The day of his arrival in Moscow, I quickly learned how badly the U.S. government wanted to apprehend him. NBC News reported he had stolen sixty gigabytes of secret information, and seemed intent on making it public.[20] As a former contractor with the National Security Agency, he had detailed information about the NSA's methods and technologies for gathering intelligence. If he told the Russians what he knew, he could cause even more damage to our security than he already had. So the stakes for getting him to return home were high.

As a first move, we invalidated Snowden's passport. Exactly when this decision took effect remains uncertain to me. Some say his passport was revoked before he boarded the Aeroflot flight in Hong Kong. Others say the annulment only showed up in the computer system after he had taken off. In either scenario, however, we created an excuse for the Russians to hold him in Moscow. They reported to the world that the evil Americans had violated Snowden's constitutional rights by revoking his passport, and therefore the law-abiding Russian government could not let him board his connecting flight to Cuba.

Of course, that was nonsense. The Russian government could easily have let him board that flight. They have a track record of violating international rules and norms when they decide it is in the national interest. Maybe the Cubans didn't want him — at the time, they were engaged in secret negotiations with our government to normalize relations — but why would the Russian government care about Cuban concerns? Clearly, the Russians wanted to keep Snowden around for a while. They had made contact with him in Hong Kong.[21] We would have done the same thing if a Russian intelligence officer had landed at Dulles.

Snowden was a public relations disaster for us, and a public relations coup for the Russians. Because of Snowden's revelations about NSA programs for gathering signals intelligence inside the United States, the whole world now focused on violations of human rights of Americans, not Putin's human rights abuses. Of course, the Russian government regularly monitored the telephone conversations and emails of Russian citizens. Snowden probably knew a great deal about these Russian operations, but only revealed information about the NSA's surveillance program. He claimed (in my view misleadingly, though with just enough facts to sound credible) that the U.S. government was spying on its citizens and thereby violating the U.S. Constitution, a message that his Russian hosts wanted the world to hear clearly and often, so they made sure Snowden had easy access to international media outlets.[22] Snowden also revealed information about the NSA's efforts to monitor the telephone conversations of our allies. The Russian government loved watching us squirm while trying to explain why we were tapping Angela Merkel's cell phone.

Initially, the Russian government kept Snowden at the airport in legal limbo, not granting him asylum, but also not handing him over to us. Whether Snowden actually spent several weeks in the hotel airport remains unknown.[23] Hundreds of journalists poked around that hotel, but no one ever saw him. We wondered if the Russian government had Snowden somewhere else and already had secured his cooperation. Such worries compelled us into a full-court press to try to negotiate his release to us.

Nearly every day, or so it seemed, for several weeks, we held a Deputies Committee meeting on Snowden, chaired by Deputy National Security Advisor Lisa Monaco at the White House. The number twos at State, Defense, Justice, FBI,

CIA, NSA, and several other agencies convened in the afternoon to discuss strategies for getting Snowden back, meaning I had to go back to the embassy every night around one or two in the morning to join my colleagues in these conversations through a secure video link. By the end of the summer, I was exhausted.

It quickly became clear that our best channel of communication for dealing with this crisis was between the FBI and FSB, Mueller and Bortnikov; Snowden was an intelligence, not a diplomatic, issue. Mueller and Bortnikov spoke frequently. Mueller's main "ask" was always for Russia to hand Snowden over to us. We did not have an extradition treaty with Russia, but did on occasion ship back to Russia criminals we had arrested; Moscow had done the same for us. We argued that Snowden was accused of breaking our laws. By holding him, Russia was interfering in our internal legal affairs.

Our second ask was to get Snowden on a plane out of Russia. The destination did not matter. We determined that almost any country in the world was a better place for him to be than Russia. Moreover, we looked at most routes out of Moscow, and decided that we could detain him almost anywhere that had a direct flight from the city. Ideally, we wanted him to land in a country with which we had an extradition treaty, but we weren't going to be picky about it. Even Cuba would be better than Russia.

Bortnikov pretended to take Mueller's concerns seriously. After weeks of placing multiple phone calls, however, the Russians never responded positively to any of our requests. They were playing us.

We at the embassy tried to contact Snowden directly. At a minimum, we wanted to confirm he was not being held against his will. Our consular officers routinely meet with Americans held in Russian jails, to ensure their proper treatment and access to legal assistance. We framed our offer to meet with Snowden in similar terms. He refused to engage. We also sent messages to him through his Russian lawyer, Anatoly Kucherena. Through this channel and others, we tried to make the case to Snowden that his cause, his reputation, and his long-term future were best served by going home and defending himself in a court of law, judged by a jury of his peers. He could use the trial to make all of his arguments directly to the American people. He risked jail time, but he had many supporters in the United States. He would be a national hero to many. Living out the rest of his days in Moscow made that less likely. His surveillance concerns were not in keeping with a country that grossly violated its citizens' privacy rights. It would be like fighting for civil rights in the United States in the 1960s while living in exile in apartheid South Africa. Back home, the Department of Justice engaged with Snowden's American lawyers to promise him a fair trial, "full due process and protections," and no special tribunals. Snowden said no.

The last drama twisted around Snowden's legal status in Russia. We warned the Russian government that if it granted Snowden asylum, Obama would not attend the Moscow summit planned for two days in September. When I presented this threat to Ushakov, he was exasperated. In his view, we had serious issues to discuss at this summit — Iran, Syria, missile defense — that should not be held hostage to a minor dispute over one defector.

I suspect that Putin believed we weren't serious about pulling out of the summit. Such an action hadn't happened in fifty years.[24] This was Obama, after all. He always wanted to talk, always wanted to engage, or so they might have reasoned. The Russian government thought we were bluffing. During one meeting with Ushakov, I discussed plans for the summit, even while reminding him that Obama would not participate if Russia granted Snowden asylum. Ushakov asked me if I was representing the "official" U.S. position, because other senior officials, according to him, had been more equivocal in recent phone calls and meetings about this threat. I confirmed without equivocation that this was our policy. Yuri chuckled, saying I was the only one in our government who talked in such blunt terms. As when Fradkov had called me a "fighter," I wasn't sure Ushakov meant his remark as a compliment, but I decided to take it that way.

Putin eventually made his decision. He gave Snowden asylum, offered him a place to live in Moscow, and found him a job. To this day, Snowden remains in Russia.

Before the Snowden drama, our government was divided on the issue of the utility of a summit. But Putin's decision on Snowden generated quick consensus in Washington. The Principals Committee, led by Susan Rice, took the decision to cancel the summit. And then Obama went one step further. He decided that there was no need to request a formal meeting with Putin in St. Petersburg in September on the sidelines of the G-20 summit. The president would instead "catch up" with Putin during a coffee break. I was surprised by this last decision by the president. Although disappointed to miss my chance to host the president in Moscow, I supported the decision to cancel the summit. But I fully expected that Obama and Putin would meet formally in St. Petersburg. We had organized a meeting between Obama and Russian presidents at every major multilateral gathering over the last five years. Now Obama would be coming to Russia and *not* meet with the Russian president?

I was instructed to deliver this news to Ushakov. As I got in my car to do my now weekly commute to see him, I worried how he would react. We had reached a new low. Yuri expressed surprise about our decision to cancel the summit, although I am sure he already knew of our decision, since we discussed our frustrations with the Russian government's handling of Snowden on open phone lines. But he sounded genuinely perturbed by our decision not to schedule a bilateral meeting in St. Petersburg, telling me that was a big mistake. But then he turned philosophical. "We will survive," he predicted, with a sheepish grin. "We will live." Those were very Russian expressions. I appreciated Yuri's sensibilities.

As I sat in traffic, inching my way back to the embassy, I decided that day that my work in Russia was over. It would be no tragedy if I left later that year. Back in January 2009, I had joined the Obama team to *do* things, to reset relations with Russia, to close out accounts, to deepen cooperation between our two governments and societies. In early 2011, I agreed to go to Moscow to continue to do things. Now, however, we weren't doing much at all. We were barely even talking. It was time to go home.

I also believed that the Snowden affair marked the end of one of the most confrontational periods in U.S.-Russian history. I thought then that we had hit the bottom and therefore there was nowhere to go but up, though it would take years to climb out of the hole we were in. I was wrong. Things could, and did, get worse.

22

GOING HOME

In the spring of 2011, when my family and I agreed to go to Moscow, I made very clear to all at the White House that I was signing up for a two-year tour. That was my deal with my older son, Cole. We agreed that we would return to Stanford in time for him to start his sophomore year at Gunn High School in Palo Alto. After five years away, going home in the fall of 2013 would give him three years in one place to reconnect to California, settle into a new social scene, and get ready for college. Everyone in the administration who needed to know was in agreement.

At some point after our first year in Moscow, our family revisited this decision one last time. Despite all the negative dramas in U.S.-Russia relations, I loved the job as ambassador. I enjoyed the challenge of negotiating outcomes, and the honor of representing my country. I also felt like I was getting better at it. Even with the Reset over, the diversity of the portfolio made the work interesting — hosting NBA stars, the CEO of eBay, Herbie Hancock, Henry Kissinger, and the Chicago Symphony Orchestra at my "house"; engaging with the Russian and American business communities; sending analytic cables back home; briefing the president, the secretary of state, the national security advisor, and the CIA director for their meetings with Russian officials. It was a challenging but interesting job. And our sons were learning a ton, not just in school, of course, but from the experience of living abroad and representing America. Instead of driving to Burlingame, my older son was flying to Bucharest for a basketball tournament! So . . . maybe we should stay? Cole could finish high school at the Anglo-American School.

Throughout the spring and summer of 2013, Donna and I debated the pros and cons. From a beach resort in Eilat, Israel, over spring break, our family together drafted two lists for Moscow and California. In the pro-Moscow column, she wrote "basketball, Pietro [our chef!], house, travel, Dad's career, VIP, No driving, Olympics, Thailand, and Africa," the latter two being future vacation destination spots. In the pro-CA column, she wrote, "Weather, America, Stanford B-Ball, more spontaneity, more money, better food, no bodyguards, Cole driving, Mom's job, more time with kids, closer to family." It was a hard decision.

When I seriously broached with Cole the idea of staying, however, I heard no enthusiasm. Cole enjoyed our setup in Russia — no cleaning, no washing dishes, living in a mansion, as well as living in Russia. But he also had been away from home for five years. The very idea of home was beginning to fade in his memories. I worried about him having no place to call home if we stayed in Moscow another three years. That concern bothered him too. And I had promised him only two years away, and we were now on year five. So eventually, we decided to stick to the original game plan. I didn't want to damage my relationship with my son for a few more years in Moscow as ambassador. Most Russians, and maybe some Americans, would find it hard to believe that I would make such a career decision for family reasons. But anyone who knew Cole and me personally understood. That's all that mattered.

Had we been doing more productive things in U.S.-Russia relations, I would have made the argument to stay more forcefully. In the spring of 2011, Obama rightly argued that we were making history with Russia, transforming a deeply troubled relationship into a cooperative one. The first three years of the Reset were extremely productive, one of the most cooperative eras in U.S.-Russian relations ever. In that moment, I could not have left the team. But 2013 had none of the optimism of 2011, none of that momentum. The Snowden affair felt like the last straw — we were not moving anywhere good in the bilateral relationship anytime soon. I felt like I was adding value in managing the difficult issues, but that was not enough to justify living five years in Russia. I did not want "to be" an ambassador. I wanted "to do" things that would make Americans and Russians better off. If that aspiration was no longer tenable, then there was not a compelling reason to stay.

On July 21, 2013, I wrote to Denis McDonough and Susan Rice to tell them of our family decision. Susan responded, "Bottom line was Boss very cool, very appreciative of you and very supportive. DM may try to convince you to come to DC." DM — Denis McDonough — at the time Obama's chief of staff, did try to get me back to Washington to serve as the under secretary of defense for policy. I was intrigued, as that job would have given me the opportunity to expand my portfolio beyond the Russia account. But no one in my family wanted to move back to D.C. In reaffirming the original plan, Denis replied that the president's view was that I had earned my freedom to return to California, provided they could find ways to keep me involved in Obama's agenda as an outside advisor. So that was it. My family would return to Stanford in August 2013 to start school; I would stay on for the summit we were planning for President Obama in Moscow in the first week in September and then rejoin my family soon thereafter.

Washington kept asking me to stay on just a little bit longer, so I ended up agreeing to stay through the end of the year. As December approached, my bosses at both the White House and State Department urged me to stay through the Winter Olympics. Obama's name was invoked. I eventually relented, but then realized there would always be cause for another delay. Already someone had floated the idea of me staying through September. The family could spend the summer back in Moscow! they proposed cheerfully. Eventually I just said no. Seven months away from my family was all I could take. A year apart seemed impossible. So to create a credible commitment to leave, I announced publicly my departure on February 4, 2014, fittingly on Twitter and on my LiveJournal blog.[1] Prompted by Howard Solomon, the head of our political section who has a PhD in Russian literature, I borrowed from Pushkin to title my last blog "It's Time, My Friend, It's Time."

In the blog, I wrote, "I will leave Russia reluctantly. I love this job. It is a tremendous honor to represent my country here . . . And my departure will mark the end of more than five years of working for President Obama and his administration (or seven years if you count my work as an unpaid advisor on his campaign). This is hard, really hard." In this same blog I rehearsed the achievements of the Reset — New START, the Northern Distribution Network (NDN), the Bilateral Presidential Commission (BPC), sanctions on Iran, Russian membership in the World Trade Organization, expanded trade and investment between our two countries, our new visa regime, and the removal and destruction of Syria's chemical weapons, as well as a dozen lesser "deliverables." We had done some big things together with Russia during my time in government. But I also hinted that the best was behind us. In the current era, I explained, a lot of my work as ambassador was managing difficult issues in U.S.-Russia relations, be it the Russian order to close USAID, the adoption ban for Americans, or responding to false claims in the Russian media about American desires to foment revolution in the state. On major policy issues, we always seemed to be playing defense.

At the same time, I reminded my Russian readers about all the good times I had had as a guest in their country. I recalled my standing-room-only sessions at American Corners in Yekaterinburg, Vladivostok, Volgograd, St. Petersburg, and Moscow, at which thousands of Russians showed up to engage with me on everything from Syria to my broken finger. I reminisced about dancing at Spaso House, watching Maestros Spivakov and Gergiev make their magic with their orchestras, or sitting in the front row at the Bolshoi and absorbing the magnificence of the best ballet dancers in the world. I wanted Russians to know that I loved their country, even if I disagreed with some of the policies of their government. More parochially, I highlighted our internal achievements and innovations as an embassy. We had accepted Clinton's challenge to deploy social media for public diplomacy, and had substantial Twitter and Facebook followings by the end of my stay. During my time in Moscow, we also started the Spaso Innovation Series to bridge the gap between Silicon Valley and Moscow, set up student lectures (in Russian and with slides!), inaugurated peer-to-peer dialogue grants, reduced considerably the time to get a visa, and more generally, through a variety of mechanisms, tried to be open, direct, and engaged with Russian society. At Spaso House, we broke with diplomatic traditions and served hot dogs and popcorn on occasion, opened our doors to Russian citizens who had never visited the mansion before, and held virtual seminars with futurists from Palo Alto. We had hosted twenty-two thousand guests in two years. I felt proud of what we had achieved during our time in Russia.

While innovating, we did not neglect our traditional, core responsibilities as a diplomatic mission in one of the most important countries in the world. That was documented in a comprehensive assessment of our activities conducted by the State Department's Office of Inspector General during my tenure in Russia, a review that occurs at embassies around the world every five years. Political appointees at other embassies warned me about the dreaded OIG. They did not like "us" — that is, political appointees — and gave our professional staff license to go after us, or so the lore went. So I was especially pleased with our good grades from OIG.[2] We could tweet and two-step, but also write excellent cables, issue visas, and conduct traditional diplomacy all at the same time. I was pleased with all that we had done, but also realistic about what little more could be achieved.

In Russia, some state-controlled media outlets explained that Obama fired me because I had failed to foment revolution. As United Russia MP Yevgeny Fyodorov explained, "They sent him here to destroy Putin and to organize Orange Revolution here . . . and he failed."[3] Another commentator concluded that I had fallen short in forging unity among liberals, nationalists, and fascists in order to produce mass disorder, and therefore had to be sent home.[4] My replacement, John Tefft, had previously served as ambassador in two "color revolution" countries, Georgia and Ukraine. After John was announced, a new story line unfolded — Obama was doubling down on his strategy of fomenting revolution, replacing the weak and friendly McFaul with an experienced, hard-nosed professional revolutionary.

Other Russian commentators claimed that I was removed because I had failed to improve relations between Russia and the United States during my time in Moscow. For instance, Pavel Zolotarev, deputy director of the Institute of the USA and Canada at the Russian Academy of Sciences, argued that I did a "fantastic job in fulfilling my responsibilities at the White House," but didn't have enough diplomatic experience," to be successful in Moscow.[5] The downward turn in U.S.-Russia relations was my fault.

Both of these contradictory explanations were silly. Of course, Obama did not send me to Russia to organize a revolution. Nor did I single-handedly ruin Russian-American relations. (And how could my mission have been revolution against the Kremlin and improved relations with the Kremlin at the same time?) Forces much bigger than me were driving our relations in a negative direction. One Russian friend suggested to me that the outsize role assigned to me in ruining relations was an indirect compliment, since it meant that the Russians believed I had more power than I really did. No one was blaming Russia's ambassador to the United States, Sergey Kislyak, for the downward turn because no one believed Kislyak had any real influence on policy.

I was most affected by the reaction of ordinary Russian citizens. My Twitter feed and Facebook page filled up with tens of thousands of heartfelt, emotional messages expressing disappointment at my departure, and appreciation for the unique, open, and engaging way that I had performed my role as ambassador. Even some of our critics seemed genuinely sad to see me go. Most shockingly, Russian nationalist leader Vladimir Zhirinovsky called me one of the greatest U.S. ambassadors ever, adding, "He deeply loves Russia and speaks perfect Russian. He didn't want to act in the same way as the American ambassador in Kiev."[6] And the latter reason, in his view, was why Washington was replacing me. My love for Russia had gotten in the way of fulfilling my mission of regime change! As liberal politician Leonid Gozman wrote in an article around the time of my departure, "They hated him not because he is an enemy of Russia, but because he is a friend."[7]

I deliberately delayed the announcement of my departure until just a few weeks before I left, so as to minimize my lame-duck status. Nonetheless, I enjoyed a series of memorable parting celebrations. In our last one-on-one meeting, Foreign Minister Lavrov was friendly, playful, and loose. We talked about some of the big issues, but mostly just bantered informally. Lavrov and my mother are the only two people in the world who refer to me as "Michael." There was something about that strange pairing that I found appealing. I think "Sergey" liked me more as a White House advisor than as an ambassador. I didn't always play by the rules of diplomatic decorum. I know he didn't like it. That was fair. But I in turn disliked how he'd gone out of his way to make my work more difficult. I also could not respect the way that his government had treated my family. Was it really necessary to send intelligence officers to my son's soccer games? Did they really have to tail my sons' car on their way to school? Are those the actions of a "great power"? Of course, Lavrov had no control over these actions, but it was still his government. I did wonder what he thought about it all. I never asked.

Putin's foreign affairs advisor Ushakov was warm and cordial in our last meeting. We had a long history. That counted for something. His boss really didn't like me, but Ushakov understood the role I played on the Obama team. "Ush," as we called him, was always asking questions about American electoral politics, personalities, and informal networks between American politicians, so he had a better feel than most about the special circumstances that landed me in Moscow. He seemed sorry to see me go, but of course I can't say for sure how he felt. He's a "professional" diplomat, after all. That was never one of my strong suits.

My final lunch with Deputy Foreign Minister Sergey Ryabkov was an elaborate affair at the MFA's beautiful nineteenth-century mansion. We drank vodka, exchanged gifts, reminisced about the early days of the Reset, and laughed about the difficult issues we'd had to manage during my time in Moscow. By this time in U.S.-Russia relations, gallows humor was an essential ingredient in our conversations. Sergey was a true partner. Like me, he believed in the promise of the Reset, and lamented the negative turn in our bilateral relations. He also seemed to want the best for me. That day, Sergey gave me a welcome parting gift — a document allowing us to begin construction of a new chancellery on our embassy compound, a project I had been trying to get off the ground for two years — one last deliverable!

I also had a memorable farewell dinner with Igor Shuvalov, at a supermodern, hip restaurant called Turandot. He obviously was a regular there. We sat for several hours in a private room, reviewing where things stood between the United States and Russia, arguing about which leader — Putin or Obama — was most responsible for the negative turn in our relations, and searching for creative ways to get things back on track. Shuvalov was a problem solver, always results-oriented. I considered myself to be the same kind of bureaucrat, though obviously not at his level regarding either efficiency or power. That night, however, we didn't get anything done, but instead engaged in a long, even philosophical discussion about the relationship between democracy and development. Igor was always urging me to take the long view, to understand that Russia was on the right path, to see Putin as a necessary, transitional figure between dictatorship and democracy, between communism and capitalism. He still worried about the critics of market reforms who had not adjusted to capitalism — pensioners, state employees, or more generally, those still desiring a paternalistic state. Old communist ideas, reinvigorated with a new nationalism, still threatened Putin's modernizing trajectory, so Igor argued. He and his like-minded colleagues in the government shared some of the values of opposition liberals. And to his credit, Shuvalov did show up at some signature liberal gatherings, including making an appearance at the annual celebration of the Echo of Moscow radio station, hosted by the station's editor in chief, Alexei Venediktov. Shuvalov had much more in common with Venediktov than with Communist Party leader Gennady Zyuganov. I was just not sure his strategy for pursuing these liberal values was succeeding.

It was an old and familiar dilemma for Russian reformers who made the decision to work in the government. Ever since 1990, when the "liberals"—economists, human rights activists, democratic champions—had joined forces with former Communist Party apparatchik Boris Yeltsin to challenge the Soviet communist regime, they had been struggling to find the proper relationship with the state. For market reformers like Gaidar and Chubais, the choice was obvious back in 1991. They had to join the government and push their market ideas from within the state. If they did not, they believed that reforms would not occur and the Communists would return to power. Over time, the ethical baggage of remaining in a corrupt, increasingly undemocratic state would compel some of these reformers to peel off, to either join the opposition or retire. During the Putin era, many were simply pushed out. But Shuvalov had survived. However constrained, he still believed that his years of work for reform inside the government was still more valuable than orating for five minutes at a demonstration outside of government. His critics would remind me about how rich Shuvalov had become. He was corrupt, they charged, and that's why he stayed. I was not so sure. He was wealthy, but he claimed to me to have made his fortune before joining the government. The harder question was whether he was doing any good as an insider, or helping to enable bad actors in the government through his work with them. In the end, I decided that I was not qualified to pass judgment. On a much smaller scale, I too had compromised some of my values and championed policies that my government supported even when I personally did not. Each time I lost a policy debate, I had to decide whether to stay in the government or resign. For five years, I decided to stay, believing that I was making a contribution for good that outweighed the compromises to enable or support bad policies or decisions. Shuvalov, it seemed to me, faced the same set of dilemmas.

I also was treated to some heartwarming send-offs from other diplomats and nongovernmental officials. One was especially meaningful, a party for me hosted by an old friend, Vygaudas Ušackas, then the EU ambassador to Russia. After a private concert, we sat down to a dinner at which I was feted with numerous heartwarming toasts. One, from a European ambassador of a "frontline" state, was especially moving. He thanked me for saying the things that other ambassadors from smaller countries thought but did not dare to say. That was American leadership, he said. That meant a lot to me.

Of course, we organized one last blowout party at Spaso House. Several hundred people came — businesspeople, journalists, diplomats, civil society activists, a handful of government officials, and yes, "the opposition." I was especially pleased to see Duma deputy Leonid Kalashnikov, a member of the Communist Party. He had attended the infamous roundtable of opposition leaders we'd hosted for Bill Burns on the second day of my service as ambassador. And here he was, showing his support on one of my last days. Classy. He proved to me that we can differ in our views while working together cordially. Some liberal opposition leaders also attended the gathering, even though their association with me increasingly had created problems for them. I enjoyed a bear hug and warm, clever banter from my old friend Boris Nemtsov. He felt compelled to rail against the Obama administration from time to time for not doing enough for Russian democracy, but he never made it personal. In fact, he said to me that evening, "It will be great to get the old activist Mike back." I replied that I too looked forward to doing what I could to keep engaged in the fight for Russian democracy. It was the last time I saw Boris. A year later he was assassinated — gunned down from behind just a few hundred yards from the Kremlin.

Those final days in Russia brought home to me the incredible contrasts and contradictions the country has always posed for me.

My family returned to Moscow twice during my final seven months as ambassador, first for Thanksgiving and then Christmas, our first Christmas not celebrated in Bozeman, Montana. For my sons and my mother, that was a huge sacrifice. They endured the trip from Los Angeles to Moscow on a nonstop Transaero flight with nothing but old Soviet films for entertainment, to spend Christmas with me in cold, dark Russia. True love. Knowing it would be our last time together in Russia in a long while, we went all out. We dined at our favorite Uzbek place on Novy Arbat, Teremok on Old Arbat, and made one last visit to

Druzhba, the smoky, mediocre Chinese restaurant my younger son loved. We attended Maestro Spivakov's New Year's concert, a most Russian event. And Donna and I splurged and bought front-row seats to see *The Nutcracker* at the Bolshoi. The ticket prices were outrageous, but the performance was also over the top. Russians can do some things better than anyone else in the world. Staging that performance at the Bolshoi during the holiday season is one of them.

As we left the theater that evening, however, I was reminded of another side of Russia. Three cars were waiting for us, along with a cluster of nervous-looking Russian and American bodyguards. My security detail was expanded that evening and for the entire time of my family's visit because of information that had come in about another death threat against me. Of course, I never discussed these kinds of issues with my sons. But the additional cars and bodyguards assigned to me made it obvious that something was going on. I admit that the threat was unnerving. It was one dimension of life in Russia that I knew I would not miss.

Once I relented to Washington's request for me to stay through the Sochi Winter Olympics, I decided to embrace the opportunity and enjoy the games. Putin, his government, and his business friends did organize an incredible party. Allegedly, the event cost $50 billion; many Russian billionaires had to pitch in. Like most major Russian projects, they needed a big push at the end to complete the construction of all the facilities, since most were built from scratch. And, of course, they didn't finish some planned projects — including the hotel intended to house the American government delegation. Visa, thankfully, didn't use its quota of rooms and signed them over to us.

Before the games began, Western journalists wrote several stories about yellow water, packs of stray dogs roaming the streets, and defective bathroom doors. (On opening night, I too found myself briefly locked in a bathroom at the stadium.) But I was impressed with the opposite — at how much the Russians got done, at how beautiful the facilities were, and how well the games came off. There were hiccups throughout, including most famously one of the five Olympic rings failing to open during a segment of the opening ceremony. But for the most part they pulled it off. They could justifiably be proud of their efforts.

Most importantly, there were no terrorist attacks during the event. For our government, Sochi was first and foremost a security challenge. We set up a temporary satellite consulate in the city, staffed by over a hundred people. We put in place the means to assist Americans with lost passports and medical emergencies, but deployed most of our resources to detect and respond to a terrorist attack, including a plan to evacuate all Americans if needed. Thankfully, our security team spent most of their time at the Sochi Olympics completely bored, sucking down bad coffee and staring at their computer screens.

Athletic competition supplies the core drama of any Olympics, but hosts don't plop down billions just for sports. Like leaders hosting other Olympics, Putin had a message for the world — Russia is back. Russia is a great power, a modern, critical, integrated member of the international community. Our hosts wanted all of their visitors to feel welcome. They hired thousands of enthusiastic college-age volunteers, fluent in English and decked out in colorful coats. Most American visitors were pleasantly surprised by the hospitality and friendliness of their Russian hosts. Despite all the tensions in U.S.-Russian relations at the time, our athletes, officials, and tourists encountered only warmth and goodwill in the Olympic Village. The Russians had kindled the Olympic spirit. Putin even stopped by the American House one night and put on a charm offensive.

I personally had the same experience. Hundreds of Russians recognized me, wanted to take a photo with me, or get an autograph. Everyone was so friendly; all of the earlier dramas of being chased around by neo-nationalists and *NTV-shniki* seemed like a long time ago. Sochi provided the perfect end to my tour in Russia.

We Americans injected some politics into the games as well. President Obama did not lead our delegation to the opening or closing ceremonies. That governments even send delegations to these ceremonies seems a little odd to me. If the Olympics are supposed to be nonpolitical affairs, then why do they maintain the ritual of government delegations coming to opening and closing ceremonies? Obama did not attend the Winter Games in Vancouver in 2010 or the Summer Games in London in 2012 — hosted by two close allies of the United States — so no one should have been surprised that he did not make an appearance in Sochi, especially after having canceled the Moscow summit the previous fall. The Russian government, however, did not understand. They pleaded with me to get Obama to come. When I conveyed the president's regrets, they were deeply disappointed.

They then became downright angry when we announced the composition of our delegations to the opening and closing ceremonies, which included some gay athletes, including tennis star Billie Jean King, Olympic gold medalist figure skater Brian Boitano, and bronze medalist hockey player Caitlin Cahow. In June of the previous year, Putin had signed the so-called anti-gay propaganda law, which prohibited exposing children to any materials portraying nontraditional sexual relationships, which in essence criminalized a lot of advocacy in defense of LGBTQ rights. We criticized the legislation, and took several other symbolic measures. I met with several leaders of LGBTQ activist organizations to express solidarity. In August 2013, I hosted a reception for the U.S. team competing in Moscow in the world track and field championships, and invited the leaders of a Russian gay athletes club. U.S. athlete Nick Symmonds even dedicated his silver medal in the 800 meters to Russian gay athletes, stating eloquently, "Whether you're gay, straight, black, white, we all deserve the same rights. If there's anything I can do to champion the cause and further it, I will, shy of getting arrested," and then adding, "I do have respect for this nation. I disagree with their rules."[8] The following year, soon after the Sochi Olympics, our embassy offered its basketball court as one of the venues for the Gay Games in Russia, since it was a safe place to play matches.[9]

However, our decision to invite gay American athletes to join our opening and closing ceremony delegations was our boldest move. The Russian gay community loved it. The Russian government hated it, as did some in our government, arguing that we were politicizing a sporting event. I supported this White House decision. We could not just close our eyes to these disturbing antidemocratic, inhumane trends inside Russia. As we discussed internally, some of my colleagues back in Washington reminded us of the world's failure to criticize their hosts at the 1936 Olympic Games in Berlin. I did not accept that analogy, since Putinism in 2014 was not comparable to Nazism in 1936. But the analogy forced us to consider Putin's political agenda in hosting the games. He wanted the world to celebrate him and Russia. We wanted to celebrate Russia, but not Putin's policies. The composition of our delegation, we believed, achieved that delicate balancing act.

• • •

The last evening that our opening-ceremonies delegation was in Sochi, we decided to celebrate our good work (of mostly watching sports!). My embassy team tracked down some vodka and caviar and proceeded to the hotel bar to toast our last day together. At one point in our celebration, the waiter approached the head of our delegation, former secretary of homeland security Janet Napolitano, and reported that an unnamed guest at the hotel had offered to buy the whole table champagne. When our delegation heard the news, they roared with gratitude. I was uneasy, though, about accepting an expensive gift from an anonymous source. After all, we were sharing our hotel with the NBC staff covering the games; I didn't want to risk our team being the subject of a potentially compromising news story. To the outcries of my fellow delegates, I told the waiter that we would have to decline. A few minutes later, Mikhail Leontiev came marching over, pleading with me to accept his offering. Leontiev was the television commentator who had run the hit job on me on Channel One on my first day on the job. He was the one who had told all of Russia that I was a revolutionary — a usurper — sent by Obama to overthrow Putin. Now, on one of my last days as ambassador, he wanted to buy me champagne. Leontiev was insulted by my refusal and urged me to reconsider. He obviously had indulged in a little of his own champagne drinking that evening. He explained that he was just trying to be friendly. I gently reminded him of all the false news reports he had broadcast about me and my country over the last two years, which hadn't seemed too friendly to me. He brushed it all off, telling me not to take things so personally. We were all just doing our jobs, he insisted. He then went on to tell me about the terrific trip he and his daughter had recently taken to the United States. Such a fantastic country, he gushed — he who had devoted dozens of his television shows to trashing America.

The exchange left me depressed. I had seen and heard it too many times over the last two years. "It's all a game." "We don't believe any of this propaganda. Neither should you." "It's what we say and do to survive in this system." They called it pragmatism — a mind-set that allowed them to reconcile the deep contradictions in their lives. I called it cynicism. I understood it, but also felt fortunate not to have to grapple with these ethical dilemmas. By 2014, large segments of Russian elite society had become deeply cynical. No amount of understanding, no amount of appreciation of Russian culture and history, no amount of engagement was ever going to help me close my disagreement with these cynics. That evening, I finally came to closure about my decision to leave this job. My work with this regime was done.

On February 26, 2014, I met with Deputy Chief of Mission Sheila Gwaltney at Spaso House and signed the "Certificate of Transfer," putting her in charge of the embassy. I said one final goodbye to the Spaso House staff and my senior leadership team at the embassy, and got into my black Cadillac for one last time, the American flag prominently displayed. After one last set of goodbyes and a photo with my bodyguards, I went through customs and was done—done as the U.S. ambassador, done as a member of the Obama team, and done with Reset—not just the past five years of U.S. policy but the previous thirty years of believing in and working toward closer ties between our two countries. And maybe I was done with Russia. Since 1983, the longest I had ever been out of the USSR or Russia was three years. As I left this time, I wondered when I'd be back.

The day I landed in San Francisco, the world learned that Putin had invaded Ukraine.

23

ANNEXATION AND WAR IN UKRAINE

The conflict that would redefine an entire region — and permanently end the Reset — began while I was on the plane home from Moscow. Putin first annexed Crimea on March 14, 2014, then doubled down in support of the separatist movement in eastern Ukraine.

The invasion and the escalation of violence may have come as a shock to many, but the political fight over Ukraine had been percolating during my entire time as ambassador. If Iran, Syria, arms control, and human rights issues topped our list of agenda items with Russia while I was ambassador, Putin's top foreign policy priority during these same years was something completely different: the creation of a Eurasian Economic Union (EEU), Russia's counter to the European Union. From Putin's point of view, Berlin anchored and dominated the economies of Europe through the EU; in turn, Moscow could and should do the same with the states of the former Soviet Union (excluding the Baltic states, which had already joined the EU).

Moscow's efforts to lasso the states that had gained independence after the collapse of the Soviet Union and corral them into the EEU rarely made the radar of our top foreign policy makers, and Putin wanted to keep it that way. From his perspective, the EEU was none of our business. Internally, we debated how coercive were the terms of membership in this Kremlin-created club. Given the shared histories and geography of former Soviet states, an international organization that increased trade and investment between these countries could make economic sense. However, Russia's hegemonic position within the EEU allowed

Moscow to dictate, not negotiate, the rules of this new organization, while Russia's imperial traditions raised suspicions in our government about the virtues of the new Kremlin initiative. In December 2012, Secretary Clinton bluntly and unexpectedly labeled the effort "a move to re-Sovietize the region"[1] — a comment for which Lavrov berated me when I saw him shortly thereafter. Although her quip went too far, it did clarify our position internally. We did not have the focus or means to actively resist the EEU, but we were also not going to celebrate its development.

Belarus and Kazakhstan were logical candidates for membership in the EEU, as they already had joined a customs union with Russia. But the real prize was Ukraine, with its nearly forty-five million consumers for Russian imports, an economy that dwarfed the economies of Belarus and Kazakhstan combined, and a wealth of opportunities for Russian investment. Ukraine's membership was the linchpin to the EEU's future. Ukrainian president Viktor Yanukovych flirted with joining Putin's club, but also negotiated with officials from Brussels about signing an association agreement with the European Union. The latter infuriated Putin. He had helped Yanukovych become president; now was his moment to return the favor. Because he believed that Ukrainians and Russians were "fundamentally a single people," Putin scoffed that all the "emotional and turbulent discussions" about signing an association agreement would eventually give way to more rational, natural economic ties between the two Slavic nations.[2]

With increased intensity in the fall of 2013, the Russian government berated the Europeans and Americans for attempting to peel Ukraine away from Russia's orbit. Putin urged Yanukovych to consider the negative economic repercussions of uniting with Europe, asking rhetorically, "How would Ukraine benefit from joining the EU? Open markets? Well, this would make the economy more liberal. But I have no idea whether Ukraine's economy can cope with such liberalism."[3] Russian officials also warned Yanukovych that the Europeans were not serious about seeing Ukraine join their club. They'd be waiting to join for decades, just like the Turks.

In my official talking points about these dueling courtships, I reminded my Russian counterparts that the United States was not a member of the EU and played no role in these negotiations. I also emphasized that these competing offers need not be zero-sum; many countries, including the United States, entered into multiple trade organizations and treaties. So could Ukraine. We also suggested that Ukraine could become more prosperous through closer association with EU members (even without EU membership), which would create more opportunities for Russian investment and trade with Ukraine. Everyone could win.

During a one-on-one conversation on this subject, First Deputy Prime Minister Igor Shuvalov remarked that he could not tell whether I sincerely believed in this win-win optimism, or was just reading my Washington-generated talking points designed to trick Russia. But Putin was having none of it. Moscow was going to play hardball. The battle for Ukraine was a zero-sum contest, with winners and losers, and Putin was determined to win.

Analytically, our team at the embassy agreed with Shuvalov. The stakes were higher for him; Putin needed Ukraine for his EEU to succeed, while Ukraine held no similar value for the European Union. What EU countries, after all, really wanted Ukraine to join? The EU faced enough problems already, without tackling the difficult, controversial project of adding another fifty million relatively poor people into its ranks. Yanukovych also faced reelection in 2015. Signing an association agreement would require Ukraine to undertake unpopular reforms just a year before the election. He would never do that. The EU also was insisting that Yanukovych release from prison Yulia Tymoshenko, the former prime minister, whom everyone in the West considered a political prisoner. That was a high price for the insecure Ukrainian president on the eve of a presidential election. And Moscow offered Yanukovych serious money not to sign the EU agreement—money that he needed for his reelection campaign. The endgame seemed clear: Yanukovych would continue the bidding war to secure as much financial support as he could from Moscow while postponing signing the association agreement until a later date, sometime after his reelection.

Those Europeans pressing Yanukovych to sign the association agreement had a different assessment.[4] They believed that Yanukovych saw the benefits of choosing their rich European Union over Putin's beleaguered Eurasian Economic Union.[5] To make things easier for the Ukrainian president, the EU negotiators eventually agreed to Tymoshenko's release for medical treatment in Germany, with the assumption that she would never return to Ukraine, a face-saving compromise. Especially after this final compromise, EU negotiators pressed for a signing date of November 28, 2013, at the EU's Eastern Partnership

Summit in Vilnius, Lithuania. I, along with several others in our government, still worried that the slippery Yanukovych would renege, but also didn't think it would be a major tragedy if he postponed signing the agreement. Everything in the EU and diplomacy more generally, after all, is about process. The end of one failed negotiation is simply the opportunity to start a new one; no big deal.

A week before the Vilnius summit, Yanukovych announced that he was not ready to sign, asserting that there were "several crucial steps left to be made."[6] Our European partners were upset, and made a last-minute attempt in Vilnius to cajole Yanukovych to reconsider.[7] Those in our government following these negotiations also were disheartened, but I don't recall any senior U.S. official losing much sleep over the signing delay. We had dozens of foreign policy issues of higher importance on the agenda at the time.

Some days later, I met with First Deputy Prime Minister Shuvalov on another matter, but seized the opportunity to congratulate him on Russia's diplomatic victory. That's what diplomats do. In defeat, we congratulate the victor. Shuvalov lamented the price tag — $15 billion — but explained that the boss wanted this one badly.[8] At the time, I thought the Russians had won the battle but would still lose the war. After presidential elections, Ukraine would be back at the negotiating table, or so I believed at the time. Time was on our side.

Others in Kyiv were not so willing to wait. To me, Yanukovych's decision was just a bump in the long road toward Ukrainian integration into Europe. To the young, principled Ukrainian journalist Mustafa Nayyem, Yanukovych's decision amounted to treason. That day, Nayyem stopped reporting the news and began to make it. On Facebook, he blasted Yanukovych for betraying the European aspirations of the Ukrainian people and called upon his virtual friends to "be serious . . . If you really want to do something, don't just 'like' this post. Write that you are ready, and we can try to start something." In a later Facebook post, he called upon his readers to "meet at 10:30 p.m. near the monument to independence in the middle of Maidan."[9] Others made similar pleas, prompting thousands and then tens of thousands to show up. Within days, a massive crowd occupied the square, in what looked like a replay of the Orange Revolution a decade earlier. Just as in the Arab world in 2011 and Russia in 2011–12, democratic activists were taking a stand — and in so doing, inadvertently and unintentionally injecting themselves into U.S.-Russia relations.

In response, Yanukovych vacillated between conciliation and confrontation, at first ignoring the protesters in the hopes that the winter cold would chase them off the streets, but then ordering the police to clear the square. When images of police beating peaceful protesters triggered even bigger crowds, Yanukovych backed off—at least for a time. After the New Year, he reverted to confrontation, floating legislation that made blockading public buildings punishable by up to five years in prison and allowed for the arrest of protesters wearing masks or helmets.[10] Just a few weeks later, however, his government agreed to give amnesty to all demonstrators who vacated occupied government buildings and stopped blocking streets.

The Kremlin grew exasperated with Yanukovych's vacillations. They wanted their guy in Kyiv—the guy they had helped bring to power in 2010—to act decisively. During a meeting with Lavrov in February 2014, he asked me hypothetically what the United States government would do if protesters occupied a federal building in downtown Washington. He didn't wait to hear the end of my convoluted answer. I recall him saying bluntly, You would use force to remove them. He was probably right.

Our administration advocated a different approach to the standoff in Ukraine. We engaged with both the government and the demonstrators to try to find a peaceful way to defuse the crisis. In Washington, Vice President Biden called Yanukovych several times during those volatile days, urging him to resist using force and instead negotiate with the protesters. On the ground in Kyiv, our new ambassador, Geoffrey Pyatt, maintained close contact with leaders in the government and on the street. Assistant Secretary of State for Europe and Eurasian Affairs Victoria Nuland also traveled to Kyiv to conduct shuttle diplomacy between the government officials and the demonstrators. American and European diplomats worked in tandem to try to defuse the standoff. Our aim was not regime change but a peaceful resolution to the crisis.

Their efforts became more urgent after violence erupted between Ukrainian police and protesters on February 18, 2014. To this day, the facts of who did what to whom and when remain murky. We know that pro-Yanukovych snipers shot and killed innocent protesters. We also know that the clashes left several police officers dead.[11] We do not know for sure who shot first. Ukrainian authorities estimated that thirty-nine had been killed; the opposition put the death toll closer to one hundred.[12] Obama issued a statement the next day, condemning "in the strongest terms the violence that's taking place."[13] The president explained that the U.S. had been deeply engaged "with our European partners as well as the Ukrainian government and the opposition to try to ensure that that violence ends."[14] By adding that phrase about the opposition, we implied subtly that both sides had contributed to this tragedy. But Obama placed most of the blame on Yanukovych: "We hold the Ukrainian government primarily responsible for making sure that it is dealing with peaceful protesters in an appropriate way, that the Ukrainian people are able to assemble and speak freely about their interests without fear of repression."[15] We then promised to work with both sides to end the violence.

A few days later, on February 21, I was back in Sochi for the closing ceremony of the Winter Olympics when my BlackBerry lit up with good news: President Yanukovych and the opposition had signed an agreement. Specifically, after meeting overnight with European officials, the president had agreed to an accord to resolve the political crisis with three opposition leaders: Vitali Klitschko, Oleh Tyahnybok, and Arseniy Yatsenyuk. These signatories agreed to "refrain from the use of violence"[16] and have their supporters hand over all illegal weapons within twenty-four hours, and in return the Ukrainian government was to lift the state of emergency. The agreement also called for the restoration of the 2004 constitution, which limited the powers of the president, and for an early presidential election no later than December 2014. Signatories expressed "their intention to create a coalition and form a national unity government" within twelve days of the agreement's signing. Three European foreign ministers—Radek Sikorski from Poland, Frank-Walter Steinmeier from Germany, and Laurent Fabius from France—also signed the agreement, in an effort to bolster its legitimacy.

Vladimir Lukin, Putin's ombudsman for human rights, represented Russia at the mediation efforts. I thought Putin's decision to send Lukin in Lavrov's place was an odd choice, but better than having no Russian representation. When Lukin didn't sign the agreement like the other foreign ministers, however, I became worried. Maybe Putin was not going to support the accord because he had other plans?

Several hours after signing the agreement, Yanukovych fled Kyiv, bouncing around between several cities in Ukraine before ending up in the provincial city of Rostov, Russia. In defending his decision to leave the country, Yanukovych said that he feared for his life.

By a margin of 328–0, Ukraine's parliament, the Rada, voted to impeach Yanukovych for fleeing the country, and then designated its newly elected speaker, Oleksandr Turchynov, as interim president until a new election could be held in May. Ukraine's constitution had no provision for this impeachment vote, but Rada members argued they had no choice. Yanukovych could not govern Ukraine from Rostov. A few months later, on May 25, 2014, Ukrainians went to the polls and elected Petro Poroshenko as president.

Many in our government, as well as our allies in European capitals, celebrated Yanukovych's departure. The protesters had won. Having squandered opportunities after independence in 1991, and again after the Orange Revolution in 2004, Ukraine now had a third chance to build a democracy with a firm orientation toward Europe.

I normally admire such displays of popular power. In my White House office, we celebrated the fall of Mubarak in Egypt back in 2011. This moment in Ukrainian history, however, felt precarious. I exchanged upbeat emails with a few government counterparts and admired the courage of the Maidan demonstrators, some of whom I knew personally. But I also was confused. As I watched Yanukovych's press conference in Rostov from my hotel room in Sochi, I wondered why the Ukrainian president had not ensconced himself somewhere in eastern Ukraine and campaigned to consolidate his authority as the country's elected president. Why did he not go to Moscow? Why was he hiding out in sleepy Rostov? But my biggest worry was Putin. There was no way the Russian president was going to stand passively on the sidelines as his man in Kyiv was taken down by pro-European protesters in events reminiscent of the Orange Revolution. He was going to strike back. While I was still in Sochi, I got together with Deputy Secretary of State Bill Burns, who was leading the U.S. delegation to the closing ceremony, to compare notes. We agreed that we should brace ourselves for a strong Putin reaction. We reached out to the Kremlin to try to arrange a meeting between Burns and Putin, but the Kremlin predictably refused our request. We considered suggesting another Obama-Putin phone call, but no one could figure out a script. What would Obama say? "Sorry"?

We did not have to wait long for Putin's response. The Kremlin immediately denounced the new Ukrainian government's members as illegitimate usurpers. Then Russia annexed Crimea. For a few days, Putin successfully kept the initial phases of this operation secret. Within our government, however, we already had discussed this scenario. On February 23, 2014, National Security Advisor Susan Rice decided to publicly discourage Russian intervention, warning that "it would be a grave mistake" for Putin to send soldiers into Ukraine.[17] Obviously, Rice wouldn't have made that statement unless she knew Putin was making preparations to do exactly that—send soldiers to Ukraine. A few days later, confused reports appeared in the media about "little green men"—Russian-speaking armed soldiers taking up positions at strategic places on the peninsula. The soldiers, who lacked insignia, first took control of the local parliament and then began seizing other strategic facilities. Ukrainian military personnel quickly dispersed. For the new leadership in Kyiv, it was a humiliating defeat. As usual, Putin had struck back in an asymmetric way.

At a news conference in Moscow in early March, Putin still denied that the soldiers on the ground in Crimea were Russian soldiers, calling them "local self-defense units."[18] At the same conference, however, he asserted that any Russian involvement in Crimea would be a "humanitarian mission" to protect ethnic Russians, and that if Russia were to use force, it would be "in full compliance with general norms of international law."[19]

Referring to Western powers' reaction to Russia's actions, Putin argued, "We are often told our actions are illegitimate, but when I ask, 'Do you think everything you do is legitimate?' they say 'yes.' Then, I have to recall the actions of the United States in Afghanistan, Iraq and Libya, where they either acted without any UN sanctions or completely distorted the content of such resolutions, as was the case with Libya."[20] It was the classic whataboutism I endured every night on Twitter from pro-Kremlin bloggers and bots. On April 17, 2014, two months after the operation began, Putin finally admitted, "Crimean self-defense forces were of course backed by Russian servicemen."[21]

I have often wondered about Putin's decision making on the day that Yanukovych fled Ukraine. He most likely was surrounded by a small group of his closest friends and advisors, mostly former or current intelligence officers. No one from his economic team would have been present. It was the last days of the Sochi Olympics, so Putin and his inner circle must have been all pumped up on nationalism; Russia was leading the medal count (with, as we know now, some major help from a government-organized, performance-enhancing-drugs program). Putin must have been fuming, blaming us for the revolution that ousted

Yanukovych; he said as much subsequently. His media channels explicitly labeled the U.S. as the real plotter of the coup, with the Ukrainian people just acting upon our instructions.[22] I am guessing he believed the CIA had once again overthrown an anti-American regime, just as he thought they had in Tunisia, Egypt, and Libya in 2011, and tried to do against him in 2011–12. After we allegedly toppled his ally in Ukraine, Putin was done worrying about what we thought of him or how we could cooperate on other issues. In his view, we were now seizing control of a country of vital strategic importance to Russia, a country that Putin didn't even think should exist independent of Russia. Ukraine would become a member of NATO — or so Putin irrationally might have believed — and thereby threaten many Russian interests, including Russia's naval base in Crimea. As he stated flippantly a few weeks later, "I simply cannot imagine that we would travel to Sevastopol to visit NATO sailors."[23] He was going to strike back boldly, even if it meant violating one of the most important norms of the post–World War II order.

Contrary to Putin's accusations, the Obama administration did not organize the Maidan protests. Ukrainians did that alone. And we did not seek Yanukovych's overthrow; rather, we tried until the very last hours to forge a deal between the president and the protesters; even after Yanukovych had massacred dozens of his citizens, we were still negotiating with him. But none of these facts mattered to Putin. Even after Russia's military occupation of Crimea, Secretary Kerry held several rounds of talks with Lavrov to try to head off a complete annexation through a referendum. But Putin wanted no compromises. The fight for Ukraine was another zero-sum struggle with the West. If he was going to lose his man in Kyiv, Putin was determined to win somewhere else in Ukraine.

On March 16, 2014, the occupying authorities held a referendum that asked the residents of Crimea if they wanted to secede from Ukraine and join Russia. To shape the electoral outcome, all Ukrainian television channels were taken off the air, while stories of neo-Nazi putschists in Kyiv bombarded television viewers from Russian channels, as did the specter of future violence against ethnic Russians if the Ukrainian authorities came back to power on the peninsula.[24] Armed Russian soldiers filled the streets and stood guard at major government buildings. Russia claimed that 83.1 percent of registered Crimean voters turned out for this referendum; 96.77 percent voted to join Russia.[25] At the time, Crimea's ethnic Russian population was estimated to be around 60 percent, meaning that huge majorities of ethnic Ukrainian and Tatar citizens of Crimea allegedly voted in support of annexation as well.

Curiously, Putin's own organization, the President of Russia's Council on Civil Society and Human Rights, published radically different results on its website, reporting that turnout was well below 50 percent and that only around half of voters supported "unification," the Russian euphemism for annexation.[26] But this report did not stay on the website for very long. Later that day, Putin formally recognized the referendum results, and a few days later Crimea and Sevastopol were formally incorporated into Russia. Putin emotionally exclaimed, "After a long, hard and exhausting voyage, Crimea and Sevastopol are returning to their harbour, to their native shores, to their home port, to Russia!"[27]

Only a day out of government when the world learned of Putin's annexation campaign, I was still in close contact with many senior officials in the Obama administration, and discussed in detail our possible responses. One first decision was whether I should return to Moscow until a new ambassador had been appointed. I considered the possibility, but quickly decided that my presence in Moscow at this juncture would make no difference. As an outsider now, I did try to help shape the Obama administration's response to Russia's annexation of Crimea, coming down firmly in favor of a strong and comprehensive reaction. This moment was not just another bump in the road in U.S.-Russia relations, like Serbia in 1999 or the Iraq War in 2003. Putin's intervention in Ukraine was even qualitatively different from Russia's invasion of Georgia in August 2008; there was no debate about who started this confrontation, as there had been during the 2008 Russian-Georgian war. Moreover, Russia had never recognized Abkhazia and South Ossetia as part of Georgia, while the Kremlin, including Putin himself on the record, had acknowledged the legitimacy of Ukraine's borders. As Putin stated in 2008, "Crimea is not a disputed territory . . . Russia has long recognized the borders of modern-day Ukraine."[28] This was the first time territory had been annexed in Europe since World War II. We had to respond vigorously, both to defend the norms of the international system and because of our obligations to Ukraine. In 1994 the United States, Russia, and the United Kingdom had signed the Budapest Memorandum, which committed signatories to respect Ukrainian territorial integrity in return for Ukraine's denuclearization. The memorandum was not a treaty, and obviously, Russia had violated its terms, not us. Nonetheless, a weak response to Russian aggression would weaken America's credibility both with Ukraine and with other countries with which we sought similar nonproliferation agreements. If we responded tepidly, we were inviting more Russian aggression, both in Ukraine and maybe against our NATO allies, Estonia, Latvia, and Lithuania. We had to make deterrence credible.

In close cooperation with German chancellor Angela Merkel, President Obama galvanized the international community to respond quickly and decisively. At the United Nations, most countries joined the United States in denouncing the annexation, a level of international support that the Bush administration did not achieve when responding to the Russian invasion of Georgia in August 2008. In a vote of one hundred in favor to eleven against, with fifty-eight abstaining, the UN General Assembly adopted a resolution declaring the Crimea referendum invalid.[29] Russia, of course, vetoed a similar resolution in the UN Security Council. But thirteen other Security Council members voted in favor, with only China abstaining. Obama and the other G-8 leaders also voted to kick Russia out of this international club, which Gorbachev and Yeltsin had fought so hard to join.[30]

The most coercive and controversial action was the imposition of comprehensive sanctions against Russian individuals and Russian companies involved in the military operation in Crimea — an action I fully supported. In a memo I wrote to Dan Fried, then our point person on sanctions at the State Department, I expressed my belief that we had to make Putin pay a price for invading Ukraine, even if "at this stage in the crisis, sanctions [would be] unlikely to reverse a previous decision made by Putin." In this memo, which I circulated to others in the Obama administration, I proposed we impose sanctions against individuals and entities that fell into the following categories: "(1) those directly responsible for this military operation; (2) political leaders who supported the war and who supported democratic crackdown to silence war critics; (3) chief propagandists for the war; and (4) heads of state-owned enterprises, or companies closely tied to the regime that indirectly financially support Putin and this war effort." I lobbied against sanctioning individuals and entities in the following categories: "(1) government officials with whom the Obama administration has to continue to engage on matters of importance to U.S. foreign policy;

(2) political leaders who are critical of the war; (3) leaders of private companies who do not appear to be supporting the war."

My list of suggested individuals and companies was obviously longer and deeper than the categories that my former colleagues in the Obama administration would eventually target through sanctions.[31] On the outside again, I could articulate grandiose objectives and goals, without trying to devise or implement a strategy to achieve them. I remember how I had hated reading such essays from our critics in the op-ed pages while in government — all goals and no strategies for reaching them. Now, I was back in that ivory-tower crowd. I didn't like the feeling. I wished that I were back in the government.

Those closest to the issue in the Obama administration wanted to move forward with comprehensive sanctions. But American sanctions alone would not credibly deter further Russian military actions. The Europeans had to be on board. So this transatlantic debate was resolved in favor of incrementalism. On March 20, 2014, the administration imposed sanctions on twenty officials, including senior Russian government officials and "cronies who hold significant resources and influence in the Russian system."[32] In announcing this first round of sanctions, the White House threatened to "impose additional costs," but did not specify whether these new sanctions would be levied in response to further Russian aggression or in the absence of Russian withdrawal from Crimea. Speaking in Brussels, Obama explained these sanctions as an effort to defend not just Ukraine, but also the postwar international norms and treaties that had served the world so well: "If we defined our interests narrowly, if we applied a cold-hearted calculus, we might decide to look the other way. Our economy is not deeply integrated with Ukraine's. Our people and our homeland face no direct threat from the invasion of Crimea. Our own borders are not threatened by Russia's annexation. But that kind of casual indifference would ignore the lessons that are written in the cemeteries of this continent. It would allow the old way of doing things to regain a foothold in this young century."[33]

Moscow reacted defiantly. "We warned many times that the use of sanction instruments is double-edged and will bounce back on the United States," read a statement by the Russian Ministry of Foreign Affairs.[34] The same day, Russia issued travel bans on nine U.S. lawmakers and officials.[35] I was especially disappointed to see my former White House colleague Ben Rhodes on that list; back in 2009 Ben had been a vital champion of the Reset. Now he was banned from traveling to Russia. I was thankful not to see my own name on the list—that would come later.

In public, I applauded the Obama administration's decision, coordinated with our European allies, to sanction Russian individuals and companies. It was a bold move; the Bush administration had not sanctioned a single Russian person or company after the Russian invasion of Georgia in August 2008.[36] In private, I urged my friends and colleagues in the Obama administration to do more: not only to punish bad Russian behavior but to deter further aggression. I worried that Putin's appetite for revenge against the new Ukrainian government would not be satiated by Crimean annexation alone.

Obama rightly diagnosed the magnitude of the moment. Putin's "old way of doing things" was destabilizing, both for Ukraine and for the international system. That said, Putin was not making rational cost-benefit calculations at that moment; his choices were emotional and revenge-driven. In retrospect, I am not sure that a different response from the United States and the international community would have worked any better. But what is certain is that the initial round of sanctions was insufficient to deter Putin from seeking more. Pleased with the results in Crimea, Putin decided to green-light a complex plan to seize Novorossiya, or New Russia, a vast region from eastern Ukraine to Odessa on the Black Sea. As justification, Putin argued that he was simply protecting the rights of ethnic Russians in eastern Ukraine, asserting, "We must do everything to help these people to protect their rights and independently determine their own destiny."[37] By invoking Novorossiya, a territory that had been incorporated into Ukraine after the Bolshevik Revolution, Putin was deploying the same rationale for Russian interference in Ukraine that he had used for annexing Crimea. As Putin explained, "It's New Russia. Kharkiv, Luhansk, Donetsk, Odessa were not part of Ukraine in czarist times; they were transferred in 1920. Why? God knows."[38] Similar to Crimea, Putin seemed committed to right this communist-era wrong and restore Russian imperial borders.

The Kremlin provided money, weapons, commanders, and even soldiers to separatist proxies in eastern Ukraine.[39] These Kremlin agents achieved quick victories against Ukraine's military and local militias, seizing government buildings in the eastern cities of Luhansk, Donetsk, and Kharkiv in early April 2014. Ukrainian authorities were able to regain control of buildings in Kharkiv, but administrative buildings in Luhansk and Donetsk remained under the control of pro-Russian groups.[40] A month later, Russia used the same playbook in eastern Ukraine as in Crimea, holding "self-rule" referenda on secession (though not, in this case, "unification" with Russia). According to election organizers, 89 percent of those who cast ballots in the Donetsk region and 96 percent of voters in neighboring Luhansk voted to leave Ukraine.[41] The official numbers clashed radically with public opinion polls. In a Pew Research Center survey administered in mid-April in eastern Ukraine, 70 percent of respondents said they wished to remain part of a united Ukraine, while only 18 percent were in favor of secession.[42] A poll released by the Kyiv International Institute of Sociology based on data gathered in early to mid-April similarly found that only 15 percent of respondents were in favor of seceding and joining Russia; 70 percent were against.[43]

The referenda resolved nothing. Conventional fighting continued. Pro-Russian protests erupted sporadically in several cities, sometimes with tragic results. In Odessa, for example, three dozen pro-Russian activists trapped in a locked government building that had been set on fire were shot at as they tried to escape through the windows.[44] Ethnic Russian militias also committed crimes against innocent ethnic Ukrainians. Ukraine was drifting toward civil war.

On July 17, 2014, the conflict internationalized even further. Russian-supported separatists or Russian soldiers — the details remain murky — shot down Kuala Lumpur–bound Malaysia Airlines Flight 17 in eastern Ukraine, killing all 298 people on board. Although the actual shooter has not been identified, the Russian government clearly supplied the rocket used in this attack.[45] Putin denied responsibility, and instead argued, "The government over whose territory it occurred is responsible for this terrible tragedy."[46] Putin put no pressure on his proxies in eastern Ukraine to cooperate with an international investigation — infuriating the Australians, the Dutch, the Malaysians, and other members of the international community who had lost nationals on board the plane. The shooting down of MH17 focused greater world attention on the conflict in eastern Ukraine, prompting the West to react with greater vigor, including new sanctions.[47]

Putin's covert operation to seize Crimea began quietly before I left Moscow, but the world began to learn about Russian intervention in Ukraine the day I arrived home. After seven months of "commuting" from Moscow to Palo Alto,

I was thrilled to be reunited with my family, but also deeply depressed to watch our relations with Russia now take an even sharper turn for the worse. And by "our" I mean not only the Obama administration but also every administration since Reagan. Putin's decision to annex Ukrainian territory was a clear breach of the most basic of international norms. If there were a Ten Commandments of international behavior, "Thou shalt not annex the territory of thy neighbor" would be at the top of the list. The project of democratic development inside Russia had ended long ago, punctuated by two years of growing autocracy during Putin's third term as president. Now, the project of Russian integration with the West, already deeply damaged, also came to a halt. The reset in relations between Washington and Moscow, first started by Reagan and Gorbachev, and then resuscitated one last time by Obama and Medvedev, was dead, too. Over the last thirty years, I had believed in the possibility of Russian democracy and integration with the West. As an NGO activist and then as ambassador I had not only believed in these goals but worked to achieve them. And now we were done. Our efforts had failed.

I was not a Cold Warrior itching to get back into the arena with a Kremlin adversary; ever since my high school days debating Jackson-Vanik, I had held a different aspiration for our bilateral relationship. But Putin's actions in Ukraine compelled the United States and Europe to pivot to a fundamentally different strategy for managing relations with Russia. We were not returning to a Cold War, but we were entering a new confrontational era, a hot peace. On March 23, 2014, I spelled out my proposed strategy in an essay for the *New York Times* titled "Confronting Putin's Russia." I deliberately echoed some, though not all, of the themes of containment codified a half century earlier by George Kennan. In my diagnosis of the problem, I summarized, in effect, the argument of this book: "The decision by President Vladimir V. Putin of Russia to annex Crimea ended the post–Cold War era in Europe. Since the late Gorbachev-Reagan years, the era was defined by zigzags of cooperation and disputes between Russia and the West, but always with an underlying sense that Russia was gradually joining the international order. No more."[48] Instead of hoping that Putin would eventually come back to his senses and seek again to integrate with the West, I declared that this project was over. In its place, I recommended a strategy of selective containment and selective engagement, an update of the strategy I'd proposed in 2012 of selective engagement and selective disengagement. In this new tragic moment in U.S.-Russian relations, we had to punish Russian bad behavior with coercive responses, such as sanctions and isolation; develop meaningful methods of deterrence, like strengthening NATO; while continuing to engage with proponents of democracy, including Ukrainian allies and Russian society. Of course, as all American administrations had done during the Cold War, we would have to work with Putin during this new era of hot peace when our vital interests overlapped. But that seemed increasingly difficult. Still, like Kennan, I was confident that victory was certain—but not anytime soon. To see the fruits of this new strategy would require patience.

The evening my proposal for a grand new strategy for dealing with Russia appeared on the *New York Times* website, I felt like I was departing from long-held convictions. Before, I had believed in engagement and the promise of Russian integration into the West. I had believed in the possibility of Russian democracy. Now, those things felt illusory and unachievable, at least for the foreseeable future. I felt like I was closing a chapter—a thirty-year chapter—of my life. I knew that critics in both America and Russia would lambaste me for advocating dangerous, Cold War–era policies. But I was not the one who pivoted: Putin did, and we had to respond.

24

THE END OF RESETS (FOR NOW)

At the Munich Security Conference in February 2016, Russian prime minister Dmitry Medvedev made a frightening pronouncement. "We are rapidly rolling into a period of a new cold war," he stated. "I am sometimes confused: Is this 2016 or 1962?"[1] Medvedev deliberately invoked 1962 because that is when the Soviet Union and the United States came closest to blowing up the world in a nuclear war. As I listened to his speech, I decided that he was exaggerating — relations between Russia and the United States were not as bad as 1962 — but one most certainly had to dig deep back into the Cold War to find an analogous era of confrontation. If it was not a return to the Cold War, it was most certainly a new hot peace. U.S.-Russia relations in 2016 were not that much different from U.S.-Soviet relations in 1979, when I first started to think about our bilateral ties.

Later that day, in the crowded hallways of the Hotel Bayerischer Hof, I stepped out of the way to let Medvedev and his trailing entourage pass. Catching my eye, he stopped to say hello and catch up. He asked me about life at Stanford, a place he'd visited in 2010, during the glory days of the U.S.-Russia Reset. He was very friendly, no hostile echoes of the Cuban missile crisis. My guess is that he looks back on the early years of the Reset fondly. Those were my best days in government, and likely his as well.

A senior U.S. government official was standing next to me as Medvedev stopped to chat. After the Russian prime minister went on his way, my colleague asked me, "Who was that?"

My Munich encounters with Medvedev — his foreboding speech, his warm chat with me, and his new anonymity — made me wonder what had happened to the Reset. Only six years earlier, Obama and Medvedev stood together on the world stage, working together to get some big things done, and now we were back to a full-throttle confrontation. Six short years ago, most Americans had a favorable view of Russia, and most Russians had a positive view of the United States. Those numbers had now reversed. Six years ago, Medvedev was an important partner to President Obama; now, he had all but faded to obscurity. All of these changes had happened so fast, within the span of my service in the U.S. government.

But it was not only Obama's Reset that was over. Medvedev's allusions to the Kennedy-Khrushchev standoff underscored that the positive shift in relations initiated by Reagan and Gorbachev was also a thing of the past. Not just during Obama's time at the White House, but for thirty years, American presidents — Democrats and Republicans alike — had pursued a policy of supporting political and economic reform inside Russia as well as integrating Russia into the West. In 2016, that strategy seemed to have failed completely. Those goals of internal democracy and external integration now seemed too naive and unrealistic to warrant ever pursuing again. Why? What went wrong?

And what did I do wrong? Since an interest in the Soviet Union had first been kindled in me as a high school student in 1979, I had believed that greater engagement with Russians could bring our two countries closer together. In the 1990s I not only believed, but also actively participated, in helping Russians develop democratic institutions and closer relations with the West. I joined the Obama administration in large part to take another run at these grand objectives. Without question, I had failed to achieve them. Was I at fault for pursuing the wrong strategies for obtaining these outcomes? Or was I wrong for believing in these goals in the first place?

The debate about the failure of the Reset leads in three possible directions: the structure of international relations between great powers, our foreign policies, and Russian domestic politics.[2] Each explanation has merit. But the third — the impact of Russian domestic politics in shaping Russian foreign policy toward us — explains the most. Russia's failure to consolidate democracy or integrate into the West is first and foremost a story about Russians, not us. But it's not solely a story about decisions and politics inside Russia. The other two factors — balance-of-power politics and American actions — also played a role.

For thousands of years, great powers have risen and great powers have fallen. As they do so, they sometimes brush up against each other, often resulting in war, and in Europe in the twentieth century, in global war.[3] As nations rise and fall, borders between states also change. In Europe over the last thousand years, changes in the balance of power between the great powers have altered borders continuously.

For scholars and politicians focused on the rise and fall of great powers, the dynamics that we are witnessing today between Russia and the United States, and Russia and the West more generally, are just normal balance-of-power politics between great powers. From this line of analysis, Reagan liked Gorbachev and Clinton liked Yeltsin not because these leaders shared American values or interests, but because these Russian leaders were weak. They did what we told them to do, because they had no other choice. They had no power to resist our hegemony: our expansion of NATO, bombing of Serbia, or invasion of Iraq. As Russia recovered from economic depression and the state collapse of the 1990s, the Kremlin acquired the means to push back. The rise of Russian power produced "natural" corrections to the temporary imbalances created by the Soviet Union's implosion. Russia's annexation of Crimea reflects this new distribution of power. From this analytic perspective, leaders and their policies within states do not matter. Nor do interest groups or the kind of institutional arrangements within states. As one of the world's most important realist theorists, John Mearsheimer, has written, "Washington may not like Moscow's position, but it should understand the logic behind it. This is Geopolitics 101."[4] Similarly, as Stephen Kinzer has argued, Russian "actions in Ukraine do not constitute a radical departure from international norms. In fact they are quite the opposite: the re-enacting of a historical pattern that is as old as empires."[5]

The distribution of power in the international system — specifically the rise in Russian power over the last two decades — certainly helps explain in part current tensions in U.S.-Russian relations. If Russia did not have the capacity to impact American interests, or those of our allies, conflict would be negligible.[6] Russian military power deployed against Ukraine punctuated the end of

Obama's Reset and the Western strategy of Russian integration started thirty years ago. Without that power — that military capability — this crisis would not have happened. No one is worried about Moldova invading neighbors or disrupting the international balance of power, because Moldova does not have the power to do so. Power matters. Russia also has tremendous capabilities regarding cyber hacking and theft, international media outreach, and social media resources. The Kremlin had the means to influence the U.S. presidential election; most countries in the world do not have such capabilities.

But was conflict inevitable, just because Russia had become more powerful? My answer is no.

First, other great powers have risen in the international system and not produced conflict with the United States. After the end of World War II, Japan reemerged as a great power without open hostilities with America. The same can be said for Germany's rise again after World War II to become the most powerful country in Europe. Spain, Italy, and South Korea also have become more powerful in the last several decades, but do not threaten us. Poland is a much more powerful country today than thirty years ago, yet few worry about Warsaw ordering troops into neighboring countries.

Second, the most recent downward spiral in relations between the United States and Russia occurred rapidly, within the span of just a few years, well after Russia had reemerged as a great power. In Prague in 2010, we were clinking champagne glasses with Medvedev, celebrating our close ties, and imagining even deeper cooperation. Only two years later, Putin was calling Obama and the United States Russia's greatest enemy. Incremental changes in Russia's military or economic power did not precipitate that radical and rapid change in our relations in just a few years. Other factors — more proximate factors — have to be added to the equation.

A second kind of explanation focuses on the specific foreign policies of states and specifically American foreign policy — and not just the balance of power between states — to explain our new era of confrontation.[7] Regarding Russian-American relations, this kind of argument comes in two opposite varieties: we did too much and we did too little.

The most prevalent of these two theories blames the United States for pushing too hard on Russia for too long and too close to Russia's core national interests. This line of argument maintains that American presidents and their foreign policy advisors took advantage of Russian weakness to press for regime change inside Russia and ignore Russian national interests internationally, forcing Moscow to push back.[8] This explanation casts the United States as both an imperial and an ideological power, pushing its normative agenda of democracy and capitalism. Many Russians, including eventually Vladimir Putin, blamed the United States for the economic chaos and international marginalization that ensued in Russia during these years. The United States forced Russians to endure the hardships of shock therapy and, according to this line of thinking, exported democracy to a Russian society that did not want it. What the United States really wanted was not a vibrant economy or functioning democracy, but a weak Russia, so this critique contends. With Russia weak, the United States and its allies could expand NATO, attack Serbia, build missile defenses, invade Iraq, and foment revolution against regimes in countries of vital interest to Moscow, including Serbia, Georgia, and Ukraine.

Again, like the balance-of-power theory, some dimensions of this explanation ring true. The 1990s were a tough decade for Russians. The economy contracted for several years and then collapsed in August 1998.[9] Democracy did not take root and did not deliver on what mattered most to Russian citizens. On the international stage, Russia was weak in the 1990s and had few means to counter American actions. Moreover, in the margins, some U.S. actions influenced the course of Russian internal reforms in the 1990s.

But the phrase "in the margins" is the key one. No American policies or actions — be it NDI seminars about party building, technical assistance for privatization, Clinton's support for Yeltsin in 1996, or USAID grants to Russian NGOs in the Obama era — determined the outcome of Russian internal reforms. Russians themselves did that. Those who blame the United States grossly overstate American influence. It is the most imperial of claims to argue that we dictated the kind and pace of reform inside Russia in the 1990s or at any other time. Both at the time and today, I too wish we had done more. Writing as early as the summer of 1990, I believed that a successful transition to democracy and markets in Russia would make closer, cooperative relations with the United States more likely, while economic and social troubles could fuel law-and-order appeals from nationalist demagogues.[10] It's a theme I echoed many times for the next two decades. More American focus on strengthening democratic and market institutions might have made a difference. Other postcommunist countries, which made the transition to democracy and capitalism more quickly and more successfully, are now some of America's strongest allies in Europe. To argue that we did too much to promote internal change in Russia, therefore, strikes me as ahistorical.

Some U.S. foreign policy decisions in the 1990s and 2000s also triggered tensions in U.S.-Russian relations. Both Yeltsin and Putin reacted negatively to NATO expansion (although Putin initially entertained the idea of Russia joining the alliance). Likewise, the NATO military campaign against Serbia in 1999 and Bush's decision to invade Iraq in 2003 generated new tensions in Russian-American relations. Putin also berated Bush and his administration for supporting color revolutions in Georgia in 2003 and Ukraine in 2004, which he believed threatened core Russian interests.

Yet, these American foreign policy decisions, both real and perceived, cannot be cited as the source of our current conflict with Russia for one major reason — the successful cooperation between Russia and the United States during the Reset, from 2009 to 2011. After our alleged meddling in Russian reforms in the 1990s, rounds of NATO expansion into Eastern Europe, American-led interventions in Serbia and Iraq, and these color revolutions, U.S.-Russian relations experienced an extraordinary amount of cooperation during the early Obama-Medvedev years. We collaborated on some very big win-win outcomes: the New START treaty, new sanctions on Iran, the expansion of the Northern Distribution Network, Russian membership in the WTO, increased trade and investment, a new user-friendly visa regime, no conflicts between our two countries in the Caucasus or Central Asia, and Russian support for our military intervention in Libya. During this Reset era, earlier contentious issues such as NATO expansion or missile defense faded. As Medvedev said of the Reset in March 2012 after his last presidential bilateral meeting with Obama in Seoul, "We probably enjoyed the best level of relations between the United States and Russia during those three years than ever during the previous decades."[11] In private with Obama, Medvedev was even more effusive, the exact opposite of his dire tone at the Munich Security Conference four years later. The effect of this era of cooperation between Obama and Medvedev influenced societal attitudes. At the height of the Reset, roughly 60 percent of Russians held a favorable view of the United States and vice versa; the same number of Americans had adopted a new, favorable attitude toward Russia. All of these cooperative outcomes for American and Russian foreign policies, as well as this new support for closer ties between our two societies, occurred after NATO expansion, after the wars in Serbia and Iraq, and after the color revolutions. These factors, therefore, cannot be cited to explain the current era of confrontation. Other factors — factors that came later — must be added to the explanation.

Maybe American actions after 2012 contributed to the end the Reset? Did the Obama administration pursue new policies in 2012 that precipitated confrontation with Russia? I have thought hard about this possible explanation, since I would have been directly responsible. My answer is mostly no. We did take some actions that strained relations, such as signing into law the Magnitsky Act, failing to reach a deal on missile defense, and criticizing antidemocratic behavior by the Kremlin. Likewise our actions in Libya produced unintended consequences that heightened tensions with Moscow. But none of these actions were fundamental departures from the Reset. They were all manageable hiccups, bumps in the road of cooperation, had both countries desired to maintain the Reset's momentum. But in 2012, one side did not.

Another line of reasoning still focused on U.S. foreign policy decisions as the cause of conflict also blames the United States for the deterioration in U.S.-Russia relations, but highlights our weakness, not our aggression. According to this view, the Reset created the permissive conditions for Putin's invasion of Ukraine: Obama showed weakness, and Putin took advantage. As then Speaker of the House John Boehner asserted, "When you look at this chaos that's going on, does anybody think that Vladimir Putin would have gone into Crimea had George W. Bush been president of the United States? No! Even Putin is smart enough to know that Bush would have punched him in the nose in about 10 seconds."[12] Other commentators even suggested a direct connection between Obama's backing down on his threat to use force against Assad after he used chemical weapons, and Putin's decision to annex Crimea.[13] Obama undermined American credibility by backing away from his own red line. Because of Obama's weakness, Putin thought he could do what he wanted, where and when he wanted.

I have a different view. Correlation is not causation. Our Reset in relations with Russia occurred before Putin invaded Ukraine, but did not give him the green light to do so. Our relations with Russia soured two years before Putin annexed Ukraine. And no American president, Reset or not, would have threatened to go to war with Russia over Ukraine. At the time of Russian annexation of Crimea, our bilateral relations were already in a downward spiral. To win reelection in 2012 and marginalize his domestic opponents, Putin needed the United States as an enemy again. He rejected deeper cooperation with us. As a result, our administration pivoted to a more confrontational policy after President Putin had rebuffed our attempts to engage with him. We downgraded and slow-rolled discussions about missile defense, signed into law and then implemented the Magnitsky Act, canceled the Moscow summit in September 2013, criticized Putin's growing autocratic tendencies, chastised Russia for blocking United Nations resolutions on Syria, and sent American delegations to the Sochi Olympics to convey a message about our support for LBGTQ rights in response to Russia's "anti-gay propaganda" law. None of these policies sent a message of acquiescence to Putin about his repression at home or his more belligerent foreign policies abroad.

To be sure, our new confrontational policy toward the Kremlin did not prevent Putin from invading Ukraine, but tragically no president — Democratic or Republican — has ever succeeded in deterring Russian intervention in its neighborhood, either directly or by proxy, for the last seventy years. Contrary to Speaker Boehner's jibe, President Bush did not punch Putin in the nose after Russia invaded Georgia in August 2008, even though they were sitting in the stands together just rows apart at the Olympics in Beijing at the time. In fact, Bush did much less in response to Putin's invasion of Georgia than Obama did in response to Putin's intervention in Ukraine in 2014. The Bush administration did deploy the USS *McFaul* (named in honor of a Navy SEAL of no relation to me) off the coast of Georgia and sent massive humanitarian and economic assistance to our partners in Tbilisi. But President Bush and his team did not sanction any Russian officials or companies, did not push for Russia to be expelled from the G-8, did not lobby for new American or NATO troop deployments in allied countries most threatened by Russia, and decided not to provide sophisticated lethal weapons to the Georgian military. After deciding to invade Iraq in 2003, President Bush cannot be accused of being afraid to use military force. But his hawkish reputation did not influence the Kremlin's decision to invade Georgia in 2008.

Likewise, few would consider President Ronald Reagan as being conciliatory or weak toward Moscow. But when Polish general Jaruzelski colluded with the Soviet leader Leonid Brezhnev to impose martial law in 1981, Reagan had no power to prevent this brutal use of force. And this tragic list goes on: President Carter failed to deter Brezhnev from invading Afghanistan in 1979, President Johnson could not impede Soviet intervention in Czechoslovakia in 1968, and President Eisenhower did not stop Soviet tanks from rolling into Hungary in 1956, even though he embraced a policy of "rollback" of communism. Whether Democrat or Republican, "strong" or "weak," American presidents tragically have been consistent in their inability to deter Russian intervention in that part of the world.

There was one possible policy not pursued by us that might have deterred Russia's invasion of Ukraine: NATO membership for Ukraine. Had Ukraine been a NATO member in 2014, I doubt that Putin would have tried to annex Crimea or support separatist movements in eastern Ukraine. This outcome, however, had no chance of being realized while I was in the government. At the NATO summit in Bucharest in 2008, most of our NATO allies made clear their lack of support for Ukrainian membership. Even the discussion between NATO leaders at this summit about offering Ukraine (and Georgia) a Membership Action Plan sparked what Secretary of State Rice described as "one of the most pointed and contentious debates with our allies that I'd ever experienced."[14] Ukrainian president Viktor Yushchenko pushed for MAP, but most Ukrainians did not support the idea.[15] After Putin's ally Viktor Yanukovych was elected Ukrainian president in 2010, even the most vehement proponents of Ukrainian membership in NATO understood that the issue was dead.

American foreign policy, whether considered too strong or too weak, was not the principal cause of changes in Russian foreign policy after 2012. Instead, the main drivers of these changes were domestic factors, specifically Putin's reactions to new domestic challengers. We had little or no influence over these Russian domestic developments.[16]

During his first two terms as president, Putin maintained high public approval ratings, in large measure because of Russia's impressive economic growth during these eight years. By 2011, however, when he launched his electoral campaign to win a third term as president, his popularity had fallen.[17] Putin expected his return to the Kremlin would be welcome news. But for many Russians, including some serving in the Medvedev administration at the time, the initial reaction to his third run at the presidency was indifference. The depth of dissatisfaction among Russia's elite produced massive demonstrations on the streets of Moscow, St. Petersburg, and other large cities in response to compelling evidence of falsification in the December 2011 parliamentary election. These protesters initially focused on electoral irregularities but eventually pivoted to a grander indictment of the Russian political system and Putin personally.[18] The last time so many Russians had taken to the streets was 1991, the year the Soviet Union disintegrated. Fueling anxiety in Russian government circles was the success of mobilized crowds in bringing down dictators in the Arab world. The slowdown in economic growth was also adding to Putin's woes at the time.

The global economic meltdown in 2008 hit Russia hard as demand for its primary export — oil — fell precipitously, and global energy prices collapsed. In 2009 the Russian economy contracted by 8 percent, and grew at around 4 percent in the three years before Putin's 2012 presidential campaign.[19] That would be a great growth rate for the American economy, but it was far from the huge Russian increases of pre-recession years. Putin had struck an implicit bargain with his citizens before: rising wages and economic development in return for political passivity. That deal did not seem as attractive in 2011 compared to the previous decade. The growing middle class in Russia's largest cities also was making demands beyond bread and butter issues. They wanted accountable government — democracy.

As discussed earlier, Putin bitterly resented these activists. In his mind, he had made them rich, and now they were turning against him. But in a year when crowds in squares were toppling regimes, he also feared them. To be elected a third time as president of Russia in 2012, he needed a new argument. In the face of growing social mobilization and protest, he revived an old Soviet-era argument as his new source of legitimacy — defense of the motherland against the evil West, and especially the imperial, conniving, threatening United States. Putin, his aides, and his media outlets accused the leaders of Russian demonstrations of being American agents, traitors from the so-called fifth column. We were no longer Reset partners, but revolutionary fomenters, usurpers, enemies of the nation.

Putin's assault on the opposition did stop after his election victory. His government arrested some demonstrators and held others under house arrest, including the leading opposition leader at the time, Alexei Navalny. To add even more pressure on Navalny to behave, they imprisoned his brother. The Russian government also increased constraints on nongovernmental organizations and independent media, forcing some activists and journalists into exile. Eventually, large protests stopped occurring, in part because of these arrests and in part because the government introduced substantial fines for participation.

Putin did not have to take these draconian measures against his opponents. He chose to do so. When faced with similar societal challenges, Gorbachev chose a different path. Even President Medvedev embarked upon a more cooperative strategy for responding to demonstrators in 2011 and 2012.[20] Once back in the Kremlin, Putin interrupted abruptly and decisively Medvedev's conciliatory steps, and adopted a more confrontational, repressive approach. In the short run, it worked.

As part of his strategy for delegitimizing his opponents, Putin and his government described them as American agents — traitors. We were once again portrayed as the enemy. Putin's pivot against us made sustaining the Reset impossible. If we were the enemy, Putin could not appear to be too friendly. But his anti-American rhetoric was not just for winning votes or disparaging the opposition. Putin believed that we were out to get him. In his view, we always sought to overthrow regimes that we didn't like. He believed that we funded color revolutions in Eastern Europe, supported uprisings in the Middle East, and now were targeting him. His desire to cooperate with us on most issues, therefore, waned.

We could not force him to continue with the Reset — to work with us on Syria, arms control, or Edward Snowden. During these tumultuous two years, however, we did find common ground on some select issues, including a few big economic deals like the Rosneft-ExxonMobil joint venture, genuine counterterrorist cooperation while investigating the Boston Marathon bombing perpetrators, the Syrian chemical weapons deal, and most importantly the P5+1 negotiations with Iran.

Bilateral relations took an even sharper turn for the worse, however, after President Yanukovych fled Ukraine in February 2014, and Putin in response invaded Ukraine. Just as he accused us of supporting Russian revolutionaries against him, Putin blamed the United States again for supporting Ukrainian revolutionaries. We tried desperately to mediate a deal between the Ukrainian government and the opposition, but when these talks failed and Yanukovych fled the country, Putin believed that we had double-crossed him. In reaction, he struck back, first by annexing Crimea and then by intervening in eastern Ukraine. If the Reset had been interrupted in 2012, it was buried for good in 2014. And it was not just our most recent attempt at Reset that was buried, but the decades-long American policy of seeking cooperation with Moscow and integrating Russia into the West first started by Ronald Reagan and Mikhail Gorbachev in the late 1980s.

To focus on Russian domestic politics and Putin's personal role as the drivers of new confrontation with the United States does not necessarily exonerate the Obama administration completely. Maybe we played a role in exacerbating domestic tensions inside Russia? Putin is not the only analyst who has asserted that we played a direct role in fomenting popular mobilization against his regime. The claim needs to be examined.

Despite Russian government claims to the contrary, the United States played no direct role in sparking protests in 2011. Neither Navalny nor Nemtsov were waiting for Clinton's "signal" before taking action. People did not gather on the streets that rainy evening of December 5 in response to anything we said. The Obama administration purposely kept its distance from the protests. We believed that close association with us would only discredit the democratic opposition movement. As ambassador, I never once attended a political rally in Russia, and never once met with Navalny. Nor did we fund the political opposition. This Russian government accusation is bogus propaganda.

But are we guilty of helping to create the conditions for those demonstrations or the momentum for democratic change in Russia during the three years of the Reset? That's a harder question. Of course, Russians — nongovernmental actors from Russian society — drove the drama around democratic change in Russia in those Reset years. They acted independently. In the margins, though, maybe our policies played a small role. Putin is mostly wrong, but maybe a little bit right, in blaming us for promoting democracy and defending human rights in Russia.

At the highest levels, Medvedev wanted Obama's respect, and moving in a democratic direction was part of earning that respect. Medvedev cared deeply about the Reset. It was one of his greatest achievements. The Russian president also valued his personal relationship with his American counterpart. Medvedev wanted Obama to believe that he, too, was a progressive — a new, young, post–Cold War leader. Together, he believed, they could change U.S.-Russia relations fundamentally. Making Russia more democratic was part of the vision. So when Magnitsky was killed, Medvedev didn't just dig in and defend his government but initiated reform — partial and incremental, but reform nonetheless. When Obama pressed Medvedev about the need for greater political openness to spur postindustrial growth, Medvedev concurred. And when protests erupted in December 2011, Medvedev engaged opposition leaders and introduced reforms, such as the reinstatement of gubernatorial elections and an easing of party registration rules.[21] Repression came later, when Putin returned to power. Of course, Obama's personal relationship with Medvedev was not the main reason Medvedev supported political reforms. And Medvedev's efforts regarding democratization were minimal. He said a lot about freedom, the rule of law, and democracy, but actually did very little to promote these ideas. (Whether he wanted to do more but couldn't, or whether he never intended to do more, is a question we often asked among ourselves at the White House. I still don't know the answer.) When he did act, however incrementally, Western reaction — Obama's reaction — might have been in the back of his mind, as Medvedev wanted to be a respected member in good standing in the Western community of states and their clubs, like the G-8. In fact, that's why his conservative critics despised him. He cared too much about his image in Washington, and not enough about his image in Bryansk. Our common concern for democratic change may have helped ultimately to weaken Medvedev's reputation with Putin.

We also sought to pull Russia toward the West. That's true. We believed close ties between Russia and the West would make both Russia and the West better off. I also personally believed at the time that such policies could — not would — create a better international environment for internal democratic change. During the first years of the Reset, I believed that coercion or confrontation with the Kremlin would not facilitate a new era of Russian democratization. For me, the analogy was Gorbachev's perestroika. During that period, Gorbachev felt emboldened to pursue democratic change at home because he faced a more benign international environment. I hoped that we could do the same with Medvedev. And maybe we did, for a while. Would those massive demonstrations in

2011 have happened under a more autocratic regime in open confrontation with the United States? Maybe not. Most certainly, our more confrontational era, especially after 2014, has not been conducive to democratic change inside Russia.

Philosophically, our administration also supported the ideas and aspirations of Russians seeking a more democratic future. That's also true. Guilty as charged. We criticized the arrests of peaceful protesters. We lamented Putin's new laws restricting the activities of nongovernmental organizations. We decried the sentences against Pussy Riot and Navalny, which we believed were politically motivated. We also provided financial and technical assistance to Russian civil society groups dedicated to democratization. In 2011, funding for democracy and election programs in Russia totaled $8.5 million; $9 million if you add in $500,000 spent by the National Endowment for Democracy. We only funded what we believed were nonpartisan groups and nonpartisan activities. Putin and his entourage, however, perceived these groups as partisan; if they supported free and fair elections, they were anti-Putin.

And yet, even while acknowledging all of these American preferences, policies, and activities, let's be humble and clear — we were marginal players in Russia's internal political dramas. Russians — in and out of government, seeking change, and in combat with those seeking change — made their own history. Despite the Kremlin's propaganda claims, the Obama administration, including me, mostly just watched from the sidelines.

And they did *make* their history, and were not simply the reflections of structural forces. Different leaders making different choices could have radically altered the course of political change inside Russia, which would in turn have altered the fate of U.S.-Russia relations. That was true during times of political change in the late 1980s and early 1990s, which brought our two countries closer. It was true again during times of attempted political change and reaction to political change in 2011–12, which ultimately pushed our two countries apart.

Ever since my junior year in high school, debating ways to increase trade with the Soviet Union, I had believed in the possibility of closer cooperation between our two countries. During my activist days of working for a democracy-promotion organization at the end of the Soviet Union, I believed in the possibility of Russian democracy and did what I could to advance its development. As the author of the Reset while working at the White House in the Obama administration, I believed that it was in the American national interest to have Russia more closely integrated into the West, and therefore recommended and helped execute polices to achieve that objective. I repeat the verb "believe" deliberately here, because I was a true believer — not just an analyst — in the possibility of closer ties between our two nations based on mutual interests and shared values. Three decades later, most of these efforts have failed. Russia is not a democracy today and Russia is not integrated into the West as deeply as many of us, both in Russia and the United States, had hoped. Putin is not interested in pursuing either democracy or Western integration today. Those projects are over for now.

I helped with some achievements along the way, which may have planted the seeds for success someday in the future. But of course I have enough humility and analytic perspective to understand that all of my efforts played only a marginal role in the larger sweep of this dramatic history, for good or for ill. I did not orchestrate the anticommunist revolution in the late 1980s and early 1990s by translating electoral laws or organizing seminars on budget reform while working at an American nongovernmental organization in Russia. And I did not precipitate confrontation between our two countries by tweeting about Navalny while serving as the U.S. ambassador to Russia. But as I conclude this book, my life's work of trying to bring our two countries closer, of trying to integrate a democratic Russia as a responsible and important stakeholder in the international community of states, seems like a failure. To punctuate the tragedy, I am now persona non grata in a country I love — a place where I have lived for years and which I studied for decades. Not since Ambassador George Kennan in 1950 has a U.S. ambassador to Russia been banned from travel to Russia.

I take solace in continuing to believe that the course of these events over the last thirty years was not all predetermined by Russian history and culture, or by the balance of power in the international system. There were moments when actors and their unique decisions mattered. There was an alternative outcome available, and so it was worthwhile to try to engage in the process of change and push history in another direction. Democracy in Russia was possible; strategic partnership between the United States and Russia was possible too.

Individuals matter. Gorbachev started the reform process inside the Soviet Union that produced democratizing changes inside Russia and closer relations with the United States. Reagan enabled those processes. Another president in the White House at the time might not have been so ready to partner with Communist Party General Secretary Gorbachev. Yeltsin continued these trajectories in partnership with Presidents Bush and Clinton. Conversely, Putin, especially in his third term, undermined democracy inside Russia and pursued foreign policies that generated renewed confrontation with the United States. In doing so, Putin was reacting to another set of individuals — protesters first in Russia and then in Ukraine. Had Russians not taken to the streets against Putin in 2011–12, U.S.-Russia relations might not have deteriorated as quickly as they did. Had Ukrainians not taken to the streets against Yanukovych in 2013–14, our current hot peace with Russia might have been avoided. But it was Putin's specific responses to these demonstrators and their actions that reverberated so negatively in our bilateral relations. Putin was not forced to crack down on Russian protesters and reinvent the United States as an enemy. Those were the choices he made. Putin was not compelled to annex Crimea and intervene in Eastern Ukraine in response to events in Kyiv. Again, those actions were his choices. Different leaders in the Kremlin at the time might have made different choices.

Critics of this explanation — and critics of me as an actor in this drama — contend that this analysis is all too convenient. Glorifying Gorbachev and Yeltsin and demonizing Putin — so they say — gives me cover for my inaccurate analysis of Russia as an academic and my poor decisions as a U.S. foreign policymaker. My failure to understand deeper structural factors at play — historical, cultural, and balance-of-power factors — allowed for my misguided belief in the possibility of Russian democracy thirty years ago and in the idea of closer relations between the United States and Russia during the Reset. Gorbachev and Yeltsin were not democrats or Westernizers, so these critics contend. They were weak leaders in office at a time of Soviet and Russian state collapse. With Russian power now stronger and the Russian state restored, Putin is a return to the Russian norm. He is an expression of Russian culture and history, and the desires of the vast majority of the Russian people, who we in the West must just learn to accept.

I take this critique seriously. After all, writing a quarter of a century ago, with the structural features and phases of the French and Bolshevik Revolutions guiding my analysis, I predicted the rise of a reactionary autocrat during the Thermidor of the Soviet Revolution.[22] I did not know his name would be Putin, but that's exactly the point. Those structural features of these earlier revolutions that produced the strongmen Napoleon and Stalin also would create the same conditions for a strongman in post-revolutionary Russia, or so I argued at the time. And that he is so popular today suggests a deep societal demand for this kind of autocratic leader, and this kind of antagonistic relationship with the United States and the West.

And yet, without denying the power of these innate forces, I still hold on to the possibility of the power of individual action. Marx was right; "Men make their own history, but they do not make it as they please; they do not make it under self-selected circumstances, but under circumstances existing already, given and transmitted from the past."[23] But when the conditions are ripe, individuals can change the course of history, for better and for ill. Had a leader other than Gorbachev come to power in the Soviet Union in 1985, the pace and course of domestic reforms could have been very different. Imagine, for instance, if a leader like Putin had been elected general secretary of the Communist Party of the Soviet Union. He would not have flirted with democratization as an instrument to stimulate economic growth. That would have been too dangerous for someone with his conservative proclivities. Even thirty years later, Putin seems comfortable with having a large portion of the economy controlled by the state. He has lamented the collapse of the Soviet Union, which he claims was not inevitable. He would have fought much harder than Gorbachev to preserve it. With a different leader in the Kremlin in 1985, warming of relations with the United States also could have been delayed or not happened at all. At the time, history, culture, or geostrategic forces did not compel Gorbachev and Reagan to reset relations. Two men did.

Yeltsin also played a unique role. His ideological commitment to Soviet dissolution, anticommunism, and closer ties to the West shaped Russian history in the 1990s. A more anti-Western nationalist leader easily could have emerged to capture opposition sentiment at the time. One such leader — Zhirinovsky — seemed to be on the rise in the early 1990s. Moreover, had the coup attempt in August 1991 succeeded — and it could have succeeded — the course of Russian reform and U.S.-Russia relations would have been very different in the 1990s.

Likewise, Putin's accidental ascendency to power in Russia played a unique and consequential role in shaping Russia's political system and relations with the United States. Had Yeltsin selected Boris Nemtsov as his successor, Russian democracy might have survived, and relations with the United States and the West more generally would have deepened. I knew Boris well. He would never have cracked down on Russia's opposition; he would never have annexed Crimea.

But could Boris Nemtsov, or someone like him, have been elected in Russia? By the time of his tragic assassination on February 27, 2015, Nemtsov had been reduced by Putin and his regime to a marginal opposition figure, helping to create the impression that people like him were out of touch with Russian society. In the 1990s, however, Nemtsov was a charismatic and popular leader, first in Nizhny Novgorod and later on the national stage as first deputy prime minister, who Yeltsin hinted at times could be his heir. He was not an extreme liberal, beyond the normative bounds of Russian voters. And remember, it was Yeltsin, not the Russian people, who selected Putin to be the next president of Russia. Had Yeltsin selected Nemtsov in 1999, he would have won election in 2000, and Russia's internal and external trajectories would have been very different.

The hardest question is whether Putin represents a return to the norm in Russian and Soviet history. Or is he an interregnum — the last forceful, successful, but ultimately fading expression of the Soviet regime — in the trajectory toward a new, more open, democratic order in Russia and a new, closer relationship with the West, of which Gorbachev, Yeltsin, and Nemtsov were the (flawed, weak, and only partially successful) founders. Only three decades into Russia's current revolutionary transformation, it remains too early to judge.

When pressed, I remain optimistic about Russia's long-term trajectory and our future with Russia. I want to believe that data and intuition inform my analysis. I fear sometimes that my Montana optimism coupled with my normative commitment to democracy might be clouding my judgment. When push comes to shove, however, I am more a social scientist than an ideologue. Data matter. Comparative historical analysis illuminates.

I find it hard to believe that Russia will defy the odds of modernization. Over a half century ago, in one of the seminal articles of political science, my former colleague at Stanford, Seymour Martin Lipset, showed a positive relationship between economic development and democracy. "The more well-to-do a nation," he argued, "the greater the chances that it will sustain democracy."[24] Russia today is a wealthy nation. Russian society is also very educated, urbanized, industrialized — other attributes of modernization that foster and sustain democratic development. In Europe, Asia, and Latin America, countries that modernized eventually consolidated democratic political systems, though rarely smoothly. Many scholars even predict a similar fate for China. Why should Russia be any different? Over the long haul, it seems unlikely that Russia will defy these world historical trends.[25]

Moreover, Putin's political system lacks many attributes that have helped sustain autocracies in other countries. Unlike China, Russia has no strong ruling political party. Russia no longer has a monarchy. Some regimes, both old and new, have relied on religious authority to justify dictatorship. That option is not available in modern Russia. Russia's current political regime is highly personalized, anchored by one charismatic leader. He cannot rule forever. Few viable candidates have emerged who could replace Putin and maintain this system.

Anecdotally, I have met too many Russians who believe in democracy, believe that Russia is a European country, and believe that closer relations between our two countries will make us all better off. I met some of those people as a student in the Soviet Union in the early 1980s, people who loved Led Zeppelin just as much as I did, wanted to pursue their own loves in life, not ones dictated by the state, and did not see me as an enemy, despite what Soviet television preached at them every night. I then met hundreds of Russians, and marched with tens of thousands of Russians, in the final year of the Soviet Union, who had embraced democracy as their ideology of opposition to the Soviet regime. We were ideological allies, not separated by different histories, geographies, or passport colors. Some of these people then joined the first government in a newly independent Russia. They faced enormous challenges. They made mistakes. Some sold out. But some remained true to those ideals that first united us. They still live in Russia. These veterans from Russia's first civil society awakening were then joined by tens of thousands of younger, more affluent, more educated Russians, who took to the streets in 2011–12 and again in 2017 to advocate for basic democratic rights. Those who resisted the August 1991 coup were idealists, motivated by beliefs more than anything else. Those who demonstrated in the Putin era also embraced democratic ideals, but now had pragmatic economic motivations for demanding a more constrained, accountable government. Those preferences have not changed. These people are quiet now. They are demobilized and demoralized. Some have emigrated. But most of them remain in Russia, waiting.

A democratic Russia will not automatically seek or obtain closer relations with the United States. The United States and Russia are two great powers that have interests all over the world. Sometimes those interests will clash, no matter what system governs inside Russia. But I continue to believe that a more democratic Russia is more likely to develop closer ties with the West, and with the United States in particular. All of the democracies in the world today enjoy close relations with the United States; many are our closest allies. Conversely, all of America's enemies — both past and present — have been autocracies.[26]

My prediction is that the next reset in relations between the United States and Russia will occur, again as a result of political change inside Russia. When such change will come is impossible to predict. But it seems impossible to imagine that it will never happen. Which is the more fanciful prediction — that Putinism will last another fifty years, or that Russia's current political system will experience some change in the next half century? I am still convinced that Russia will one day consolidate democracy and that the United States and Russia will be allies. I just do not know when that "one day" will come.

EPILOGUE: TRUMP AND PUTIN

For a brief interlude during the first year of the Trump presidency, it seemed like the hot peace between the United States and Russia of the late Obama-Putin years might be cooling. Candidate and then president Donald J. Trump heaped praise on Russian president Vladimir Putin, calling him a first-class leader. Trump signaled his readiness to forget about those actions of Putin's that had sparked confrontation with the United States and the West in the second term of the Obama administration. Annexation of Crimea, military intervention in Ukraine and Syria, further rollback of democratic practices inside Russia, and interference in the 2016 U.S. presidential election didn't seem to trouble the new American president. He just wanted to get along with Putin. In response, Putin expressed his desire to work with Trump, hoping that the new American president might deliver on the series of concessions to Russia that Trump had discussed as a candidate.

Less than a year into the Trump presidency, the Putin-Trump "bromance" had faded. Trump did not deliver on his more audacious pro-Russia campaign promises, such as lifting sanctions, looking into recognizing Crimea as a part of Russia, or withdrawing support for NATO. He may well have wanted to do these things, but his own foreign policy team, the U.S. Congress, and American public opinion were united against him. Americans have not been willing to forget about past Russian behavior. In turn, Putin took very few actions to improve relations with the United States. The disappointing, lamentable, confrontational dynamics in Russian-American relations prompted by Putin's pivot away from the Reset in 2012 continued in the Trump-Putin era.

When my former boss Hillary Clinton prepared to run for president, she and her closest foreign policy advisors worried about her vulnerabilities regarding our Russia policy. Clinton handed Lavrov the reset button at their first meeting in Geneva in 2009, forever identifying her with the policy. The conventional Republican critique of Obama's (and therefore Clinton's) approach toward Russia was that we had been too weak on Putin. Reset was a failed policy that demanded correction. I made my first (and last) appearance on Sean Hannity's radio show not long after Putin's intervention in Ukraine. Back then, Hannity insisted that America needed a new Republican president who would stand up to Putin. Likewise, Senator John McCain stated bluntly that Obama's failure to deter Putin from invading Ukraine "has made America look weak."[1] Future presidential candidate Senator Ted Cruz argued the same: "In Ukraine, we are seeing the direct consequences of a failure in American leaders . . . [Putin] has taken the measure of President Obama and has determined that he has nothing to fear from the United States, and that is why he is proceeding with impunity."[2] Another Republican presidential candidate, Marco Rubio, called Putin "a gangster and thug" and promised, if elected, tougher actions against Russia.[3]

Given my new leadership responsibilities at Stanford and my work as a news analyst with NBC News, I declined an opportunity to take a formal role in the Clinton campaign. Informally, however, I talked with my former government colleagues in major policy roles on the campaign, including Jake Sullivan and his deputy for foreign policy issues, Laura Rosenberger, about how to address the coming challenge on Russia. In policy memos drafted for the Clinton campaign, the approach was threefold: (1) defend the results of the Reset; (2) applaud Obama's coercive responses to Putin's invasion of Ukraine; and (3) pledge to do more. "Obama plus" meant greater resources and soldiers for NATO in the Baltics, more economic and military support for Ukraine including defensive weapons, new sanctions on Russia, the creation of no-fly zones in Syria, stronger support for democracy and human rights, and more expansive efforts to push back on Russian propaganda around the world.[4]

As the Republican field grew crowded, every single candidate advocated a get-tough approach on Russia, except one: Donald Trump. Instead, Trump uttered phrases and policy stances on Russia that radically departed from Republican foreign policy traditions. Rather than contain or punish Russia, Trump promised to work with Putin. As he told one campaign crowd, "We're going to have a great relationship with Putin and Russia."[5] He praised Putin effusively throughout the campaign, calling him "brilliant"[6] and a "genius."[7] Trump went out of his way to claim Putin's support. Hearing that Putin had spoken favorably about him to the Russian press, he said, "When people call you brilliant, it's always good, especially when the person heads up Russia."[8] Trump would return the compliment: "I will tell you in terms of leadership he is getting an 'A,' and our president is not doing so well."[9] When asked about possible American recognition of Russia's annexation of Crimea or lifting sanctions, Trump promised to look into it.[10] Trump also denigrated the United States' NATO allies, calling the alliance obsolete and threatening not to honor our treaty commitments if our allies did not provide more for their own defense.[11] On the campaign trail, Trump never once mentioned human rights abuses or democratic erosion in Russia. On the contrary, he seemed to admire Putin's alleged strongman ways. Most shockingly, Trump assigned moral equivalency to the United States and Russia. When asked by Joe Scarborough on MSNBC's *Morning Joe* about the killing of journalists and opposition leaders in Russia, Trump countered, "Well, I think our country does plenty of killing also, Joe."[12] When given the chance to correct the record a year later on Bill O'Reilly's television show on Fox, a very friendly venue, President-elect Trump instead doubled down. "We have a lot of killers . . . you think our country is so innocent?"[13]

Clinton did not have to defend the Reset. Instead, she had to explain why Trump was the one who was too soft on Russia. Speaking at a public policy event at Stanford, Clinton exclaimed, "It'll be like Christmas in the Kremlin" if Trump won the presidency.[14]

Logically, then, Putin supported Trump. You don't need a PhD in Russian studies to understand why Putin would prefer Trump to Clinton. What an unexpected windfall to have an American presidential candidate willing to consider recognizing Crimean "unification" with Russia, lifting sanctions, weakening NATO, and avoiding any mention of democratic values. More generally, Trump promised to pull back on American commitments around the world and radically disrupt politics at home, actions that would weaken the United States, another Putin objective. As the unclassified report by the Office of the Director of National Intelligence on Russian involvement in the 2016 presidential election stated unequivocally, "We assess Putin, his advisers, and the Russian Government developed a clear preference for President-elect Trump over Secretary Clinton."[15]

In addition to supporting pro-Kremlin policies, Trump's ideological orientation overlapped with many Putin ideas. For years, Putin railed against the liberal world order, and what he called decadent Western cultural trends. Trump did the same. Putin chastised American interventionism and hegemony. So too did Trump. Putin and his government had cultivated xenophobic, nationalist, conservative allies throughout Europe, including Marine Le Pen in France, Nigel Farage in the United Kingdom, and Viktor Orbán in Hungary.[16] To the Kremlin, Trump seemed to represent another ideological ally in this transnational movement against globalism, liberalism, and multilateralism. Self-proclaimed populist, nationalist ideologues on Trump's team, such as Steve Bannon, perceived even deeper philosophical connections to like-minded Russian thinkers.[17] Alleged defenders of the white, Christian world against Islam and China worked in or hovered around both the Kremlin and the Trump campaign.[18]

Putin also despised Clinton. As Clinton herself observed, "Our relationship has been sour for a long time."[19] Clinton stood for the exact opposite of Trump: no recognition ever of Crimean annexation, continued sanctions until Russia withdrew from Ukraine, strengthening of NATO, and a firm commitment to defending human rights. Clinton also categorically rejected the construct of a white, Christian world against the rest. And she knew Putin blamed her personally for sparking protests against the Russian government in December 2011. Putin wanted revenge. Putin also wanted to strike back for what he saw as American efforts to discredit Russia's image by revealing Russia's doping operation for Olympic athletes, and exposing massive Russian corruption, including by those close to Putin, in the publication of the Panama Papers.[20] The Russian president had no interest in having to deal with Clinton again for four more years.

Putin made his preferences known. He called Trump a "vivid" or "colorful" person (*yarkii chelovek*), which the Russian news agency, Interfax, translated as "brilliant." When pressed to explain if this was a statement of support, Putin replied, "I said in passing that Trump is a vivid personality. Is he not? He is . . . However what I definitely note and what I definitely welcome . . . is that Mr. Trump said that he is ready for the full-scale restoration of Russian-U.S. relations . . . We all welcome this!"[21] Every Russian media outlet owned or controlled by the Kremlin praised Trump and ridiculed Clinton. Most vocal in support of Trump was Dmitry Kiselyov, the host of one of Russia's most popular news programs on the state-controlled television channel Rossiya 1. He lauded Trump as strong, principled, and consistent, predicting that with Trump there would be no more "lies and other nonsense packed in beautiful words" (such as "democracy" and "human rights").[22] Pro-Putin Russian parliamentarians also endorsed Trump. Vitaly Milonov, a member of parliament from Putin's party, asserted, "I officially state that Hillary Clinton is a cursed witch. Therefore, even such a funny dude as Donald Trump looks more adequate."[23] Likewise, Russian senator Alexey Pushkov expressed his open support for Trump, claiming that he was less ideological and more businesslike than Obama or Bush.[24] In parallel, communicators of the official Kremlin line made clear their disdain for Clinton, calling her unstable, corrupt, absentminded, "another blond," racist, unattractive, and ill.[25]

Putin and his surrogates did not just root from the sidelines for their preferred candidate, but actively intervened in the American electoral process to help Trump win. More generally, Russian actions during the 2016 U.S. presidential election aimed to sow doubt about the integrity of the American electoral process and democratic institutions more broadly.

Most audaciously, cyber criminals affiliated with Russian intelligence agencies — code-named Cozy Bear and Fancy Bear by our intelligence community — stole private data from the Democratic National Committee and Clinton campaign chairman John Podesta. A private cybersecurity company, CrowdStrike, confirmed the Russian break-in and made its results public in a report called *Bears in the Midst* in June 2016, although the U.S. government had detected this Russian hacking operation long before.[26] On the eve of the Democratic National Convention in July 2016, WikiLeaks published roughly twenty thousand emails and several thousand attached files stolen from the DNC.[27] Throughout the summer and fall, WikiLeaks and DCLeaks.com continued to publish more data, including emails from Podesta, to achieve the maximum negative effect on Clinton's campaign.[28] The American intelligence community later revealed that these Russian agents conducted operations against both Republicans and Democrats, but only data stolen from the Democratic side were published.[29]

On October 7, 2016, a joint statement issued by the Department of Homeland Security and the Office of the Director of National Intelligence stated without caveat or qualification, "The U.S. Intelligence Community (USIC) is confident that the Russian Government directed the recent compromises of emails from US persons and institutions, including from US political organizations. The recent disclosures of alleged hacked emails on sites like DCLeaks.com and WikiLeaks and by the Guccifer 2.0 online persona are consistent with the methods and motivations of Russian-directed efforts. These thefts and disclosures are intended to interfere with the US election process."[30] (To distract attention away from this U.S. government intelligence assessment, WikiLeaks started publishing emails stolen from Podesta within one hour after this statement's release.) After the election, the Office of the Director of National Intelligence issued a more comprehensive report, supported by all other relevant intelligence agencies, with this conclusion:

> Russian efforts to influence the 2016 U.S. presidential election represent the most recent expression of Moscow's longstanding desire to undermine the U.S.-led liberal democratic order, but these activities demonstrated a significant escalation in directness, level of activity, and scope of effort compared to previous operations. We assess Russian President Vladimir Putin ordered an influence campaign in 2016 aimed at the U.S. presidential election. Russia's goals were to undermine public faith in the U.S. democratic process, denigrate Secretary Clinton, and harm her electability and potential presidency. We further assess Putin and the Russian Government developed a clear preference for President-elect Trump.[31]

As a result of the publication of embarrassing emails, DNC chair Debbie Wasserman Schultz was forced to resign. Recriminations within the Democratic Party ensued, as supporters of Senator Bernie Sanders — Clinton's main opponent in the primaries — accused the Democratic Party of unfairly supporting Clinton based on what they read in these emails. Clinton's image was damaged by daily media coverage of the emails, and Bernie supporters became more enraged at Clinton.

In addition to stealing and publishing private information, Russian media outlets in the United States, including the Russian state-controlled television and social media company RT and the Russian government's print and radio organizaton, Sputnik, openly supported Trump and campaigned against Clinton. RT, for instance, ran stories (recirculated on social media) claiming that Obama and Clinton created the terrorist organization ISIS; that the Clinton Foundation paid for Chelsea Clinton's wedding; and that Clinton needed to be brought to justice for her various criminal activities.[32] Sputnik echoed similar themes while also promoting Trump attacks against Clinton, including on Twitter, unabashedly using the hashtag #CrookedHillary. RT had established a strong presence on YouTube, where several million people viewed anti-Clinton clips.[33] Russian bloggers and bots on Twitter and elsewhere, often disguised as Americans, frequently amplified pro-Trump and anti-Clinton stories, including disinformation.[34] By saturating social media platforms with retweets, repostings, and likes, Russian actors aimed to influence the algorithms used by social media companies to determine what readers and viewers saw. Senator Mark Warner, the vice chairman of the Senate Select Committee on Intelligence, charged with investigating these activities, drew this conclusion: "This Russian propaganda on steroids was designed to poison the national conversation in America. The Russians employed thousands of paid internet trolls, fake accounts, and automated bots to push out disinformation and fake news at a high volume. These stories, fake news and disinformation was then hyped by the American media echo chamber and our own social media networks to reach tens of millions of Americans."[35] As Clinton remembers, "Russian trolls posted stories about how I was a murderer, money launderer, and secretly had Parkinson's disease."[36] Russian social media activists and their bots targeted potential Trump voters in swing states.[37] The Internet Research Agency, an entity whose owner the U.S. intelligence community has described as "a close Putin ally with ties to Russian intelligence," also purchased ads on Facebook in support of Trump.[38] IRA and other Russian actors also made advertising purchases on several different social media platforms. Russian actors operating on these platforms even helped to arrange political demonstrations inside the United States. These Russian efforts were multipronged and mostly undetected during the election campaign, benefiting from the absence of regulation or transparency regarding political and campaign messaging on social media platforms.

The objective of Russian efforts in U.S. media was not only to support Trump and damage Clinton, but also to undermine the truth more generally.[39] This Russian assault on American independent media echoed similar arguments advanced by Trump as well as "alt-right" publications such as Breitbart News and Infowars. The attacks damaged Clinton, helped Trump, and raised doubts about independent media, damaging American democracy in ways that lingered well beyond the 2016 presidential election. Russian activities on social media platforms continued after the election.

Neither the U.S. government nor American social media companies did anything to stop these Russian activities during the 2016 campaign. American law prohibits foreigners from contributing financially to candidates, but still allowed Russian government proxies to provide in-kind assistance by buying Facebook ads, broadcasting partisan stories on cable television and YouTube, and disseminating #CrookedHillary and other negative hashtags.

Nor during the campaign was there a huge public demand to discuss these Russian efforts to influence our elections. On August 17, 2016, I published a *Washington Post* column titled "Why Putin Wants a Trump Victory (So Much He Might Even Be Trying to Help Him)."[40] I discussed in detail the motivations behind the Russian theft and publication of DNC emails as well as Russia's comprehensive propaganda efforts to support Trump. Back then, however, many journalists, commentators, and national security experts remained skeptical about such theories. Some reporters went out of their way to dismiss these allegations as a Clinton campaign tall tale, designed to distract from the real "news" contained in the published DNC emails. Media coverage of the emails, in part because they reminded voters of Clinton's earlier email scandal, vastly exceeded attention devoted to Russian interference.[41] I was being "alarmist," one D.C. mandarin schooled me, because of my difficult times in Russia. Clinton surrogates on television experienced the same ridicule. As Clinton writes in her 2017 memoir regarding her campaign manager Robby Mook's media appearances on this subject, "Robby was treated like a kook."[42] On social media, the reaction to me when discussing these matters was explosive, especially from Trump and Putin supporters who lambasted my conspiratorial "propaganda." These virtual attacks reminded me of the same vitriol I'd been subjected to when criticizing the Russian invasion of Ukraine in 2014, only this time Americans were in the lead. And as during my time as ambassador, sometimes the attacks became threats, spilling out of the virtual and into the real world. One antagonist was particularly persistent and belligerent, calling me frequently to charge that I was undermining the Republic by making such statements in print and on television and therefore had to be stopped. If one of Putin's aims was to spark tensions within the United States, he was achieving his goals. I felt it personally.

During the campaign and even after the election, even when presented with the firm assessment of the U.S. intelligence community that Russia was behind the email hacking and other propaganda efforts, Donald Trump continued to push back. Weeks before leaving office, the Obama administration sanctioned more Russian officials in response to this meddling campaign. (It also expelled two dozen Russian diplomats in response to Kremlin harassment of American diplomats working in Russia.) Nonetheless, Trump argued repeatedly that it was impossible to know who was responsible and raised the issue of the U.S. intelligence failure in Iraq. The passion of his denials, reinforced by misleading statements about Russian contacts from his campaign team, fueled suspicion about possible collusion between Trump campaign officials and Russian agents during the election. For instance, Donald Trump Jr., Trump's oldest son, received an offer to obtain compromising information on Clinton as "part of Russia and its government's support for Mr. Trump."[43] Such alleged *kompromat* could only have been obtained by Russian intelligence services. In response to this enticing proposal from Moscow, delivered by email through a British intermediary, Donald Jr. replied, "If it's what you say I love it."[44] A Russian delegation with ties to Russian intelligence subsequently traveled to Trump Tower to meet with Donald Jr. and two of the most senior officials in the campaign at the time, Paul Manafort, the chairman of the campaign, and Trump's son-in-law and close advisor, Jared Kushner.[45] What quid pro quo was offered or agreed upon in that meeting remains unclear. That the meeting happened at all shows the Trump campaign's intent to collaborate with Russian officials in their shared objective of defeating Clinton.

Numerous other meetings between Russian emissaries and Trump campaign officials occurred during the 2016 presidential election and transition. Well before the election was over, the FBI began investigating these ties, looking for evidence of collusion, conspiracy, and blackmail. Manafort had deep, long-standing ties with pro-Putin officials and businessmen as a result of his work for Ukrainian president Viktor Yanukovych, ties that included financial transactions measured in the tens of millions of dollars. His relations with Russia became so controversial that he was forced to resign as Trump's campaign chair in August. Trump's senior foreign policy advisor on the campaign and subsequent first national security advisor, Michael Flynn, also had cultivated relationships with Russian officials, including appearing for a hefty fee at RT's tenth-anniversary celebration in Moscow, at which he was seated next to Putin. During the campaign and transition, Flynn also had numerous conversations with Sergey Kislyak, Russia's ambassador to the United States. Because he failed to disclose these Russian contacts properly — apparently lying about them to Vice President Mike Pence — Flynn, too, was forced to resign.[46] He later confessed to lying to the FBI about the substance of these conversations with Kislyak, a criminal offense. Another Trump confidant, Roger Stone, seemed to have inside information about the timing of the publication of Podesta's emails. Several other figures in the campaign and the Trump organization, including Jared Kushner, also arranged meetings with Russian officials and businesspeople in 2016.[47] Concluding that any of these contacts were improper is premature. But as this book goes to press, the investigations into these ties being conducted by Special Counsel Robert Mueller on behalf of the Justice Department and the Senate and House Intelligence Committees as well as the Senate Judiciary Committee remain under way.

In the weeks and days before the 2016 presidential election, Russian cyber actors even explored tampering with the electoral machinery used for tracking registered voters and counting votes. As then White House chief of staff Denis McDonough recalls, Russia's cyber probes were an attack on the "heart of our system . . . We set out from a first-order principle that required us to defend the integrity of the vote."[48] Initially, the FBI had identified Russian efforts to break into election systems in twenty-one states,[49] but subsequent reporting has suggested that the number may have been as high as thirty-nine.[50] The Obama administration worried that Russian hackers would tamper with voter-registration lists and even vote tabulations, actions that would call into question the integrity of the election. At the time, Trump was warning Americans that the vote would be rigged. Only the smallest of disruptions would have triggered a major crisis in confidence in the integrity of the voting process. Members of the Obama administration were so concerned about Russian interference on Election Day that they pursued a series of measures to deter Russian action. They consulted with state election officials to encourage them to enhance their cybersecurity.[51] They made preparations to attack Russia in the cyber world as well as implement crippling economic sanctions should something happen. And in his final meeting with Putin as president at a G-20 meeting in Hangzhou, China, Obama warned the Russian leader that he knew what Russia was doing in the U.S. elections and had better knock it off or face serious consequences. Thankfully, Putin did not use Russia's full cyber capacities on our Election Day.

And then Donald Trump won.

Russians close to the Kremlin celebrated Trump's unexpected election victory. "Our Trump" or #TrumpNash, as they tweeted, had just been elected president of the United States.[52] As Andrei Soldatov and Irina Borogan report, "Trump's victory was met with jubilation in Moscow. Parties were given, and in the State Duma, champagne bottles were popped."[53] Few in Moscow expected him to win, but many Russians wanted him to win. At the peak of his popularity, 70 percent of Russians held a favorable view of Trump, a much higher percentage than Trump ever achieved among Americans.[54] Russian officials looked forward to launching a new era in Russian-American relations, now that Trump would abandon Obama's allegedly flawed policies toward Russia and make the proper corrections to build a positive relationship with Putin.[55]

Did Putin make Trump president? Of course not. American voters did that. Did Putin help Trump win? Maybe. Answering this latter question definitively is more difficult, since isolating the impact of Russian efforts to influence the outcome of the 2016 presidential election is a very difficult social science problem. Obviously, many factors combined to deliver Trump's victory in the Electoral College, most of which had nothing to do with Russian actions.[56] In the margins, however, Russian interference played a role, and this election was won in the margins. Trump triumphed in the Electoral College by winning Michigan, Pennsylvania, and Wisconsin by a combined total of less than seventy-eight thousand votes. Postelection surveys revealed that 12 percent of Sanders's supporters in the Democratic primaries voted for Trump in the general election.[57] Voter turnout among Democrats also was lower in 2016 than in 2012 in key swing states. Votes cast for third-party candidates were significantly higher in 2016 than in 2012, including in swing states.[58] The publication of DNC and Podesta emails, as well as Russian anti-Clinton messaging campaigns in U.S. media (both broadcast and online) contributed to these outcomes. Right up until Election Day, Russians bought social media ads targeted at voters in swing states, while WikiLeaks continued to publish information stolen by Russian agents, fueling the damaging association between Clinton and emails. As Clinton remembers, "The WikiLeaks email dumps kept coming and coming. It was like Chinese water torture. No single day was that bad, but it added up, and we could never get past it."[59] That Trump consistently mentioned WikiLeaks — by one count more than 160 times in the last month of the campaign — suggests that he knew the electoral power of this Putin gift.[60]

Before being fired by Trump, FBI director James Comey testified, "They [the Russians] were unusually loud in their intervention."[61] In violating U.S. sovereignty during one of America's most sacred acts as a democracy, Putin did not seem to care if Russian actions were exposed. Officially, of course, Putin repeated, "We have never meddled in the domestic affairs and never will." But when pressed about the theft of DNC and Podesta emails after the election, Putin cynically joked that maybe it was done by "patriotically minded" Russians.[62] Putin also employed whataboutism, deflecting criticism by making a counteraccusation. When confronted about Russian interference in the American presidential election, Putin responded, "The United States, everywhere, all over the world, is actively interfering in electoral campaigns in other countries,"[63] as if to justify Russian intervention in the American election. In another interview, Putin defensively countered, "When . . . we are told: 'Do not meddle in our affairs. Mind your own business. This is how we do things,' we feel like saying: 'Well then, do not meddle in our affairs.'"[64] Some pro-Kremlin pundits even seemed proud of their achievement. And why not? As *Washington Post* investigative journalists Greg Miller, Ellen Nakashima, and Adam Entous concluded, "Russia's interference was the crime of the century, an unprecedented and largely successful destabilizing attack on American democracy."[65]

Putin did not receive the payoff he expected from Trump's victory. As President-elect Trump assembled his foreign policy team, he invited into his administration several national security experts with more skeptical views of the

Russian leader and his policies. He nominated my Hoover Institution colleague General James Mattis, no pushover where Putin is concerned, to serve as his secretary of defense. After three weeks in the White House, Trump replaced his pro-Russian national security advisor, General Michael Flynn, with General H. R. McMaster, whose last job in the U.S. Army before moving over to the White House was crafting a new military strategy to deter Russian aggression in Europe.[66] At the United Nations, Trump's new ambassador, Nikki Haley, quickly established a rhetorical track record of talking tough on Russia. In a sharp departure from her boss in the White House, Ambassador Haley stated boldly, "Everybody knows that Russia meddled in our elections."[67] She also lambasted Russian indifference toward Assad's use of chemical weapons in Syria, and she spoke repeatedly in support of Ukrainian sovereignty.[68] Trump's choice as secretary of state, Rex Tillerson, expressed more conciliatory views regarding Putin and Russia. As the former CEO of ExxonMobil, Tillerson penned a joint venture with Rosneft, Russia's largest oil company, worth potentially hundreds of billions of dollars. Putin was intimately involved in crafting this deal. Tillerson never supported sanctions against Russia in response to Russia's invasion of Ukraine and instead pressed for moving forward with Russia, despite past transgressions.[69] Steve Bannon, Trump's chief strategist at the White House, also favored closer ties with Russia, but he was forced to resign after just several months in the job. On balance, the Trump national security team during their first year in office demonstrated more continuity with than disruption of Obama's approach to Russia.

The one dissident on Russia on the Trump national security team was Trump himself. With more consistency than on almost any other foreign policy issue, Trump continued to insist that the United States needed a better relationship with Russia; that he personally desired closer ties with Putin; and that Russia didn't interfere in our election. Trump defended Putin and his commitment to work with the Kremlin even as the congressional and Justice Department investigations probing ties between Russian interlocutors and Trump, his children, and his advisors expanded.

President Trump's first meeting with a Russian official took place on May 10, 2017, when Foreign Minister Lavrov visited the Oval Office. I watched that event closely, as I had organized the first Lavrov meeting with Obama at the White House in May 2009. There were many contrasts between the two encounters.

Most notably, Trump's White House team did not allow American reporters to cover the event, but did invite a Russian photographer from TASS (a Russian government news agency) to snap photos of Trump and Lavrov sharing hearty laughs. They did not look like they were discussing Russia's invasion of Ukraine, Crimean annexation, attacks on Russian opposition leaders, the obliteration of the Syrian city of Aleppo, or Russian violations of American sovereignty in the 2016 presidential election. Instead, Trump blasted FBI director Comey as a "nut job," and provided Lavrov with secret information provided to the United States by Israel.[70] In his press conference, Lavrov seemed very pleased with his meeting with the new American president.

Trump then went out of his way to accommodate Putin when they finally met for the first time, on July 7, 2009, on the sidelines of the G-20 summit in Munich, Germany. Again I watched closely, as I had arranged the first meeting between Obama and President Medvedev on the sidelines of a G-20 summit in London in April 2009. This sit-down was also different. Even the format was unorthodox. Trump allowed Secretary of State Tillerson alone to attend the meeting, leaving out of the room his national security advisor, H. R. McMaster, and his senior Russian advisor at the National Security Council, Fiona Hill. When I worked at the White House, I attended every single meeting that Obama had with a Russian official. The content, as best we can guess from the readout provided by Tillerson, was also unique. Trump did not express his personal disapproval of Putin's meddling in the U.S. presidential campaign, but raised "the concerns of the American people regarding Russian interference in the 2016 election."[71] When Putin pushed back, saying he'd done nothing, the discussion ended. As Tillerson later explained, "I think what the two Presidents, I think rightly, focused on is how do we move forward; how do we move forward from here. Because it's not clear to me that we will ever come to some agreed-upon resolution of that question between the two nations."[72] Later that evening, at the dinner for G-20 leaders, Trump spent an inordinate amount of time talking to Putin confidentially, prompting others at the dinner to leak information about their off-the-record chat.[73]

President Trump is quick to criticize. In his first year in office, he berated not only his Democratic critics and the media but also American allies, Republican leaders in Congress, members of his cabinet, and senior officials on his White House staff. Putin, however, got a pass. President Trump continued to deny and downplay Russian interference in the 2016 presidential election, fearing that recognition of Russian involvement would undermine the legitimacy of his electoral victory. He repeatedly claimed, for instance, as he insisted in this February 16, 2017, tweet, that the story of Russian meddling had been generated by the Democrats: "The Democrats had to come up with a story as to why they lost the election, and so badly (306), so they made up a story—RUSSIA. Fake news!"[74]

Trump's obsession with winning favor from the Russian president fueled yet more speculation about collusion during the presidential election.[75] Trump critics claimed that this praise of Putin was the payoff for Russian help in 2016. Others asserted that the Russian government had collected compromising material on Trump during his trips to Russia, including his last visit in 2013 to attend his Miss Universe pageant.[76] A third theory focused on financial ties and even money laundering—that Trump, his sons, and his associates had done shady, secretive business deals with Russian oligarchs, which allowed the Kremlin to blackmail the American president.[77] None of these allegations have been proven. But again, as this book goes to press, congressional and Justice Department investigations continue to examine all of these hypotheses.

Trump's eagerness to appease Putin worried congressional leaders, as well as other American national security officials in and outside government, including me. Trump's hints at lifting sanctions—without spelling out what the United States or allies would get in return—were particularly troubling. By 2017, a large majority of Americans involved in security issues saw Putin as acting well out of the bounds of the norms, rules, and treaties of the international system. Putin annexed territory in Ukraine, a shocking violation of international law that had not occurred in Europe since World War II. Putin authorized the carpet bombing of Syrian cities, including the ancient city of Aleppo, to shore up a ruthless dictator who had slaughtered hundreds of thousands and displaced millions more of his own citizens. Putin intervened dramatically in our election, another outlandish first.[78] Putin's aggressive behavior, therefore, had to be contained, not rewarded. Fearing Trump's conciliatory impulses—some even called these tendencies appeasement[79]—the Republican-controlled Congress tied the president's hands by passing with veto-proof majorities the Countering America's Adversaries Through Sanctions Act. This law put in place some new penalties against Russians and made it difficult for the executive branch to lift existing sanctions without congressional approval. Trump reluctantly signed into law what he called "seriously flawed" legislation.[80]

Days before Trump signed this new sanctions law, the Russian Ministry of Foreign Affairs instructed the State Department to reduce the staff at the Moscow Embassy and three consulates by 755 people. Putin had delayed reaction to Obama's expulsion of three dozen Russian diplomats in December 2016 as punishment for interfering in the U.S. presidential election and harassing American diplomats serving in Russia. Putin waited because Trump pledged to reverse Obama's actions. Six months into the Trump presidency, however, Putin had concluded that the new American president could not reverse Obama's expulsions or lift sanctions. So he struck back, in characteristically asymmetrical fashion.

Putin's action dismayed and angered me. I knew many of these embassy employees, including the Russian nationals who were going to lose their jobs and have little chance of finding new work in Russia. They would be labeled traitors for working with us. Almost immediately after the Russian declaration, I heard from a former Russian employee in Moscow asking me to help him obtain a visa to immigrate to the United States. He did not want to leave his homeland, family, and friends, but felt he had no choice. It was disheartening. Not even during the most contentious days of the Cold War had the Kremlin taken such actions against our diplomats.

Trump did not try to negotiate a smaller staff reduction. He expressed no solidarity with those serving him and our country inside Russia, but instead flippantly applauded Putin, "because we're trying to cut down our payroll, and as far as I'm concerned, I'm very thankful that he let go of a large number of people because now we have a smaller payroll."[81] He proved unwilling to stand up for his own team serving his administration and our country in Russia. Thankfully, the State Department eventually responded by forcing the Russian Ministry of Foreign Affairs to close its consulate in San Francisco as well as two other diplomatic annexes.

For all his conciliatory gestures, Trump won no favors from Moscow. After he signed the sanctions law, Russian prime minister Medvedev ridiculed Trump in public, writing on his Facebook page that the new legislation "ends hopes for improving our relations with the new U.S. administration. Second, it is a declaration of a full-fledged economic war on Russia. Third, the Trump administration has shown its total weakness by handing over executive power to Congress in the most humiliating way."[82] Other Russian officials and Kremlin supporters berated America's alleged "deep state," which had boxed in the maverick, antiestablishment president.[83] As Putin himself argued, "I have already spoken to three U.S. Presidents. They come and go, but politics stay the same at all times. Do you know why? Because of the powerful bureaucracy."[84] Putin also blamed the Democrats — the losers in the last presidential election — for blocking Trump's laudable efforts to normalize relations with Russia.[85] That Trump could not overcome these Democrats, especially when Republicans controlled both the Senate and the House — underscored for Moscow Trump's ineptitude. Expectations in Moscow about Trump faded fast; the official Kremlin line on Trump was reversed. The same television commentator Kiselyov who had praised Trump during the campaign now, mere months into his presidency, compared him to North Korea's Kim Jong-un in his impulsivity and unpredictability.[86] We Americans were the enemy again. Russian citizens as well quickly withdrew their support for the new American president.[87]

In his first year in office, Trump failed to articulate a coherent strategy for dealing with Russia. While clear about his desire to befriend Putin, Trump never elucidated the foreign policy objectives he sought to achieve while engaging Russia. "Good relations" should never be the goal of U.S. foreign policy toward Russia or any country. Diplomacy is not a popularity contest. Rather, better engagement always must be used as one of many possible means to advance American security, prosperity, and values goals. Trump confused means and ends in formulating his ideas about Russia policy, defining "better relations" as his main objective instead of articulating concrete security and economic objectives of benefit to the American people. The result was unimpressive.

But Putin did little to help the cause either. The central contentious issue in U.S.-Russian relations during my last years in government — how to respond to protest movements in the Arab Spring, Russia, and Ukraine — had subsided by January 2017. Putin should have been pleased. And yet, he took no initiatives to signal credibly his interest in better relations with the United States. For instance, he could have lifted the ban on adoptions for Russian orphans, proposed a new initiative on trade and investment, or signaled a desire to have the Trump administration more involved in negotiations regarding Ukraine. He took none of these actions. The Kremlin's only cooperative initiative with the new team in Washington concerned the creation of safe zones in Syria, but even this gesture reflected Russia's victories in Syria and not a conciliatory gesture toward the United States.[88] More dramatic were Putin's negative signals: the expulsion of several hundred American employees from the U.S. Embassy and consulates in Russia, arming the Taliban terrorists fighting our soldiers in Afghanistan, threatening vaguely "to eliminate this threat" if Sweden joined NATO, and new instances of cyberattacks and meddling in other elections.[89] Instead of cooperation, preserving the United States as an enemy — especially just months before his next reelection bid in March 2018 — seemed like Putin's more important objective.

Irrespective of who was in the White House, Putin in 2016 was no longer seeking to join our Western clubs, or play by the rules of the international order. Instead of acting as a responsible stakeholder in the existing international system, Putin has pushed to undermine this system, which he believes was erected and dominated by the United States to serve American national interests. Invasion and annexation of Ukraine, military intervention in Syria to shore up the world's worst dictator, and audacious interference in the American presidential election are but three in a series of Putin's foreign actions that demonstrate his fearlessness in challenging the norms of the liberal world order — in challenging us. Strikingly, Putin has remained unrepentant about his foreign policy actions in Ukraine, Syria, or the United States. He also seems to care little what outsiders think about his undemocratic practices at home. Instead, Putin, his government, and his media outlets have continued to blame the United States for Russian domestic problems. America is still looking for ways to overthrow the Putin regime, or so the Kremlin continues to assert, just as it did during my tenure in Russia as ambassador. When the American Embassy announced that it had to discontinue issuing tourist visas for several days in August 2017 to meet the reduction in staff mandated by the Kremlin, Foreign Minister Lavrov fantasized, "The American authors of these decisions have come up with another attempt to stir up discontent among Russian citizens about the actions of the Russian authorities. It's a well known logic . . . and this is the logic of those who organize color revolutions."[90] Delaying Russian tourist trips to Disneyland was yet another clever scheme of ours to overthrow Putin! Lavrov's quip was further proof that the Russian government's obsession with what it saw as American efforts to spur revolution in the state had not ended with my departure. Most amazingly, Lavrov's press spokesperson blamed me personally for the U.S. government decision on visas, as if I were continuing to pull strings at the embassy from seven thousand miles away in California. (A Russian television series called *Sleepers*, in which the United States is plotting to overthrow the Russian government, began airing in the fall of 2017. In the show, the U.S. ambassador is blond; his name is Michael. They haven't forgotten about me yet!) Kremlin paranoia about America did not end just because Trump was elected president.

Even without a breakthrough on sanctions relief or recognition of Crimea as part of Russia, the first year of the Trump presidency did serve Putin's interests. First, America's image as a democracy had been damaged. Trump's constant discussion of "fake news" confirmed a Putin claim about American media. Deadly clashes between neo-Nazis and protesters in Charlottesville, Virginia, in August 2017 underscored another of Putin's favorite claims: America has no right to criticize Russia, as America has its own problems with the practice of democracy. Trump's disrespect for the rule of law and obsession with a deep state underscored Russian government themes about the facade of American democracy championed on Russian state-controlled media outlets for years. Trump's loose commitment to the truth (by the summer of 2017, more than 60 percent of Americans thought that the president was not honest)[91] further reinforced the argument that Americans were no different from anyone else.[92] Few abroad, including Russian democratic activists or government officials, remain inspired by American democracy today. Some American commentators even expressed worries that the United States was becoming an autocratic regime.

My own view is that American democratic institutions remain resilient and will survive Trump.[93] But the growing perception that American democracy is no different from or better than any other political system in the world will linger throughout the Trump presidency and maybe beyond. All of these disturbing challenges to American democracy help Putin make his case that the United States is not exceptional, that American democracy is no different from Russia's political system, and that we have no moral authority to preach to other countries about their behavior.

American leadership in the world as well as the perception of the positive role of American global leadership have also diminished in the first year of the

Trump era. That too serves Putin's interests. Trump's decision to withdraw from the Trans-Pacific Partnership (TPP), the Paris Climate Accords (the United Nations Framework Convention on Climate Change), and maybe even the nuclear agreement with Iran has undercut America's global stature. Trump's protestations about supporting American allies have reduced our stature in Europe and Asia. Trump's refusal to ever discuss democracy or mention human rights means that the United States is no longer perceived as the leader of the free world. American retrenchment has created opportunities for Putin to fill the leadership void.

In addition, Trump's erratic foreign policy pronouncements have undermined America's reputation around the world, allowing Putin to look like the rational one in the eyes of many. Globally, trust in the American president declined dramatically, on average by three times, between the last year of the Obama administration and Trump's first year in office. People in only two countries — Israel and Russia — had a more favorable view of Trump than Obama. Around the world, trust in Putin is low — only a quarter of respondents in thirty-seven countries' polls say that they have confidence in the Russian president — but Trump's standing is even lower.[94] Most shockingly, in 2017 citizens in nine American allies — Japan, South Korea, France, Germany, Greece, Hungary, Italy, Spain, and Turkey — trusted Putin more than Trump to do the right thing in international affairs.[95]

The Kremlin initially intervened in the U.S. election with the objectives of damaging Clinton and disrupting the American democratic process. It could not have anticipated a Trump victory; Trump himself did not expect to win. Nonetheless, the results — even if unintended — have clearly benefited Putin, at least in the short run, and not the United States. At the same time, the American backlash against Russia's meddling in the 2016 presidential election has begun. The U.S. Congress has frozen in sanctions, probably for a long time. Several congressional and judicial investigations, as well as the Department of Justice inquiry, continue. Americans have grown even warier of Putin and his intentions. American commitments to strengthening NATO have grown. In the long run, Putin's election meddling may produce more permanent damage to U.S.-Russian relations than any short-term upside for the Kremlin. The hot peace, tragically but perhaps necessarily, seems here to stay.

ACKNOWLEDGMENTS

I have written many academic and policy-oriented books, but never anything like this one. I don't think that I have ever used the word "I" until writing *From Cold War to Hot Peace*. For pushing me to have the courage to write such a book — at the same time historical, analytic, and personal — I want to thank Tina Bennett, my friend and agent at William Morris Endeavor (whom I first met when I was her TA for a Stanford summer program in Krakow, Poland, in the 1980s). Without her vision, I either would not have written another book at all or defaulted back to my academic ways. Whether this genre works or not for the reader is entirely on me. But I am satisfied with the product, and thank Tina first and foremost for the prod, guidance, and confidence.

I also want to thank Bruce Nichols, my editor at Houghton Mifflin Harcourt, for believing immediately in this project and guiding its development from beginning to end. He too embraced this mix of voices in the book and encouraged me to write the story the way I wanted to tell it. I also thank him for making sure that the book did not extend to 800 pages! In addition, I received terrific editorial assistance from Ben Hyman and Melissa Dobson, and I am grateful to Ivy Givens for her help with the photos.

I also am deeply indebted to John Powers, director for access management at the National Security Council at the White House. John arranged for me to review my own archival material from my time at the National Security Council, and then he and his team reviewed the manuscript not once but twice to make sure no classified material seeped from my memory into these pages. He also ran point for getting clearances from the State Department. Some sections had to be cut — saved for future historians to reveal — but I am deeply appreciative for John's work and guidance throughout the process.

At Stanford, a terrific team of research assistants, who sometimes doubled as editors as well, helped to produce this book. I am grateful to Breana Dinh, Michael Goldfien, Will Roth, Paul Shields, and Henry Ulmer for their research work. Anna Coll and Alice Underwood engaged in both extraordinary research and editorial assistance. In addition, Anna undertook the difficult job of protecting my time so that I could actually write this book. Magdalena Fitipaldi helped with some major administrative lifts to get this project over the finish line. And as full-time research assistants, polite but tough critics, interested analysts, and just overall great collaborators on this book, I am especially grateful to Sam Rebo and Anya Shkurko for their tremendous efforts on this project. Sam was there in the beginning, helping to transform disjointed chapters into narratives with endnotes. Anya was with me for the final push, adding value on everything from grammatical edits in English (though English is her second language!) to tracking down obscure Russian sources and making decisions about which photos to use. My students in PoliSci 213, a seminar I taught on U.S.-Russian relations in the spring of 2017, also provided many great comments that improved my thinking on and writing of this manuscript.

I also thank those who agreed to read parts of the book. It always amazes me when people take the time, without compensation, to read my work in draft. I owe them all, as everyone made the book better. I especially want to thank Chris Miller, Masha Lipman, Amr Hamzawy, and Kathryn Stoner, who provided detailed comments on specific chapters, as well as several people from the Obama administration who did the same on individual chapters and sections, including Ambassador Bill Burns, Tom Donilon, Ambassador Robert Ford, George Look, Denis McDonough, Stacy McTaggert, Jim Miller, Ambassador Victoria Nuland, Ben Rhodes, Ambassador Susan Rice, Brad Roberts, Laura Rosenberger, Gary Samore, Howard Solomon, Jim Timbie, Celeste Wallander, and Jeremy Weinstein. It should go without saying that none of the people are responsible for the content of this book. No one read the entire manuscript. Some may disagree with some of my conclusions. One of the pleasures of writing this book, however, was remembering the good times and great memories I shared with these government officials. I loved being on the Obama team.

And for that opportunity, I want to thank President Barack Obama. It was the greatest privilege of my professional life to work for him, first at the White House and then at the U.S. Embassy in Moscow. Thank you, Mr. President, for the chance to "dream big dreams" with you, even if our biggest dreams regarding U.S.-Russia relations were never realized.

I also want to thank my parents, Kip and Helen McFaul, who were there at the beginning of my interest in the Soviet Union, encouraging me all along the way to take intellectual chances, learn about new places even when that meant traveling to the "evil empire" for my first trip abroad, and try to make the world a better place.

My wife, Donna, signed up not just for two years in Russia or five in the Obama administration, but for thirty-four years so far with me and by association with Russia. She first came to see me in the USSR when we were in college in the spring of 1985, lived with me again in Moscow in the 1990s, and joyfully packed our bags together for our last stint in Russia earlier this decade. We have endured together the travails of KGB harassment and Putin's wrath, but also the joys of conversation, song, dance, and laughter with so many Russian friends. On all dimensions of life, but including our shared history with Russia and Russians, I am forever grateful for her encouragement, support, and equanimity.

Finally, I want to thank my two sons, Cole and Luke, for not just coming along for the ride in allowing me to pursue my ideas and passions for Russian-American relations, but also for actively participating in the process. I pulled them all out of paradise — our home on Stanford's campus — to spend three years in Washington. I then asked them to sign up for two more years with Team Obama, only now in Moscow. That is a lot to ask. For embracing the adventure with humor, a sense of national service, and sacrifice for me, I will be forever grateful. I wrote this book with them especially in mind, so that they will know what I was trying to do regarding U.S.-Russian relations, and what might still be done in the future (maybe even by one of them!).

NOTES

PROLOGUE

1. After decades of a focus on structural variables in international relations theory, the role of leaders has recently enjoyed a small renaissance. See, for instance, Giacomo Chiozza and H. E. Goemans, *Leaders and International Conflict* (Cambridge: Cambridge University Press, 2011), and Michael Horowitz, Allan Stam, and Cali Ellis, *Why Leaders Fight* (Cambridge: Cambridge University Press, 2015).
2. To hold myself accountable and avoid a revisionist history of my thinking, I have included extensive footnotes to my own writings during the period in question, allowing interested readers to check for themselves what I was saying at the time.

1. THE FIRST RESET

1. Residents of the USSR who were refused permission to emigrate, especially Jews forbidden to emigrate to Israel.
2. For a full account of Gorbachev's economic reforms, see Chris Miller, *The Struggle to Save the Soviet Economy: Mikhail Gorbachev and the Collapse of the USSR* (Chapel Hill: University of North Carolina Press, 2016).
3. Mikhail Gorbachev, "Korennoi vopros ekonomicheskoi politiki partii," excerpt from Gorbachev's speech to the CPSU Central Committee, June 11, 1985; reprinted in Gorbachev, *Gody trudnykh reshenii, 1985–1992* (Moscow: Alpha-Print, 1993), 33.
4. David F. Schmitz, *Brent Scowcroft: Internationalism and Post–Vietnam War American Foreign Policy* (Lanham, MD: Rowman & Littlefield, 2011), 116.
5. Associated Press, "Shevardnadze Resigns: Gorbachev Says He's 'Stunned,'" *Los Angeles Times,* December 20, 1990, http://articles.latimes.com/1990-12-20/news/mn-9704_1_shevardnadze-resigns.
6. Serge Schmemann, "Soviet Crackdown: Latvia; Soviet Commandos Stage Latvia Raid; At Least 5 Killed," *New York Times,* January 21, 1991, http://www.nytimes.com/1991/01/21/world/soviet-crackdown-latvia-soviet-commandos-stage-latvia-raid-at-least-5-killed.html.
7. *Forbes* estimated a turnout of 800,000 (Антон Трофимов, "15 самых знаменитых массовых акций постсоветской России," *Forbes,* February 3, 2012, http://www.forbes.ru/sobytiya-slideshow/lyudi/78962-15-samyh-znamenityh-massovyh-aktsii-postsovetskoi-rossii?photo=3); the *New York Times* estimated 100,000 (Serge Schmemann, "Soviet Crackdown: Moscow, 100,000 in Moscow Protest Crackdown," *New York Times,* January 21, 1991, http://www.nytimes.com/1991/01/21/world/soviet

-crackdown-moscow-100000-in-moscow-protest-crackdown.html), as did *Getty* (Vitaly Armand, "About 100,000 Demonstrators March on the Kremlin in Moscow on January 20, 1991," *Getty Images,* January 20, 1991, http://www.gettyimages.com/detail/news-photo/about-100-000-demonstrators-march-on-the-kremlin-in-moscow-news-photo/104790422#about-100-000-demonstrators-march-on-the-kremlin-in-moscow-on-january-picture-id104790422).

8. The full interview with Zhirinovsky appears in Michael McFaul and Sergey Markov, *The Troubled Birth of Russian Democracy* (Stanford, CA: Hoover Institution Press, 1993).
9. President George H. W. Bush, "Remarks to the Supreme Soviet of the Republic of the Ukraine," Kyiv, Soviet Union, August 1, 1991, American Presidency Project, http://www.presidency.ucsb.edu/ws/?pid=19864.

2. DEMOCRATS OF THE WORLD, UNITE!

1. "End of the Soviet Union; Text of Bush's Address to Nation on Gorbachev's Resignation," *New York Times,* December 26, 1991, http://www.nytimes.com/1991/12/26/world/end-soviet-union-text-bush-s-address-nation-gorbachev-s-resignation.html.
2. Stanley Fischer, "Russia and the Soviet Union: Then and Now," in *The Transition in Eastern Europe,* vol. 1, ed. Olivier Jean Blanchard, Kenneth A. Froot, and Jeffrey D. Sachs (Chicago: University of Chicago Press, 1994), 234; and Igor Filtochev and Roy Bradshaw, "The Soviet Hyperinflation: Its Origins and Impact Throughout the Former Republic," *Soviet Studies* 44, no. 5 (1992): 1.
3. Stanley Fischer, "Stabilization and Economic Reform in Russia," *Brookings Papers on Economic Activity* 1 (1992): 79.
4. Andrei Shleifer and Daniel Treisman, "A Normal Country: Russia After Communism," *Journal of Economic Perspectives* 19, no. 1 (Winter 2005): 153.
5. As quoted in Mahnaz Ispahani, "Briefing Paper: The Situation of Russian Democracy," NDI memo to international participants, December 3, 1991, 4.
6. As Bruce Russett has summarized, "First, democratically organized political systems in general operate under restraints that make them more peaceful in their relations with other democracies . . . Second, in the modern international system, democracies are less likely to use lethal violence toward other democracies than toward autocratically governed states or than autocratically governed states are toward each other. Furthermore, there are no clear-cut cases of sovereign stable democracies waging war with each other in the modern international system." Russett, *Grasping the Democratic Peace: Principles for a Post–Cold War World* (Princeton, NJ: Princeton University Press, 1993), 11.
7. President Bill Clinton, "A Strategic Alliance with Russian Reform," April 1, 1993, in *U.S. Department of State Dispatch* 4, no. 13 (April 5, 1993): 189–94.
8. James M. Goldgeier and Michael McFaul, *Power and Purpose: U.S. Policy Toward Russia After the Cold War* (Washington, DC: Brookings Institution Press, 2003), 93.
9. Michael McFaul, *Russia's Unfinished Revolution: Political Change from Gorbachev to Putin* (Ithaca, NY: Cornell University Press, 2001), 195.
10. Доклад Комиссии Государственной Думы Федерального Собрания Российской Федерации по дополнительному изучению и анализу событий, происходивших в городе Москве 21 сентября — 5 октября 1993 года, Москва, 1999, https://refdb.ru/look/1893418-pall.html.
11. Douglas Jehl, "Clinton Repeats Support for Yeltsin," *New York Times,* September 30, 1993, http://www.nytimes.com/1993/09/30/world/clinton-repeats-support-for-yeltsin.html.
12. Carolina Vendi Pallin, "The Russian Ministries: Tool and Insurance of Power," *Journal of Slavic Military Studies* 20, no. 1 (2007): 2.
13. Michael McFaul, "Why Russia's Politics Matter," *Foreign Affairs,* January/February 1995, 87–99.
14. Guillermo O'Donnell and Philippe C. Schmitter, *Transitions from Authoritarian Rule: Tentative Conclusions About Uncertain Democracies* (Baltimore, MD: Johns Hopkins University Press, 1986).
15. Hertsen was later renamed Bolshaya Nikitskaya Street.
16. Michael McFaul, "Nut 'N' Honey: Why Zhirinovsky Can Win," *New Republic,* February 14, 1994; "*Spektr Russkogo Fashizma*" [The Specter of Russian Fascism], *Nezavisimaya Gazeta,* November 2,

1994; and "The Specter of Russian Fascism: Recommendations for U.S. Policy," *Conversion,* no. 3, January 1994.

17. Fred Hiatt, "Alleged Swindler, Neo-Nazi Vie for Russian Voters Seeking 'New Force,'" *Washington Post,* October 21, 1994, https://www.washingtonpost.com/archive/politics/1994/10/21/alleged-swindler-neo-nazi-vie-for-russian-voters-seeking-new-force/1a70e900-e16e-42d7-991a-3aa52cf2b10c/.

3. YELTSIN'S PARTIAL REVOLUTION

1. Michael McFaul, "A Communist Rout?" *New York Times,* December 20, 1995, http://www.nytimes.com/1995/12/20/opinion/a-communist-rout.html?mcubz=3.
2. Strobe Talbott, *The Russia Hand: A Memoir of Presidential Diplomacy* (New York: Random House, 2002), 195.
3. William Neikirk and James P. Gallagher, "Smiles at the Summit," *Chicago Tribune,* April 22, 1996, http://articles.chicagotribune.com/1996-04-22/news/9604220164_1_chechnya-crisis-russian-president-boris-yeltsin-chechen.
4. Michael McFaul, *Russia's 1996 Presidential Election: The End of Polarized Politics* (Stanford, CA: Hoover Institution Press, 1997).
5. Игорь Клямкин, Александр Сегал, Леонтий Бызов, Сергей Марков, Владимир Боксер, Майкл Макфол, Василий Осташев, *Анализ электората политических сил России,* (Москва: Московский центр Карнеги, 1995).
6. Talbott, *The Russia Hand,* 195.
7. Boris Yeltsin, *Midnight Diaries* (London: Orion, 2001), 25.
8. McFaul, *Russia's Unfinished Revolution,* 292.
9. "Итоги голосования по выборам Президента Российской Федерации 1996 года," Vybory.ru, http://www.vybory.ru/spravka/results/president.php3.
10. Кирилл Родионов, "Были ли чистыми президентские выборы 1996?," Gaidar Center, July 30, 2015, http://gaidar.center/articles/transparent-election-1996.htm.
11. Валентин Михайлов, "Демократизация России: различная скорость в регионах. (Анализ выборов 1996 и 2000 гг. Место Татарстана среди субъектов РФ.)" Democracy.ru, http://www.democracy.ru/library/articles/tatarstan/page3.html.
12. Adam Przeworski, "Minimalist Conception of Democracy: A Defense," in *The Democracy Sourcebook,* ed. Robert Alan Dahl, Ian Shapiro, and José Antonio Cheibub (Cambridge, MA: MIT Press, 2003), 12.
13. Michael McFaul, published as "Russia: Still a Long Way to Go" in the *Washington Post* and "A Victory for Optimists" in the *Moscow Times. Washington Post,* July 7, 1996, https://www.washingtonpost.com/archive/opinions/1996/07/07/russia-still-a-long-way-to-go/e8658742-ff85-45d8-8c51-e13087054af2/.
14. George Kennan, as quoted in Thomas L. Friedman, "Foreign Affairs; Now a Word from X," *New York Times,* May 2, 1998, http://www.nytimes.com/1998/05/02/opinion/foreign-affairs-now-a-word-from-x.html.
15. *U.S. Policy Toward NATO Enlargement, Hearing Before the Committee on International Relations, House of Representatives, One Hundred Fourth Congress, Second Session, June 20, 1996* (Washington, DC: U.S. Government Printing Office, 1996), 31.
16. "NATO-Russia Council," website of the North Atlantic Treaty Organization, http://www.nato.int/cps/en/natohq/topics_50091.htm.
17. Nemtsov had joined the government earlier, in March 1997, as first deputy prime minister.
18. David Woodruff, *Money Unmade: Barter and the Fate of Russian Capitalism* (Ithaca, NY: Cornell University Press, 1999), xi.
19. "The Russian Crisis 1998," RaboResearch, September 16, 2013, https://economics.rabobank.com/publications/2013/september/the-russian-crisis-1998/.
20. Stanley Fischer, "The Russian Economy at the Start of 1998," International Monetary Fund, January 9, 1998, https://www.imf.org/en/News/Articles/2015/09/28/04/53/sp010998.

21. Ron Rimkus, "Russian Bond Default/Ruble Collapse," CFA Institute, May 5, 2011, https://www.econcrises.org/2016/05/05/russian-bond-defaultruble-collapse/.
22. Ibid.
23. As quoted in John Thornhill, "Opposition Grows to Yeltsin's Economic Reforms," *Irish Times,* August 19, 1998, http://www.irishtimes.com/business/opposition-grows-to-yeltsin-s-economic-reforms-1.184520. See also Joanna Chung, "Investors Are Warming to Russian Paper — FT.com," *Financial Times,* June 20, 2007, https://www.ft.com/content/ff0a2c26-1eca-11dc-bc22-000b5df10621.
24. U.S. Congress, *Russia's Road to Corruption: How the Clinton Administration Exported Government Instead of Free Enterprise and Failed the Russian People* (Washington, DC: Defense Technical Information Center, 2000).
25. John Lloyd, "The Russian Devolution, or Who Lost Russia?" *New York Times,* August 15, 1999, http://www.nytimes.com/library/magazine/home/19990815mag-russia-crisis.html.
26. "Евгений Примаков — человек, победивший дефолт 1998," Vestifinance.ru, June 29, 2015, http://www.vestifinance.ru/articles/59350.
27. Michael McFaul, "Authoritarian and Democratic Responses to Financial Meltdown in Russia," *Problems of Post-Communism,* July/August 1999, 22–32.
28. Michael McFaul, "Getting Russia Right," *Foreign Policy,* Winter 1999–2000, 58–73.
29. "NATO to Strike Yugoslavia," BBC News, March 24, 1999, http://news.bbc.co.uk/2/hi/europe/302265.stm.
30. "Конфликты между РФ и Чеченской Республикой в 1994–1996/1999–2009 годы," РИА Новости, April 18, 2011, https://ria.ru/history_spravki/20110418/365668402.html.
31. John B. Dunlop, *The Moscow Bombings of September 1999: Examinations of Russian Terrorist Attacks at the Onset of Vladimir Putin's Rule* (Stuttgart: Ibidem Press, 2014).
32. "Конфликты между РФ и Чеченской Республикой в 1994–1996/1999–2009 годы," РИА Новости, April 18, 2011, https://ria.ru/history_spravki/20110418/365668402.html.
33. "War Crimes in Chechnya and the Response of the West," Human Rights Watch, February 29, 2000, https://www.hrw.org/news/2000/02/29/war-crimes-chechnya-and-response-west.
34. Biography of Vladimir Putin, President of Russia official website, http://en.putin.kremlin.ru/bio; and Steven Lee Myers, *The New Tsar: The Rise and Reign of Vladimir Putin* (New York: Knopf, 2016), 107–11.
35. Timothy Colton and Michael McFaul, "Are Russians Undemocratic?" *Post-Soviet Affairs* 18, no. 2 (April–June 2002): 91–121.
36. For my more comprehensive assessments at the time, see the final two chapters of McFaul, *Russia's Unfinished Revolution.*

4. PUTIN'S THERMIDOR

1. This flirtation is documented in a Russian film about Nemtsov called *The Man Who Was Too Free* (Moscow: TVINDIE, 2016).
2. "Frozen Out," *Economist,* January 6, 2011, http://www.economist.com/node/17851285.
3. Mumin Shakirov, "Who Was Mister Putin? An Interview with Boris Nemtsov," *Open Democracy,* February 2, 2011, https://www.opendemocracy.net/od-russia/mumin-shakirov/who-was-mister-putin-interview-with-boris-nemtsov.
4. Steven Lee Myers, *The New Tsar: The Rise and Reign of Vladimir Putin* (New York: First Vintage, 2015), 108.
5. "The Putin Curve," *Wall Street Journal,* November 26, 2002, https://www.wsj.com/articles/SB1038271514450758308; and "Frozen Out," *Economist,* January 6, 2011, http://www.economist.com/node/17851285.
6. Ian Jeffries, *The New Russia: A Handbook of Economic and Political Developments, 1992–2000* (London: Routledge, 2002), 543.
7. "Vladimir Putin, Interview by David Frost," *Breakfast with Frost,* BBC News, March 5, 2000, http://news.bbc.co.uk/hi/english/static/audio_video/programmes/breakfast_with_frost/transcripts/putin5.mar.tx.

8. Sabina Tavernise, "Putin Critic Yields Control of His Media Empire to the Kremlin," *New York Times,* November 14, 2000, http://www.nytimes.com/2000/11/14/world/putin-critic-yields-control-of-his-media-empire-to-the-kremlin.html?ref=topics.
9. Michael McFaul, "Indifferent to Democracy," *Washington Post,* March 3, 2000, http://www.washing tonpost.com/wp-srv/WPcap/2000-03/03/065r-030300-idx.html.
10. In the 1980s those focused on such futuristic scenario-planning envisioned Japan as an eventual rival of the United States. In 1979 Harvard professor Ezra Vogel published *Japan as Number One,* in which he argued, "It is a matter of urgent national interest for Americans to confront Japanese successes." Japan eventually would translate its enormous economic power into military power, or so many believed, which in turn would create problems for the United States. But after an economic recession and slow recovery in the 1990s, the Japanese threat to U.S. security never materialized.
11. Michael McFaul and Alexandra Vacroux, "Russian Resilience as a Great Power," *Post-Soviet Affairs* 22, no. 1 (January–March 2006): 24–33.
12. Thomas Graham Jr., *Russia's Decline and Uncertain Recovery* (Washington, DC: Carnegie Endowment for International Peace, 2002).
13. Peter Baker, "The Seduction of George W. Bush," *Foreign Policy,* November 6, 2013, http://foreignpol icy.com/2013/11/06/the-seduction-of-george-w-bush/.
14. White House, Office of the Press Secretary, "Press Conference by President Bush and Russian Federation President Putin," June 16, 2001, https://georgewbush-whitehouse.archives.gov/news/re leases/2001/06/20010618.html.
15. Peter Baker, *Days of Fire: Bush and Cheney in the White House* (New York: Doubleday, 2013), 107.
16. Jane Perlez, "Cordial Rivals: How Bush and Putin Became Friends," *New York Times,* June 18, 2001, http://www.nytimes.com/2001/06/18/world/cordial-rivals-how-bush-and-putin-became-friends.html.
17. Vladimir Putin, "Why We Must Act," *New York Times,* November 14, 1999, http://www.nytimes. com/1999/11/14/opinion/why-we-must-act.html.
18. Jill Dougherty, "9/11 a 'Turning Point' for Putin," CNN, September 10, 2002, http://www.edition.cnn. com/2002/WORLD/europe/09/10/ar911.russia.putin/index.html.
19. Alexander Cooley, *Great Games, Local Rules: The New Great Power Contest in Central Asia* (New York: Oxford University Press, 2012), 53.
20. Bradley Graham and Mike Allen, "Bush to Tell Russia U.S. Will Withdraw from '72 ABM Pact," *Washington Post,* December 12, 2001, https://www.washingtonpost.com/archive/politics/2001/12/12/ bush-to-tell-russia-us-will-withdraw-from-72-abm-pact/c22d075f-d9b4-4b53-a937-d6fda1fb07f1 /?utm_term=.c5ef708229fb.
21. Terence Neilan, "Bush Pulls Out of ABM Treaty; Putin Calls Move a Mistake," *New York Times,* December 13, 2001, http://www.nytimes.com/2001/12/13/international/bush-pulls-out-of-abm-treaty -putin-calls-move-a-mistake.html.
22. "A Statement Regarding the Decision of the Administration of the United States to Withdraw from the Antiballistic Missile Treaty of 1972," President of Russia official website, December 13, 2001, http://en.kremlin.ru/events/president/transcripts/21444.
23. Igor Ivanov, "Organizing the World to Fight Terror," *New York Times,* January 27, 2002, http://www. nytimes.com/2002/01/27/opinion/organizing-the-world-to-fight-terror.html.
24. "President Bush, President Putin Discuss Joint Efforts Against Terrorism: Remarks by President Bush and President Putin in Photo Opportunity," Calgary Kananaskis, Canada, June 27, 2002, www. georgewbush-whitehouse.archives.gov/infocus.
25. Amy Woolf, *Nuclear Arms Control: The Strategic Offensive Reductions Treaty* (Washington, DC: Congressional Research Service, 2010), 1.
26. NATO-Russia Council, "About NRC," website of the North Atlantic Treaty Organization, http:// www.nato.int/nrc-website/EN/about/index.html.
27. See Timothy Colton and Michael McFaul, "America's Real Russian Allies," *Foreign Affairs,* November/December 2001, 1–13.
28. Michael McFaul, "The Precarious Peace: Domestic Politics in the Making of Russian Foreign Policy," *International Security* 22, no. 3 (Winter 1997/98): 5–35.

29. "Vladimir Putin: The NPR Interview," NPR, November 15, 2001, http://www.npr.org/news/specials/putin/nprinterview.html.
30. Mira Duric and Tom Lansford, "US-Russian Competition in the Middle East: Convergences and Divergences in Foreign Security Policy," in *Strategic Interests in the Middle East,* ed. Jack Covarrubias and Tom Lansford (Burlington, VT: Ashgate, 2007), 74.
31. "The World of 100% Election Victories," BBC News, March 11, 2014, http://www.bbc.com/news/blogs-magazine-monitor-26527422.
32. "Kyiv" is the direct transliteration from Ukrainian and is Ukraine's preferred Latinization of the spelling of its capital city. The U.S. government adopted the spelling under the Obama administration. The conventional spelling "Kiev" reflects the Russian transliteration.
33. For elaboration, see Nikolai Petrov and Andrey Ryabov, "Russia's Role in the Orange Revolution," in Anders Aslund and Michael McFaul, eds., *Revolution in Orange: The Origins of Ukraine's Democratic Breakthrough* (Washington, DC: Carnegie Endowment for International Peace, 2006), 145–64.
34. Condoleezza Rice, *Democracy: Stories from the Long Road to Freedom* (New York: Twelve, 2017), 169.
35. Robert W. Orttung, "Russia," in *Nations in Transit: Democratization from Central Europe to Eurasia,* ed. Jeanette Goehring (New York: Freedom House, 2007), 588.
36. Ibid., 579.
37. Michael McFaul, "U.S. Ignores Putin's Assault on Rights," *Los Angeles Times,* February 2, 2003, http://articles.latimes.com/2003/feb/02/opinion/op-mcfaul2.
38. Alexey Miller profile on Gazprom website, http://www.gazprom.com/about/management/board/miller/.
39. Marcel H. Van Herpen, *Putinism: The Slow Rise of a Radical Right Regime in Russia* (New York: Palgrave Macmillan, 2013), 297.
40. Anders Aslund, "Russia's New Oligarchy," *Washington Post,* December 12, 2007, http://www.washingtonpost.com/wp-dyn/content/article/2007/12/11/AR2007121101833.html.
41. Michael McFaul and Kathryn Stoner, "The Myth of the Authoritarian Model: How Putin's Crackdown Holds Russia Back," *Foreign Affairs,* January/February 2008, http://fsi.stanford.edu/sites/default/files/Myth_of_the_Authoritarian_Model.pdf.
42. "Transcript: Putin's Prepared Remarks at 43rd Munich Conference on Security Policy," *Washington Post,* February 12, 2007, http://www.washingtonpost.com/wp-dyn/content/article/2007/02/12/AR2007021200555.html.
43. Francesca Mereu, "Medvedev Taking Words from Putin's Mouth," *Moscow Times,* February 27, 2008, https://www.pressreader.com/russia/the-moscow-times/20080227/281526516753017.
44. "Russia Brushes off Western Call to Revoke Abkhaz, S. Ossetia Move," *Civil Georgia,* April 24, 2008, http://www.civil.ge/eng/article.php?id=17677.
45. Condoleezza Rice, *No Higher Honor: A Memoir of My Years in Washington* (New York: Crown, 2011), 685–86.
46. Luke Harding, "Bush Backs Ukraine and Georgia for NATO Membership," *Guardian,* April 1, 2008, http://www.theguardian.com/world/2008/apr/01/nato.georgia.
47. Michael McFaul, "U.S.-Russia Relations in the Aftermath of the Georgia Crisis," testimony to the House Committee on Foreign Affairs, September 9, 2008, http://cddrl.fsi.stanford.edu/sites/default/files/MCFAUL-Testimony-9-9-2008-FINAL.pdf.

5. CHANGE WE CAN BELIEVE IN

1. "Statements on Russia-Georgia Conflict," School of Russian and Asian Studies, September 11, 2008, http://www.sras.org/official_statements_on_russia_georgia_conflict.
2. Ibid.
3. "Gallup Daily: Election 2008," Gallup.com, http://www.gallup.com/poll/17785/election-2008.aspx.
4. Barack Obama and Joe Biden on U.S. Foreign Policy Issues, Acronym Institute, October 2008, http://www.acronym.org.uk/old/archive/docs/0810/doc10.htm.
5. McFaul, "U.S.-Russia Relations in the Aftermath of the Georgia Crisis."

6. "The First Presidential Debate," transcript, *New York Times,* September 26, 2008, http://elections.nytimes.com/2008/president/debates/transcripts/first-presidential-debate.html.
7. Dan Balz and Jon Cohen, "Economic Fears Give Obama Clear Lead over McCain in Poll," *Washington Post,* September 24, 2008, http://www.washingtonpost.com/wp-dyn/content/article/2008/09/23/AR2008092303667.html?hpid=topnews; and George F. Will, "McCain Loses His Head," *Washington Post,* September 23, 2008, http://www.washingtonpost.com/wp-dyn/content/article/2008/09/22/AR2008092202583.html.
8. "Interview with Tom Brokaw on NBC's *Meet the Press,*" December 7, 2008, http://www.presidency.ucsb.edu/ws/?pid=85042.

6. LAUNCHING THE RESET

1. See Rice, *No Higher Honor,* 685–86.
2. "Realistic Reengagement with the Soviets" is the title of chapter 25 of Shultz's memoir, *Turmoil and Triumph: My Years as Secretary of State* (New York: Scribner's, 1993).
3. Janny Scott, "Obama's Account of New York Years Often Differs from What Others Say," *New York Times,* October 30, 2007, http://www.nytimes.com/2007/10/30/us/politics/30obama.html.
4. Shultz, *Turmoil and Triumph,* 165.
5. White House, Office of the Press Secretary, "Remarks by the President at the New Economic School Graduation," July 7, 2009, https://obamawhitehouse.archives.gov/realitycheck/the-press-office/remarks-president-new-economic-school-graduation.
6. Ibid.
7. Office of the Historian, Department of State, *Foreign Relations of the United States, 1969–1976,* vol. 1, *Foundations of Foreign Policy, 1969–1972* (Washington, DC: Department of State, 2003), https://history.state.gov/historicaldocuments/frus1969-76v01/summary.
8. "Foreign Trade: Data," U.S. Census Bureau, Foreign Trade Division, https://www.census.gov/foreign-trade/balance/c5700.html.
9. "U.S. Direct Investments in China 2000–2015 | Statistic," *Statista,* https://www.statista.com/statistics/188629/united-states-direct-investments-in-china-since-2000/.
10. "Foreign Trade: Data," U.S. Census Bureau, Foreign Trade Division, https://www.census.gov/foreign-trade/balance/c5700.html.
11. "Direct Investment Position of the United States in Russia from 2000 to 2015 (in billion U.S. dollars, on a historical-cost basis)," *Statista,* https://www.statista.com/statistics/188637/united-states-direct-investments-in-russia-since-2000/.
12. "WTO: How Membership Has Changed Rules in Russia," *International Financial Law Review,* March 20, 2013, http://www.iflr.com/Article/3181873/WTO-how-membership-has-changed-rules-in-Russia.html.
13. For a review of this academic literature, see chapter 2 of Michael McFaul, *Advancing Democracy Abroad: Why We Should and How We Can* (Lanham, MD: Rowman & Littlefield, 2010).
14. Hillary Rodham Clinton, *Hard Choices* (New York: Simon & Schuster, 2014), 230.
15. "Remarks by the Vice President at the 45th Munich Conference on Security Policy," February 7, 2009, American Presidency Project, http://www.presidency.ucsb.edu/ws/index.php?pid=123108.
16. White House, Office of the Press Secretary, "Joint Statement by President Dmitry Medvedev of the Russian Federation and President Barack Obama of the United States of America," April 1, 2009, https://obamawhitehouse.archives.gov/the-press-office/joint-statement-president-dmitriy-medvedev-russian-federation-and-president-barack-.
17. Michael Schwirtz and Clifford J. Levy, "In Reversal, Kyrgyzstan Won't Close a U.S. Base," *New York Times,* June 23, 2009, http://www.nytimes.com/2009/06/24/world/asia/24base.html.
18. White House, Office of the Press Secretary, "Background Readout by Senior Administration Officials on President Obama's Meeting with Russian President Medvedev," April 1, 2009, https://obamawhitehouse.archives.gov/the-press-office/background-readout-senior-administration-officials-president-obamas-meeting-with-ru.

19. White House, Office of the Press Secretary, "Remarks by President Obama and Russian President Medvedev After Meeting," April 1, 2009, https://obamawhitehouse.archives.gov/the-press-office/remarks-president-obama-and-russian-president-medvedev-after-meeting.
20. Ibid.

7. UNIVERSAL VALUES

1. Bush's National Security Presidential Directive 58 defined ending tyranny in the world as the ultimate goal.
2. Five days before Obama's inauguration, I had mailed off to the publisher my book on this subject, *Advancing Democracy Abroad.*
3. This academic literature is reviewed in McFaul, *Advancing Democracy Abroad.*
4. "Text: Obama's Speech in Cairo," *New York Times,* June 4, 2009, http://www.nytimes.com/2009/06/04/us/politics/04obama.text.html.
5. "Text: Obama's Speech at the New Economic School," *New York Times,* July 7, 2009, http://www.nytimes.com/2009/07/07/world/europe/07prexy.text.html.
6. "Obama Ghana Speech: Full Text," *Huffpost,* August 11, 2009, http://www.huffingtonpost.com/2009/07/11/obama-ghana-speech-full-t_n_230009.html.
7. "Obama's Speech to the United Nations General Assembly," *New York Times,* September 23, 2009, http://www.nytimes.com/2009/09/24/us/politics/24prexy.text.html.
8. *National Security Strategy 2010,* National Security Strategy Archive, http://nssarchive.us/national-security-strategy-2010/.
9. Open Government Partnership website, http://www.opengovpartnership.org/about/mission-and-goals.
10. *National Security Strategy 2010.*
11. Brian Montopoli, "Full Remarks: Obama at United Nations," CBS News, September 23, 2009, http://www.cbsnews.com/news/full-remarks-obama-at-united-nations/.
12. In fact, President Bush did not use force to promote democracy, but that was a widely held perception at the time.
13. David Adesnik and Michael McFaul, "Engaging Autocratic Allies to Promote Democracy," *Washington Quarterly* 29, no. 2 (2006): 5–26; and Kathryn Stoner and Michael McFaul, eds., *Transitions to Democracy: A Comparative Perspective* (Baltimore, MD: Johns Hopkins University Press, 2013).
14. "Russian Democracy Weak, Admits Medvedev," *Sydney Morning Herald,* September 12, 2009, http://www.smh.com.au/world/russian-democracy-weak-admits-medvedev-20090911-fllm.html.
15. Shaun Walker, "Medvedev Promises New Era for Russian Democracy," *Independent,* November 13, 2009, http://www.independent.co.uk/news/world/europe/medvedev-promises-new-era-for-russian-democracy-1819927.html.
16. "Dmitry Medvedev Addressed the World Economic Forum in Davos," President of Russia official website, January 26, 2011, http://en.kremlin.ru/events/president/news/10163.
17. Francesca Mereu, "Medvedev Taking Words from Putin's Mouth," *Moscow Times,* February 27, 2008, https://www.pressreader.com/russia/the-moscow-times/20080227/281526516753017.
18. Dmitry Medvedev, "Go Russia!" President of Russia official website, September 10, 2009, http://en.kremlin.ru/events/president/news/5413.
19. Ibid.
20. "Analysis: Putin Critics Hit Back over Charge of Western Funding," Reuters, December 13, 2011, http://www.reuters.com/article/us-russia-financing-idUSTRE7BC0ZZ20111213.

8. THE FIRST (AND LAST) MOSCOW SUMMIT

1. Some referred to them as нашисты (*nashisty*), which sounds similar to *fashisty,* meaning fascists. The opposition called them Putin-Jugend, referring to the Hitler-Jugend.

2. Peter Pomerantsev, "The Hidden Author of Putinism," *Atlantic*, November 7, 2014, http://www.theatlantic.com/international/archive/2014/11/hidden-author-putinism-russia-vladislav-surkov/382489/.
3. Ibid.
4. White House, Office of the Press Secretary, "U.S. Russia Relations: 'Reset' Fact Sheet," June 24, 2010, https://obamawhitehouse.archives.gov/the-press-office/us-russia-relations-reset-fact-sheet.
5. "'Safe Haven' for Taliban, Al Qaeda in Pakistan," CBS News, August 8, 2007, http://www.cbsnews.com/news/safe-haven-for-taliban-al-qaeda-in-pakistan/. See also Clinton, *Hard Choices*, 159–73.
6. Thom Shanker and Richard Oppel Jr., "U.S. to Widen Supply Routes in Afghan War," *New York Times*, December 30, 2008, http://www.nytimes.com/2008/12/31/world/asia/31military.html\pagewanted=all&_r=0.
7. Cong. Record vol. 155, no. 174 (November 21, 2009), statement of Sen. Kyl, https://www.gpo.gov/fdsys/pkg/CREC-2009-11-21/html/CREC-2009-11-21-pt1-PgS11969.htm.
8. White House, Office of the Press Secretary, "Press Conference by President Obama and President Medvedev of Russia," July 6, 2009, https://obamawhitehouse.archives.gov/the-press-office/press-conference-president-obama-and-president-medvedev-russia.
9. Jesse Lee, "In Russia, Defining the Reset," White House blog, July 6, 2009, https://obamawhitehouse.archives.gov/blog/2009/07/06/russia-defining-reset.
10. The lead drafter of this speech was Rhodes's deputy, Terry Szuplat, an incredibly gifted speechwriter with whom I worked closely on this address, but of course others engaged closer to delivery.
11. Сергей Измалков, "В чём важность речи Обамы в РЭШ," Slon.ru, July 9, 2009, http://slon.ru/world/v_chem_vazhnost_rechi_obamy_v_resh-84463.xhtml.
12. Our partners were Sarah Mendelson at the Center for Strategic and International Studies (CSIS), Horton Beebe-Center at the Eurasia Foundation in Washington, and Andrey Kortunov at the New Eurasia Foundation in Moscow.
13. White House, Office of the Press Secretary, "Remarks by the President at Parallel Civil Society Summit," July 7, 2009, https://obamawhitehouse.archives.gov/the-press-office/remarks-president-parallel-civil-society-summit.

9. NEW START

1. White House, Office of the Press Secretary, "Remarks by President Barack Obama in Prague as Delivered," April 5, 2009, https://obamawhitehouse.archives.gov/the-press-office/remarks-president-barack-obama-prague-delivered.
2. The old START treaty did not directly verify how many warheads each missile carried, but instead established an upper limit on the number of warheads that could be deployed on each type of missile, so it was critical to monitor missile development through flight-test data as a means to estimate the number of warheads on the missile.
3. "New START," U.S. Department of State, https://www.state.gov/t/avc/newstart/index.htm.
4. Amy Woolf, "The New START Treaty: Central Limits and Key Provisions," *Current Politics and Economics of Russia, Eastern and Central Europe* 26, no. 2 (2011): 187.
5. Treaty Between the United States of America and the Russian Federation on Measures for the Further Reduction and Limitation of Strategic Offensive Arms (New START), April 8, 2010, http://www.state.gov/documents/organization/140035.pdf.
6. "New START Signed; Senate Battle Looms," Arms Control Association website, May 5, 2010, https://www.armscontrol.org/print/4211.
7. Michael Shear, "Obama, Medvedev Sign Treaty to Reduce Nuclear Weapons," *Washington Post*, April 8, 2010, http://www.washingtonpost.com/wp-dyn/content/article/2010/04/08/AR2010040801677.html.
8. Philip Stephens, "Now Obama Is a President with an Endgame," *Financial Times*, April 8, 2010, http://www.ft.com/cms/s/0/06f11ad2-4342-11df-9046-00144feab49a.html#axzz4GyhYQZCT.
9. Barron Youngsmith, "START to Finish?" *New Republic*, November 12, 2010, https://newrepublic.com/article/79148/start-treaty-obama-kyl-nuclear.

10. Ibid.
11. Peter Baker, "GOP Senators Detail Objections to Arms Treaty," *New York Times,* November 24, 2010, http://www.nytimes.com/2010/11/25/world/europe/25start.html?_r=0.
12. George Shultz, William Perry, Henry Kissinger, and Sam Nunn, "A World Free of Nuclear Weapons," *Wall Street Journal,* January 4, 2007, http://www.wsj.com/articles/SB116787515251566636.
13. Peter Baker, "Obama's Gamble on Arms Pact Pays Off," *New York Times,* December 22, 2010, http://www.nytimes.com/2010/12/23/world/23start.html.
14. Mary Beth Sheridan, "In Letter to Senate, Obama Promises That New START Treaty Won't Limit Missile Defense," *Washington Post,* December 19, 2010, http://www.washingtonpost.com/wp-dyn/content/article/2010/12/18/AR2010121802999.html.
15. Mary Beth Sheridan and William Branigin, "Senate Ratifies New U.S.-Russia Nuclear Weapons Treaty," *Washington Post,* December 22, 2010, http://www.washingtonpost.com/wp-dyn/content/article/2010/12/21/AR2010122104371.html.

10. DENYING IRAN THE BOMB

1. "Transcript: Fourth Democratic Debate," *New York Times,* July 24, 2007, http://www.nytimes.com/2007/07/24/us/politics/24transcript.html.
2. Mark Landler, *Alter Egos: Obama's Legacy, Hillary's Promise, and the Struggle over American Power* (New York: Random House, 2016), 233.
3. David Sanger, *Confront and Conceal: Obama's Secret Wars and Surprising Use of American Power* (New York: Crown, 2012), 192.
4. Arms Control Association, "Official Proposals on the Iranian Nuclear Issue, 2003–2013," January 2015, https://www.armscontrol.org/factsheets/Iran_Nuclear_Proposals.
5. Thomas Erdbrink and William Branigin, "Iran's Supreme Leader Warns Against Negotiating with U.S.," *Washington Post,* November 4, 2009, http://www.washingtonpost.com/wp-dyn/content/article/2009/11/03/AR2009110301397.html.
6. Robert Tait, "Iran Arrests 500 Activists in Wake of Election Protests," *Guardian,* June 17, 2009, http://www.theguardian.com/world/2009/jun/17/iran-election-protests-arrests. See also Simon Jeffery, "Iran Election Protests: The Dead, Jailed and Missing," *Guardian,* July 29, 2009, https://www.theguardian.com/world/blog/2009/jul/29/iran-election-protest-dead-missing.
7. U.S. Department of State, "Six-Year Anniversary of the House Arrests of Mir Hossein Mousavi, Mehdi Karroubi and Zahra Rahnavard," February 14, 2017, http://www.state.gov/r/pa/prs/ps/2017/02/267627.htm.
8. See, for example, Michael McFaul and Abbas Milani, "The Right Way to Engage Iran," *Washington Post,* December 29, 2007, http://www.washingtonpost.com/wp-dyn/content/article/2007/12/28/AR2007122802299.html.
9. White House, Office of the Press Secretary, "Remarks by President Obama and Prime Minster Berlusconi of Italy in Press Availability," June 15, 2009, https://obamawhitehouse.archives.gov/video/President-Obama-Meets-with-Italian-Prime-Minister-Berlusconi?tid=8#transcript.
10. White House, Office of the Press Secretary, "The President's Opening Remarks on Iran, with Persian Translation," June 23, 2009, https://obamawhitehouse.archives.gov/blog/2009/06/23/presidents-opening-remarks-iran-with-persian-translation.
11. White House, Office of the Press Secretary, "Remarks by President Obama and President Medvedev of Russia After Bilateral Meeting," September 23, 2009, https://obamawhitehouse.archives.gov/the-press-office/remarks-president-obama-and-president-medvedev-russia-after-bilateral-meeting.
12. "Medvedev Signals Openness to Iran Sanctions After Talks," CNN, September 23, 2009, http://www.cnn.com/2009/POLITICS/09/23/us.russia.iran/index.html?iref=nextin.
13. David Sanger and William J. Broad, "U.S. and Allies Warn Iran over Nuclear 'Deception,'" *New York Times,* September 25, 2009, http://www.nytimes.com/2009/09/26/world/middleeast/26nuke.html.
14. Sanger, *Confront and Conceal,* 182.
15. White House, Office of the Press Secretary, "Statements by President Obama and French President

Sarkozy and British Prime Minister Brown on Iranian Nuclear Facility," September 25, 2009, https://obamawhitehouse.archives.gov/the-press-office/2009/09/25/statements-president-obama-french-president-sarkozy-and-british-prime-mi.

16. Ibid.
17. Ian Traynor, Julian Borger, and Ewen MacAskill, "Obama Condemns Iran over Secret Nuclear Plant," *Guardian*, September 25, 2009, https://www.theguardian.com/world/2009/sep/25/iran-secret-underground-nuclear-plant.
18. "Statement by President of Russia Dmitry Medvedev Regarding the Situation in Iran," President of Russia official website, September 25, 2009, http://en.kremlin.ru/events/president/transcripts/48467.
19. White House, Office of the Press Secretary, "Remarks by President Obama and President Medvedev of Russia at Joint Press Conference," June 24, 2010, https://obamawhitehouse.archives.gov/realitycheck/the-press-office/remarks-president-obama-and-president-medvedev-russia-joint-press-conference.
20. "U.S. Trade in Goods with Iran," U.S. Census Bureau, Foreign Trade Division, https://www.census.gov/foreign-trade/balance/c5070.html#2011.
21. The three entities were Dmitry Mendeleev University of Chemical Technology, the Moscow Aviation Institute, and the Tula Instrument Design Bureau. See Peter Baker and David Sanger, "U.S. Makes Concessions to Russia for Iran Sanctions," *New York Times*, May 21, 2010, http://www.nytimes.com/2010/05/22/world/22sanctions.html.
22. United Nations, Department of Public Information, "Security Council Imposes Additional Sanctions on Iran, Voting 12 in Favour to 2 Against, with 1 Abstention," SC/9948, June 9, 2010, http://www.un.org/press/en/2010/sc9948.doc.htm.
23. United Nations, Security Council, Resolution 1929 (2010), June 9, 2010, https://www.un.org/press/en/2010/sc9948.doc.htm.
24. "Obama Says New U.S. Sanctions Show International Resolve in Iran Issue," CNN, July 1, 2010, http://www.cnn.com/2010/POLITICS/07/01/obama.iran.sanctions/index.html.
25. SWIFT, "SWIFT and Sanctions," November 5, 2015, https://www.swift.com/about-us/legal/compliance/swift-and-sanctions.
26. Vladimir Radyuhin, "Russia Comes Out on Top in Iran Deal," *Hindu*, May 20, 2010, http://www.thehindu.com/todays-paper/tp-opinion/russia-comes-out-on-top-in-iran-deal/article770723.ece.
27. Gary Samore, *The Iran Nuclear Deal: A Definitive Guide* (Cambridge, MA: Belfer Center for Science and International Affairs, 2015), http://belfercenter.ksg.harvard.edu/publication/25599/iran_nuclear_deal.html.

11. HARD ACCOUNTS: RUSSIA'S NEIGHBORHOOD AND MISSILE DEFENSE

1. White House, Office of the Vice President, "Remarks by the Vice President at the Munich Security Conference," February 7, 2015, https://obamawhitehouse.archives.gov/the-press-office/2015/02/07/remarks-vice-president-munich-security-conference.
2. NATO, Bucharest Summit Declaration, April 3, 2008, http://www.nato.int/cps/en/natolive/official_texts_8443.htm.
3. Rogert Gates, *Duty: Memoirs of a Secretary at War* (New York: Knopf, 2014), 399.
4. IISS Strategic Dossiers, "Iran's Ballistic Missile Capabilities: A Net Assessment," May 10, 2010, https://www.iiss.org/en/publications/strategic%20dossiers/issues/iran-39-s-ballistic-missile-capabilities-a-net-assessment-885a.
5. Jaganath Sankaran, "Missile Defense Against Iran Without Threatening Russia," Arms Control Association, November 4, 2013, https://www.armscontrol.org/act/2013_11/Missile-Defense-Against-Iran-Without-Threatening-Russia.
6. Ibid.
7. White House, Office of the Press Secretary, "Remarks by President Barack Obama in Prague as Delivered," April 5, 2009, https://obamawhitehouse.archives.gov/the-press-office/remarks-president-barack-obama-prague-delivered.

8. On Czech reluctance to build the radar, see Gates, *Duty,* 399.
9. Gates, *Duty,* 403.
10. White House, Office of the Press Secretary, "Remarks by the President on Strengthening Missile Defense in Europe," September 17, 2009, https://obamawhitehouse.archives.gov/the-press-office/remarks-president-strengthening-missile-defense-europe.
11. Gates, *Duty,* 402.
12. Peter Spiegel, "U.S. to Shelve Nuclear-Missile Shield," *Wall Street Journal,* September 17, 2009, https://www.wsj.com/articles/SB125314575889817971.
13. Gates, *Duty,* 404.
14. Russian Prime Minister Vladimir Putin, interviewed by the German ARD TV channel, August 29, 2008, "Archive of the Official Site of the 2008–2012 Prime Minister of the Russian Federation Vladimir Putin — Events," accessed March 17, 2017, http://archive.premier.gov.ru/eng/events/news/1758/.
15. White House, "The Vice President: We Stand by Ukraine," July 22, 2009, https://obamawhitehouse.archives.gov/blog/2009/07/22/vice-president-we-stand-ukraine.
16. Josh Rogin, "White House: We're Not Throwing Georgia Under the Bus," *Foreign Policy,* June 11, 2010, http://foreignpolicy.com/2010/06/11/white-house-were-not-throwing-georgia-under-the-bus/.
17. White House, Office of the Press Secretary, "U.S.-Russia Relations: 'Reset' Fact Sheet," June 24, 2010, https://obamawhitehouse.archives.gov/realitycheck/the-press-office/us-russia-relations-reset-fact-sheet.
18. Michael McFaul, "The False Promise of Autocratic Stability," *Hoover Daily Report,* September 14, 2005, http://www.hoover.org/research/false-promise-autocratic-stability.
19. Clifford Levy, "Upheaval in Kyrgyzstan Could Imperil Key U.S. Base," *New York Times,* April 7, 2010, http://www.nytimes.com/2010/04/08/world/asia/08bishkek.html?_r=0.
20. International Crisis Group, "Kyrgyzstan: Widening Ethnic Divisions in the South," March 29, 2012, https://www.crisisgroup.org/europe-central-asia/central-asia/kyrgyzstan/kyrgyzstan-widening-ethnic-divisions-south.
21. David Trilling and Chinghiz Umetov, "Kyrgyzstan: Is Putin Punishing Bakiev?" April 5, 2010, Eurasianet.org, http://www.eurasianet.org/departments/insight/articles/eav040610a.shtml.
22. White House, Office of the Press Secretary, "Remarks by President Obama and President Medvedev of Russia at Joint Press Conference," June 24, 2010, https://obamawhitehouse.archives.gov/realitycheck/the-press-office/remarks-president-obama-and-president-medvedev-russia-joint-press-conference.
23. Mark Hosenball, "Pentagon Confirms It Gave $1.4 Billion in No-Bid Fuel Contracts to Mysterious Companies," *Newsweek,* April 28, 2010, http://www.newsweek.com/pentagon-confirms-it-gave-14-billion-no-bid-fuel-contracts-mysterious-companies-217042.
24. Thom Shanker, "U.S. Gets a First Look at Russian Radar," *New York Times,* November 2, 2007, http://www.nytimes.com/2007/11/02/world/europe/02iht-missile.4.8167475.html?mcubz=1.
25. White House, Office of the Press Secretary, "Fact Sheet: President Obama's Participation in the NATO Summit Meetings in Lisbon," November 20, 2010, https://obamawhitehouse.archives.gov/the-press-office/2010/11/20/fact-sheet-president-obamas-participation-nato-summit-meetings-lisbon.

12. BURGERS AND SPIES

1. For a review of that literature, see McFaul, *Advancing Democracy Abroad,* chapter 2.
2. White House, Office of the Press Secretary, "Remarks by President Obama and President Medvedev of Russia at the U.S.-Russia Business Summit," June 24, 2010, https://obamawhitehouse.archives.gov/the-press-office/remarks-president-obama-and-president-medvedev-russia-us-russia-business-summit.
3. Ibid.
4. Hillary Clinton, "Remarks at the U.S.-Russia 'Civil Society to Civil Society' Summit," Washington, DC, June 24, 2010, https://2009-2017.state.gov/secretary/20092013clinton/rm/2010/06/143628.htm.
5. Ibid.

6. White House, Office of the Press Secretary, "Remarks by President Obama and President Medvedev of Russia at Joint Press Conference," June 24, 2010, https://obamawhitehouse.archives.gov/reality check/the-press-office/remarks-president-obama-and-president-medvedev-russia-joint-press-con ference.
7. White House, Office of the Press Secretary, "U.S.-Russia Relations: 'Reset' Fact Sheet," June 24, 2010, https://obamawhitehouse.archives.gov/the-press-office/us-russia-relations-reset-fact-sheet.
8. Michael Schwirtz, "Officer Who Exposed Russian Spies Is Sentenced in Absentia," *New York Times*, June 27, 2011, http://www.nytimes.com/2011/06/28/world/europe/28moscow.html.
9. "What the Russian Spy Case Reveals," Council on Foreign Relations, July 12, 2010, http://www.cfr.org/world/russian-spy-case-reveals/p22620.
10. Scott Shane and Benjamin Weiser, "Accused Spies Blended In, but Seemed Short on Secrets," *New York Times*, June 29, 2010, http://www.nytimes.com/2010/06/30/world/europe/30spy.html.
11. Jerry Markon and Philip Rucker, "The Suspects in a Russian Spy Ring Lived All-American Lives," *Washington Post*, June 30, 2010, http://www.washingtonpost.com/wp-dyn/content/article/2010/06/29/AR2010062905401.html.
12. "Edward Snowden: Leaks That Exposed US Spy Programme," BBC News, January 17, 2014, http://www.bbc.com/news/world-us-canada-23123964.
13. Ian Black, "Igor Sutyagin Is Odd Man Out in Spy Swap Deal," *Guardian*, August 17, 2010, https://www.theguardian.com/world/2010/aug/17/igor-sutyagin-spy-swap.
14. Tom Parfitt and Chris McGreal, "Russia and US 'Planning Spy Swap,'" *Guardian*, July 7, 2010, https://www.theguardian.com/world/2010/jul/07/russia-us-spy-igor-sutyagin.
15. Tom Parfitt, "Spy Swap Russian: I Want to Go Home," *Guardian*, August 13, 2010, http://www.theguardian.com/world/2010/aug/13/spy-swap-russian-igor-sutyagin.

13. THE ARAB SPRING, LIBYA, AND THE BEGINNING OF THE END OF THE RESET

1. Jeremy M. Sharp, *Egypt: Background and U.S. Relations* (Washington, DC: Congressional Research Service, 2012), 6.
2. Clinton, *Hard Choices*, 339–40.
3. Ibid., 341.
4. My article on the subject, coauthored with David Adesnik, is "Engaging Autocratic Allies to Promote Democracy," *Washington Quarterly* 29, no. 2 (2006), 7–26.
5. See Stoner and McFaul, *Transitions to Democracy*, for a compilation of these studies.
6. Mark Landler, "Obama Seeks Reset in Arab World, *New York Times*, May 11, 2011, http://www.nytimes.com/2011/05/12/us/politics/12prexy.html?mcubz=3.
7. White House, "Remarks by the President on Egypt," February 11, 2011, https://obamawhitehouse.archives.gov/the-press-office/2011/02/11/remarks-president-egypt.
8. Ibid.
9. King gets credit for this phrase, but the author is the nineteenth-century reformer and abolitionist Theodore Parker.
10. "Obama's Remarks on the Resignation of Mubarak," *New York Times*, February 11, 2011, http://www.nytimes.com/2011/02/12/world/middleeast/12diplo-text.html?mcubz=2.
11. In parallel, the policy planning staff at the State Department was also drafting and circulating a paper on lessons learned from other democratic transitions.
12. At the time, I was drawing heavily on Guillermo O'Donnell and Philippe Schmitter, *Transitions from Authoritarian Rule: Tentative Conclusions about Uncertain Democracies* (Baltimore: Johns Hopkins University Press, 1986).
13. H. A. Heller, "'National Dialogue' in Egypt Was Just Empty Symbolism," *National*, December 11, 2012, https://www.thenational.ae/national-dialogue-in-egypt-was-just-empty-symbolism-1.367417.
14. On "managed transition" as our preferred outcome in Syria, see Derek H. Chollet, *The Long Game: How Obama Defied Washington and Redefined America's Role in the World* (New York: PublicAffairs, 2016), chapter 5.

15. White House, Office of the Press Secretary, "Remarks of President Obama — 'A Moment of Opportunity,'" May 19, 2011, https://obamawhitehouse.archives.gov/the-press-office/2011/05/19/remarks-president-barack-obama-prepared-delivery-moment-opportunity.
16. Surprising to me, Saad supported Egyptian army chief General Abdel Fattah el-Sisi's military coup two years later, after fighting against Mubarak's dictatorship for decades.
17. Jackson Diehl, "Fulfilling the Arab Spring," *Washington Post,* April 26, 2015, https://www.washingtonpost.com/opinions/investing-in-the-legacy-of-the-arab-spring/2015/04/26/c44b1638-e9c7-11e4-9767-6276fc9b0ada_story.html?utm_term=.ae548adda9f0.
18. Michael McFaul, "Turtle Bay Tango," *Wall Street Journal,* October 14, 2003, https://www.wsj.com/articles/SB106609501293370400.
19. Gadhafi, as quoted in Maria Golovnina and Patrick Worsnip, "UN Approves Military Force; Gaddafi Threatens Rebels," Reuters, March 18, 2011, http://in.reuters.com/article/2011/03/18/idINIndia-55674620110318.
20. Gates, *Duty,* 518.
21. Ibid.
22. Gates writes in his memoir that he considered resigning over the Libya decision. See *Duty,* 522.
23. White House, "Remarks by President on Libya," March 19, 2011, https://obamawhitehouse.archives.gov/the-press-office/2011/03/19/remarks-president-libya.
24. The West only secured a UN resolution authorizing U.S. military involvement in Korea because the Soviets were boycotting the UN Security Council at the time.
25. "Байден пытался отговорить Путина от участия в выборах," http://newsland.com/news/detail/id/653351/.
26. United Nations, Security Council, "Security Council Approves 'No-Fly Zone' over Libya, Authorizing 'All Necessary Measures' to Protect Civilians, by Vote of 10 in Favour with 5 Abstentions," March 17, 2011, https://www.un.org/press/en/2011/sc10200.doc.htm.
27. Putin, as quoted in Gleb Bryanski, "Putin Likens UN Libya Resolution to Crusades," Reuters, March 11, 2011, http://www.reuters.com/article/us-libya-russia-idUSTRE72K3JR20110321.
28. Putin, as quoted in Jill Dougherty, "Putin and Medvedev Spar over Libya," CNN, March 23, 2011, http://www.cnn.com/2011/WORLD/europe/03/21/russia.leaders.libya/.
29. Ibid.
30. "Statement by Dmitry Medvedev on the Situation in Libya," President of Russia official website, March 21, 2011, http://en.kremlin.ru/events/president/news/10701.
31. Ibid.
32. Ibid.
33. It was much later that the breakdown of the Libyan state allowed for civil war to drag on for years. At the time of the fall of Gadhafi, a more stable, positive outcome seemed possible.
34. "'Reset' with U.S. Ended with Libya, Not Crimea, Putin Says," *Moscow Times,* April 17, 2014, https://themoscowtimes.com/articles/reset-with-us-ended-with-libya-not-crimea-putin-says-34134.

14. BECOMING "HIS EXCELLENCY"

1. Robert Kagan, "The Senate Passed New START. What's Next?" *Washington Post,* December 23, 2010, http://www.washingtonpost.com/wp-dyn/content/article/2010/12/22/AR2010122203206.html.
2. "The Russian Economy and U.S.-Russia Relations," conference, Peterson Institute for International Economics, Washington, DC, April 15, 2011, event recap, https://piie.com/events/russian-economy-and-us-russia-relations.
3. Josh Rogin, "McFaul: We Must Give Russia Privileged Trade Status," *Foreign Policy,* October 12, 2011.
4. "Washington's Infighting Delays an Ambassador," *Washington Post,* December 2, 2011.
5. U.S. Department of State, "Secretary Clinton Holds a Swearing-In Ceremony for Ambassador-Designate to Russia Mike McFaul," YouTube, originally posted on January 11, 2012, https://www.youtube.com/watch?v=WB2cgI6lKHc.
6. In assisting with the writing of Obama's speech in Moscow in July 2009, I had tried to use this same

phrase, "a strong, prosperous, and democratic Russia." In the editorial process, the word "democratic" got cut. I deliberately added it this time around, and did not seek anyone's permission to use the phrase.

15. PUTIN NEEDS AN ENEMY—AMERICA, OBAMA, AND ME

1. Ellen Barry, "Putin Once More Moves to Assume Top Job in Russia," *New York Times,* September 24, 2011, http://www.nytimes.com/2011/09/25/world/europe/medvedev-says-putin-will-seek-russian-presidency-in-2012.html?mcubz=2.
2. James R. Clapper, Director of National Intelligence, "Statement for the Record on the Worldwide Threat Assessment of the U.S. Intelligence Community," U.S. Senate, Select Committee on Intelligence, January 31, 2012.
3. Julie Ray and Neli Esipova, "Russian Approval of Putin Soars to Highest Level in Years," Gallup.com, July 18, 2014, http://www.gallup.com/poll/173597/russian-approval-putin-soars-highest-level-years.aspx.
4. See McFaul and Stoner, "The Myth of the Authoritarian Model."
5. Steven Lee Myers, *The New Tsar: The Rise and Reign of Vladimir Putin* (New York: Simon & Schuster, 2015), 392.
6. Dvorkovich, as quoted in Ellen Barry, "Putin Once More Moves to Assume Top Job in Russia," *New York Times,* September 24, 2011.
7. "Бывший 'кардинал' Владимира Путина вытеснен из правительства," *Русинформ,* May 8, 2013, http://rusinform.ru/inopressa/2726-byvshiy-kardinal-vladimira-putina-vytesnen-iz-pravitelstva.html; see also "Зарубежные СМИ считают отставку В.Суркова ударом по Д.Медведеву," *РБК,* May 13, 2013, http://www.rbc.ru/politics/13/05/2013/5704084c9a7947fcbd448b9c.
8. "Возможные результаты президентских выборов," Levada-Center, December 4, 2017, https://www.levada.ru/2017/12/04/vozmozhnye-rezultaty-prezidentskih-vborov-4/print/.
9. Ellen Barry, "Tens of Thousands Protest in Moscow, Russia, in Defiance of Putin," *New York Times,* December 10, 2011, http://www.nytimes.com/2011/12/11/world/europe/thousands-protest-in-moscow-russia-in-defiance-of-putin.html.
10. Myers, *The New Tsar,* 396–97.
11. Putin's remarks to the United Russia Congress, November 26, 2011, as quoted in Miriam Elder, "Vladimir Putin Rallies Obedient Crowd at Party Congress," *Guardian,* November 27, 2011, http://www.theguardian.com/world/2011/nov/27/vladimir-putin-party-congress.
12. Putin, as quoted in Steve Gutterman and Gleb Bryanski, "Putin Says U.S. Stoked Russian Protests," Reuters, December 8, 2011, http://www.reuters.com/article/us-russia-idUSTRE7B610S20111208.
13. Ibid.
14. "McFaul Arrives to Keep 'Reset' Alive," *Moscow Times,* January 15, 2012, https://themoscowtimes.com/news/mcfaul-arrives-to-keep-reset-alive-11892.
15. Андрей Васильев, "НТВ, 'нашисты' или чекисты? Кто стоит за скандальным видео," *Golos Ameriki,* January 18, 2012, http://www.golos-ameriki.ru/content/russia-opposition-video-2012-01-18-137605948/250101.html. See also "Встреча нового посла США в Москве с оппозицией: партия власти негодует, а дипломат доволен," Новости NEWSru.com, January 18, 2012, http://www.newsru.com/russia/18jan2012/mcfail.html.
16. Ibid.
17. "Обзор самых популярных любительских съемок из интернета за прошедшую неделю—Первый Канал," Channel One, January 22, 2012, http://www.1tv.ru/news/social/197093.
18. Андрей Козенко, "Посол США встретился с сопротивлением," *Газета Коммерсантъ,* January 18, 2012, http://www.kommersant.ru/doc/1853261.
19. "Специалист по демократии," Lenta.ru, January 18, 2012, https://lenta.ru/articles/2012/01/18/mcfaul/.
20. "Человек Обамы позвал россиян в Америку," *NTV,* January 14, 2012, http://www.ntv.ru/novosti/259973/.

21. "Аналитическая программа 'Однако' с Михаилом Леонтьевым," *Новости, Первый канал,* January 17, 2012, https://www.1tv.ru/news/2012-01-17/102215-analiticheskaya_programma_odnako_s_mihailom_leontievym.
22. "Осадок от первых дней пребывания Майкла Макфола останется у российской власти надолго," *Коммерсантъ,* January 20, 2012, http://www.kommersant.ru/doc/1854220. Translated from Russian by Anya Shkurko. The *Odnako* show was actually broadcast my second working day; the Monday before was an American holiday.
23. Ellen Barry, "New U.S. Envoy Steps into Glare of a Russia Eager to Find Fault," *New York Times,* January 23, 2012, http://www.nytimes.com/2012/01/24/world/europe/in-russia-new-us-envoy-mcfaul-ruffles-feathers.html.
24. "Voting LIVE: Russians to Monitor Elections Online," RT International, February 28, 2012, https://www.rt.com/politics/russia-elections-putin-cameras-407/.
25. "Дмитрий Медведев внес в Госдуму законопроект, упрощающий выдвижение кандидатов на выборах," *Свободная Пресса,* December 23, 2011, http://svpressa.ru/all/news/51271/.
26. In fact, after years of growing autocracy, this electoral "reform" had no democratizing effect at all, but instead returned to the parliament an even bigger majority for Putin.
27. Maxim Tkachenko, "Medvedev Proposes Sweeping Political Reform in Russia," CNN.com, December 22, 2011, http://www.cnn.com/2011/12/22/world/europe/russia-political-reform/.
28. Ibid.
29. Gleb Bryanski, "Russia's Putin Signs Anti-Protest Law Before Rally," Reuters, June 8, 2012, http://www.reuters.com/article/us-russia-protests-idUSBRE8570ZH20120608.
30. "Госдума: 'Взбесившийся принтер' или Россия в миниатюре?" BBC, *Русская Служба,* March 5, 2013, http://www.bbc.com/russian/russia/2013/03/130303_duma_crazy_printer.
31. "'Blasphemy Bill' Passes Duma Unanimously," *Moscow Times,* June 11, 2013, https://themoscowtimes.com/news/blasphemy-bill-passes-duma-unanimously-24862.
32. Владимир Федоренко, "В 'стоп-лист' НКО вошли Фонд Сороса и Всемирный конгресс украинцев," *Риа Новости,* July 7, 2015, https://ria.ru/society/20150707/1118850670.html.
33. See Andrew Kramer, "From Russian Health Official, Food Criticism with a Dash of Politics," *New York Times,* July 25, 2012; David Herszenhorn and Ellen Barry, "Russia Demands U.S. End Support of Democracy Groups," *New York Times,* September 18, 2012; Ellen Barry, "Stung by Criticism, Russian Lawmakers Point to Human Rights Abuses in U.S.," *New York Times,* October 22, 2012; and David Herszenhorn, "U.S.-Russian Trade Ties Face Some Political Snags," *New York Times,* February 24, 2012.
34. Вячеслав Правдзинский, "MC Faul Girls 2012," *Blog RussiaRu,* February 10, 2012, https://russiaru.net/id253480/#?/id253480/status/3de280000002f.
35. "News Conference of Vladimir Putin," President of Russia official website, December 19, 2003, http://en.kremlin.ru/events/president/news/19859.
36. Shaun Walker, "Putin's Praise for Trump May Mask 'Conflicted' Feelings, Kremlin Watchers Say," *Guardian,* August 16, 2016, https://www.theguardian.com/world/2016/aug/16/vladmir-putin-donald-trump-ties-russia-us-election?CMP=share_btn_fb.
37. I discuss all of these cases in McFaul, *Advancing Democracy Abroad.*
38. David Remnick, "Vladimir Putin's New Anti-Americanism," *New Yorker,* August 3, 2014, http://www.newyorker.com/magazine/2014/08/11/watching-eclipse.
39. Putin, as quoted in Peter Baker, "U.S.-Russian Ties Still Fall Short of 'Reset' Goal," *New York Times,* September 3, 2013, http://www.nytimes.com/2013/09/03/world/europe/us-russian-ties-still-fall-short-of-reset-goal.html.
40. "Президент России | интервью германским телеканалам АРД и ЦДФ," President of Russia official website, May 5, 2005, http://kremlin.ru/events/president/transcripts/22948.
41. Helene Cooper, "Obama Tells UN New Democracies Need Free Speech," *New York Times,* September 25, 2012, http://www.nytimes.com/2012/09/26/world/obamas-address-to-united-nations.html?mcubz=2.

16. GETTING PHYSICAL

1. Between 2011 and 2014, Rusnano's U.S. arm, Rusnano USA, invested about $1.2 billion in American companies. See Benny Evangelista, "Russia, U.S. Tied by Technology Investments," *SFGate,* April 5, 2014, http://www.sfgate.com/news/article/Russia-U-S-tied-by-technology-investments-5379796.php.
2. Arkady Ostrovsky, "Youth Group Hounds UK Moscow Ambassador," *Financial Times,* December 8, 2006, http://www.ft.com/cms/s/0/15054982-8661-11db-86d5-0000779e2340.html?ft_site=falcon&desktop=true#axzz3HbkOmUXO.
3. Office of Inspector General, U.S. Department of State and the Broadcasting Board of Governors, *Inspection of Embassy Moscow and Constituent Posts, Russia,* September 2013, https://oig.state.gov/system/files/217736.pdf.
4. Ellen Barry, "U.S. Envoy to Russia Accuses TV Station of Spying," *New York Times,* March 30, 2012, http://www.nytimes.com/2012/03/31/world/europe/russia-ambassador-michael-mcfaul-ntv-hacking.html.
5. Alexey Navalny, "Не понимаю Макфола @McFaul. У него же дипиммунитет. Он может избивать журналистов НТВ совершенно безнаказанно! Вперёд, Майк! Один за всех!" tweet, March 29, 2012, https://twitter.com/navalny/status/185403492620439552.
6. Michael McFaul, "Just Watched NTV. I Mispoke," tweet, March 29, 2012, https://twitter.com/McFaul/status/185475286408757249.
7. Jonathan Martin, "Obama Apologizes for Remark," *Politico,* March 21, 2009, http://www.politico.com/story/2009/03/obama-apologizes-for-remark-020268.
8. "Выгонят ли из Госдумы Депутатов за визит в Американское Посольство?" TV Rain, January 24, 2012, http://tvrain.ru/articles/vygonyat_li_iz_gosdumy_deputatov_za_vizit_v_amerikanskoe_posolstvo-151148/.
9. Russian businessperson, email to author, April 25, 2013.
10. Alexander Zemlianichenko, "The Unaccountable Death of Boris Nemtsov," *New Yorker,* February 26, 2016, http://www.newyorker.com/news/news-desk/the-unaccountable-death-of-boris-nemtsov.
11. Clinton, *Hard Choices,* 551.

17. PUSHBACK

1. "Trade in Goods with Russia," U.S. Census Bureau, Foreign Trade Division, https://www.census.gov/foreign-trade/balance/c4621.html.
2. Direct investment position of the United States in Russia from 2000 to 2015 (in billion U.S. dollars, on a historical-cost basis), *Statista,* http://www.statista.com/statistics/188637/united-states-direct-investments-in-russia-since-2000/.
3. Ibid.
4. To conduct such advocacy, we had to get special permission from the U.S. government officials in Washington, who rightly regulated these activities.
5. Jon Ostrower and Andy Pasztor, "Boeing, United Technologies Stockpile Titanium Parts," *Wall Street Journal,* August 7, 2014, https://www.wsj.com/articles/boeing-united-technologies-stockpiling-titanium-parts-1407441886?mg=prod/accounts-wsj.
6. Andrew E. Kramer, "Exxon Wins Prized Access to Arctic with Russia Deal," *New York Times,* August 30, 2011, http://www.nytimes.com/2011/08/31/business/global/exxon-and-rosneft-partner-in-russian-oil-deal.html.
7. Kennan, as quoted in Alexis Wichowski, "Social Diplomacy: Or How Diplomats Learned to Stop Worrying and Love the Tweet," *Foreign Affairs,* April 5, 2013, https://www.foreignaffairs.com/articles/2013-04-05/social-diplomacy.
8. Michael McFaul verified, "BTW @navalny, We Should Meet Someday," tweet, July 18, 2012, https://twitter.com/McFaul/status/225676227539787776.

9. Хахахаха is Russian for "Ha ha ha," signifying laughter. Alexey Navalny verified, "@McFaul Хахахаха. Отлично," tweet, July 19, 2012, https://twitter.com/navalny/status/225861082747650048.
10. "Mission and Activities," Perspektiva website, https://perspektiva-inva.ru/en/about/.
11. Larry Diamond, *Developing Democracy: Toward Consolidation* (Baltimore, MD: Johns Hopkins University Press, 1999).
12. Майкл Макфол, "Что делают и чего не делают США для поддержки гражданского общества," *Moskovskiy Komsomolets,* March 26, 2012, http://www.mk.ru/politics/2012/03/26/685554-chto-delayut-i-chego-ne-delayut-ssha-dlya-podderzhki-grazhdanskogo-obschestva.html.
13. Will Englund and Tara Bahrampour, "Russia's Ban on U.S. Adoptions Devastates American Families," *Washington Post,* December 28, 2012, https://www.washingtonpost.com/world/europe/russia-set-to-ban-us-adoptions/2012/12/27/fd49c542-504f-11e2-8b49-64675006147f_story.html.
14. "Amendments Planned for White-Collar Crime Laws | News," *Moscow Times,* July 8, 2010, http://www.themoscowtimes.com/sitemap/free/2010/7/article/amendments-planned-for-white-collar-crime-laws/409966.html/.
15. U.S. Embassy & Consulates in Russia website, http://russian.moscow.usembassy.gov/st_080112_russia_prosecutions_rus.html.
16. Michael McFaul verified and @USEmbRu, "Сегодняшний приговор по делу Pussy Riot выглядит несоразмерным содеянному," tweet, August 17, 2012, https://twitter.com/USEmbRu/status/236462859536719872.
17. *National Security Strategy 2010,* 6, National Security Strategy Archive, http://nssarchive.us/national-security-strategy-2010/6/.
18. Garry Kasparov, "A Shared Enemy Does Not Mean Shared Values," *Wall Street Journal,* May 12, 2013, https://www.wsj.com/articles/SB10001424127887324244304578473603662138038.
19. U.S. Embassy & Consulates in Russia website, Fact Sheet Archive, https://ru.usembassy.gov/category/fact-sheets/.
20. "'Космоскоп' на открытии выставки 'Вселенная 'Хаббла,'" *Cosmoscope,* May 13, 2012, http://www.cosmoscope.ru/news/231/.
21. "Игорь Бутман вступил в Единую Россию," *Korrespondent,* December 12, 2008, http://korrespondent.net/showbiz/music/676699-igor-butman-vstupil-v-edinuyu-rossiyu.
22. "Chicago Symphony Orchestra on Tour — Russia Italy 2012," Chicago Symphony Orchestra website, accessed January 30, 2015, https://cso.org/media/tours/RussiaItaly2012/.
23. Yelena Agamyan, "The Average Lifetime of a Soldier Coming to Stalingrad Front Was 24 Hours. Just One Day," *Siberian Times,* February 2, 2013, http://siberiantimes.com/other/others/features/the-average-lifetime-of-a-soldier-coming-to-stalingrad-front-was-24-hours-just-one-day/.

18. TWITTER AND THE TWO-STEP

1. Clinton, *Hard Choices,* 460.
2. We decided not to use VKontakte, a successful Russian copycat of Facebook, due to its violation of the intellectual property rights of several American video and music companies. Russia Internet Users, *Internet Live Stats,* http://www.internetlivestats.com/internet-users/russia/.
3. "Макфол в твиттере извинился за свой 'Ёбург,'" *Информационное Агентство европейско-азиатские новости,* 4 Июля, 2012, http://eanews.ru/news/scandals_sensations/makfol_v_tvittere_izvinilsya_za_svoy_burg/.
4. Alexey Navalny verified, "'Организовал.' Значит красивой сцены с оправданием не будет. Всем привет, кстати," tweet, July 18, 2013, https://twitter.com/navalny/status/357728359976943617.
5. Мария Климова, Сергей Смирнов, "Протесты после приговора Навальному," *Русская планета,* July 18, 2013, http://rusplt.ru/policy/prigovor_protest.html.
6. "Подсчитаны 100 процентов голосов на выборах мэра Москвы," Lenta.ru, September 9, 2013, http://lenta.ru/news/2013/09/09/hundred/.
7. MFA Russia verified, "The Foreign Ministry Is Utterly Shocked," tweet, May 28, 2012, https://twitter.com/mfa_russia/status/207171456524750848.

8. MFA Russia verified, "Ambassadors' Job, as We Understand It, Is to Improve Bilateral Ties," tweet, May 28, 2012, https://twitter.com/mfa_russia/status/207172362343415808.
9. MFA Russia verified, "Michael McFaul's Analysis Is a Deliberate Distortion," tweet, May 28, 2012, https://twitter.com/mfa_russia/status/207171509628837889.
10. Carl Bildt verified, "I See That Russia MFA Has Launched a Twitter-War," tweet, May 28, 2012, https://twitter.com/carlbildt/status/207181238434398208.
11. Jonathan Earle, "McFaul Tears Up Dance Floor at Spaso Hootenanny," *Moscow Times*, March 28, 2012, https://themoscowtimes.com/articles/mcfaul-tears-up-dance-floor-at-spaso-hootenanny-13646.
12. "Negative Views of U.S. in Russia," Pew Research Center, Global Attitudes and Trends, June 22, 2015, http://www.pewglobal.org/2015/06/23/global-publics-back-u-s-on-fighting-isis-but-are-critical-of-post-911-torture/bop-report-35/.
13. "Блогеры — с 16 по 23 Января 2013," *Медиалогия — Система Мониторинга СМИ*, accessed February 13, 2015, http://www.mlg.ru/company/pr/2328/.
14. Alicia P. Q. Wittmeyer, "The FP Twitterati 100: A Who's Who of the Foreign-Policy Twitterverse in 2013," *Foreign Policy*, August 13, 2013, http://foreignpolicy.com/2013/08/13/the-fp-twitterati-100/.
15. "Russians Consider TV the Most Trustworthy News Source," *Moscow Times*, October 23, 2015, https://themoscowtimes.com/news/russians-consider-tv-the-most-trustworthy-news-source-50444.
16. Michael Bohm, "Farewell, Ambassador McFaul," *Moscow Times*, February 14, 2014, http://old.themoscowtimes.com/mobile/opinion/article/494528.html.

19. IT TAKES TWO TO TANGO

1. Clinton, *Hard Choices*, 215.
2. Email from McFaul to NSC and State Department Officials. Subject: RE: papers 1 and 2, November 7, 2012, CLASSIFICATION: UNCLASSIFIED.
3. Blair Euteneuer, "Russia Seeks U.S. Ties, Won't Skip 'Reset,' Shuvalov Says," Bloomberg, October 4, 2011, http://www.bloomberg.com/news/articles/2011-10-04/russia-seeks-more-u-s-ties-won-t-skip-reset-shuvalov-says.
4. White House, Office of the Press Secretary, "Remarks by President Obama and President Medvedev of Russia After Bilateral Meeting," March 26, 2012, https://www.whitehouse.gov/the-press-office/2012/03/26/remarks-president-obama-and-president-medvedev-russia-after-bilateral-me.
5. Ibid.
6. Michael Paul, *Missile Defense Problems and Opportunities in NATO-Russia Relations* (Berlin: Stiftung Wissenschaft und Politik, 2012), https://www.swp-berlin.org/fileadmin/contents/products/comments/2012C19_pau.pdf.
7. White House, Office of the Press Secretary, "Remarks by the President in the State of the Union Address," February 12, 2013, https://obamawhitehouse.archives.gov/the-press-office/2013/02/12/remarks-president-state-union-address.

20. CHASING RUSSIANS, FAILING SYRIANS

1. Landler, *Alter Egos*, 215–20.
2. Mark Mazzetti, "CIA Study of Covert Aid Fueled Skepticism About Helping Syrian Rebels," *New York Times*, October 14, 2014, https://www.nytimes.com/2014/10/15/us/politics/cia-study-says-arming-rebels-seldom-works.html.
3. Clinton, *Hard Choices*, 389.
4. "Syria: Hillary Clinton Calls Russia and China 'Despicable' for Opposing UN Resolution," *Telegraph*, February 25, 2012, http://www.telegraph.co.uk/news/worldnews/middleeast/syria/9105470/Syria-Hillary-Clinton-calls-Russia-and-China-despicable-for-opposing-UN-resolution.html.
5. Barack Obama, "A Just and Lasting Peace," Nobel Lecture, December 10, 2009, https://www.nobelprize.org/nobel_prizes/peace/laureates/2009/obama-lecture_en.html.

6. Andrew Osborn, "Vladimir Putin Lashes Out at America for Killing Gaddafi and Backing Protests," *Telegraph,* December 15, 2011, http://www.telegraph.co.uk/news/worldnews/europe/russia/8958475/Vladimir-Putin-lashes-out-at-America-for-killing-Gaddafi-and-backing-protests.html.
7. Ariel Cohen, "Obama-Putin Meeting Brings Chill to Mexico," *National Interest,* June 20, 2012, http://nationalinterest.org/commentary/obama-putin-meeting-brings-chill-mexico-7089.
8. Peter Collett, "G20 Body Language Decoded: A Spring in Obama's Step but No Pat for Putin," *Guardian,* September 5, 2013, http://www.theguardian.com/world/2013/sep/05/g20-handshakes-obama-putin-cameron.
9. "Poker-Faced Meeting: Putin, Obama Avoid Pushing Sore Points," RT News, June 19, 2012, http://rt.com/news/putin-obama-g20-meeting-144/.
10. Bogdanov, as quoted in Elizabeth Kennedy and Vladimir Isachenkov, "Russia, NATO Chief Say Assad Losing Control," *Boston Globe,* December 13, 2012. https://www.bostonglobe.com/2012/12/13/russia-nato-chief-say-assad-losing-control/cTOpYUXlQGVez0Upwf6NiL/story.html.
11. Neil MacFarquhar, "Western Nations, Protesting Killings, Expel Syrian Envoys," *New York Times,* May 29, 2012, http://www.nytimes.com/2012/05/30/world/middleeast/kofi-annan-meets-with-bashar-al-assad.html?mcubz=2.
12. "Lavrov-Clinton Talks: 'Very Good Chance' of Progress on Syria in Geneva," *RT News,* June 29, 2012, https://www.rt.com/news/clinton-lavrov-syria-talks-103/.
13. United Nations, "Action Group for Syria Final Communiqué," June 30, 2012, http://www.un.org/News/dh/infocus/Syria/FinalCommuniqueActionGroupforSyria.pdf.
14. Nick Meo, "Geneva Meeting Agrees 'Transition Plan' to Syria Unity Government," *Telegraph,* June 30, 2012, http://www.telegraph.co.uk/news/worldnews/middleeast/syria/9367330/Geneva-meeting-agrees-transition-plan-to-Syria-unity-government.html.
15. U.S. Mission in Geneva, "Remarks by Secretary of State Hillary Rodham Clinton, Press Availability Following the Meeting of the Action Group for Syria," June 30, 2012, https://geneva.usmission.gov/2012/06/30/secretary-clinton-press-availability-following-the-meeting-of-the-action-group-on-syria/.
16. Clinton, *Hard Choices,* 459.
17. "Geneva Decisions on Syria Already Being Distorted — Lavrov," RT News, July 3, 2012, https://www.rt.com/politics/russia-us-syria-geneva-talks-282/.
18. Joshua Yaffa, "The Putin of Chechnya," *New Yorker,* February 8, 2016, http://www.newyorker.com/magazine/2016/02/08/putins-dragon.
19. Clinton, *Hard Choices,* 675.
20. "Kofi Annan Resigns as UN-Arab League Joint Special Envoy for Syrian Crisis," UN News Service, August 2, 2012, http://www.un.org/apps/news/story.asp?NewsID=42609#.WQdiJlPyiRs.
21. U.S. Department of State, "Remarks with Russian Foreign Minister Sergey Lavrov," May 7, 2013, https://2009-2017.state.gov/secretary/remarks/2013/05/209117.htm.
22. Ministry of Foreign Affairs of the Russian Federation, "Speech of and Answers to Questions by Russian Foreign Minister Sergey Lavrov During Joint Press Conference Summarizing the Results of Negotiations with US Secretary of State John Kerry, Moscow," May 7, 2013, http://www.mid.ru/en/vistupleniya_ministra/-/asset_publisher/MCZ7HQuMdqBY/content/id/111326.
23. Joby Warrick, "More Than 1,400 Killed in Syrian Chemical Weapons Attack, U.S. Says," *Washington Post,* August 30, 2013, http://www.washingtonpost.com/world/national-security/nearly-1500-killed-in-syrian-chemical-weapons-attack-us-says/2013/08/30/b2864662-1196-11e3-85b6-d27422650fd5_story.html. In his speech on August 30, 2013, condemning these attacks, Secretary Kerry reported the number killed was 1,429, including 426 children.
24. James Ball, "Obama Issues a 'Red Line' Warning on Chemical Weapons," *Washington Post,* August 20, 2012, https://www.washingtonpost.com/world/national-security/obama-issues-syria-red-line-warning-on-chemical-weapons/2012/08/20/ba5d26ec-eaf7-11e1-b811-09036bcb182b_story.html.
25. Chollet, *The Long Game,* 2.
26. Ibid., 3.
27. "Full Transcript: Secretary of State John Kerry's Remarks on Syria on Aug. 30," *Washington Post,*

August 30, 2013, https://www.washingtonpost.com/world/national-security/running-transcript-secretary-of-state-john-kerrys-remarks-on-syria-on-aug-30/2013/08/30/f3a63a1a-1193-11e3-85b6-d27422650fd5_story.html?utm_term=.824bcc2d8aa3.

28. Mark Landler, Jonathan Weisman, and Michael R. Gordon, "Split Senate Panel Approves Giving Obama Limited Authority on Syria," *New York Times*, September 4, 2013, http://www.nytimes.com/2013/09/05/world/middleeast/divided-senate-panel-approves-resolution-on-syria-strike.html.
29. Chollet, *The Long Game*, 15.
30. Lesley Wroughton, "As Syria War Escalates, Americans Cool to U.S. Intervention: Reuters/Ipsos Poll," Reuters, August 24, 2013, www.reuters.com/article/us-syria-crisis-usa-poll-idUSBRE97O00E20130825.
31. White House, Office of the Press Secretary, "Joint Statement on Syria," September 6, 2013, https://obamawhitehouse.archives.gov/the-press-office/2013/09/06/joint-statement-syria.
32. "Grip and Grin: Obama, Putin in Awkward Handshake," CNBC, September 5, 2013, http://www.cnbc.com/id/101011330; and "Just How Awkward Was the Obama-Putin G20 Handshake?" *Daily Intelligencer*, September 5, 2013, http://nymag.com/daily/intelligencer/2013/09/how-awkward-was-the-obama-putin-g20-handshake.html.
33. "Daily Rundown: US, Russia Relationship Difficult, Says Ambassador," NBC News, September 6, 2013, http://www.nbcnews.com/video/daily-rundown/52936374.
34. My undergraduate advisor at Stanford, Alex George, labeled this technique "forceful persuasion." See his *Forceful Persuasion: Coercive Diplomacy as an Alternative to War* (Washington, DC: United States Institute of Peace, 1992).
35. Geoff Dyer, "Kerry Comments Change Dynamic of Debate on Syria for Obama," *Financial Times*, September 9, 2013, http://www.ft.com/cms/s/0/5208110c-1999-11e3-afc2-00144feab7de.html#axzz3ZOZh1qzm.
36. Jeffrey Goldberg, "The Obama Doctrine," *Atlantic*, April 2016, http://www.theatlantic.com/magazine/archive/2016/04/the-obama-doctrine/471525/.
37. Paul Lewis, "Obama Welcomes Syria Chemical Weapons Deal but Retains Strike Options," *Guardian*, September 14, 2013, http://www.theguardian.com/world/2013/sep/14/barack-obama-syria-chemical-weapons-deal.
38. Naftali Bendavid, "Removal of Chemical Weapons from Syria Is Completed," *Wall Street Journal*, June 23, 2014, http://www.wsj.com/articles/removal-of-chemical-weapons-from-syria-is-completed-1403529356.
39. Paul F. Walker, "Syrian Chemical Weapons Destruction: Taking Stock and Looking Ahead," Arms Control Association, December 2014, https://www.armscontrol.org/ACT/2014_12/Features/Syrian-Chemical-Weapons-Destruction-Taking-Stock-And-Looking-Ahead#note1.
40. Pamela Engel, "Former U.S. Defense Secretary: Obama Hurt U.S. Credibility When He Backed Down from His Red Line on Syria," *Business Insider*, January 26, 2016, http://www.businessinsider.com/robert-gates-syria-red-line-obama-2016-1.
41. S. A. Miller, "The Knives Are Out: Panetta Eviscerates Obama's 'Red Line' Blunder on Syria," *Washington Times*, October 7, 2014, http://www.washingtontimes.com/news/2014/oct/7/panetta-decries-obama-red-line-blunder-syria/.
42. Zack Beauchamp, "How Russian Bombing Is Changing Syria's War, in 3 Maps," *Vox*, February 16, 2016, https://www.vox.com/2016/2/16/11020140/russia-syria-bombing-maps.
43. "Syria Chemical 'Attack': What We Know," BBC News, April 26, 2017, http://www.bbc.com/news/world-middle-east-39500947.
44. "Syrian Refugees: A Snapshot of the Crisis," http://syrianrefugees.eu/.
45. U.S. Department of Defense, "Operation Inherent Resolve," http://www.defense.gov/News/Special-Reports/0814_Inherent-Resolve.
46. Thomas L. Friedman, "Obama on the World," interview, *New York Times*, August 8, 2014, https://www.nytimes.com/2014/08/09/opinion/president-obama-thomas-l-friedman-iraq-and-world-affairs.html?_r=0.
47. Obama, as quoted in Goldberg, "The Obama Doctrine."

48. Landler, *Alter Egos,* 215.
49. Obama, "A Just and Lasting Peace."

21. DUELING ON HUMAN RIGHTS

1. U.S. Embassy & Consulates in Russia, "U.S.-Russia Peer-to-Peer Dialogue Program," https://ru.usembassy.gov/education-culture/us-russia-peer-peer-dialogue-program/.
2. Prague Civil Society Centre, http://www.praguecivilsociety.org/.
3. White House, "Jackson-Vanik and Russia Fact Sheet," http://georgewbush-whitehouse.archives.gov/news/releases/2001/11/20011113-16.html.
4. Browder tells the entire story in *Red Notice: How I Became Putin's No. 1 Enemy* (New York: Transworld, 2015).
5. U.S. Department of State, "2012 Investment Climate Statement: Russia," http://www.state.gov/e/eb/rls/othr/ics/2012/191223.htm.
6. Andrew Higgins, "How Moscow Uses Interpol to Pursue Its Enemies," *New York Times,* November 6, 2016, https://www.nytimes.com/2016/11/07/world/europe/how-moscow-uses-interpol-to-pursue-its-enemies.html. Browder was born in the United States but later obtained British citizenship.
7. White House, Office of the Press Secretary, "Presidential Proclamation—Suspension of Entry as Immigrants and Nonimmigrants of Persons Who Participate in Serious Human Rights and Humanitarian Law Violations and Other Abuses," August 4, 2011, https://obamawhitehouse.archives.gov/the-press-office/2011/08/04/presidential-proclamation-suspension-entry-immigrants-and-nonimmigrants-.
8. U.S. Senate, 112th Congress, 1st Session, "A Bill to Impose Sanctions on Persons Responsible for the Detention, Abuse, or Death of Sergei Magnitsky," http://www.cardin.senate.gov/imo/media/doc/mag.pdf.
9. Kathy Lally, "U.S. Puts Russian Officials on Visa Blacklist," *Washington Post,* July 25, 2011, http://www.washingtonpost.com/world/europe/us-puts-russian-officials-on-visa-blacklist/2011/07/25/gIQArcTbZI_story.html.
10. "Russia Strikes Back with Magnitsky List Response," RT News, April 13, 2013, https://www.rt.com/news/anti-magnitsky-list-russia-799/.
11. Alexandra Odynova, "State of the Wards," *Russian Life,* March/April 2013, http://www.russianlife.com/archive/article/params/Number/2560/.
12. U.S. Department of State, "Intercountry Adoption," http://travel.state.gov/content/adoptionsabroad/en/about-us/statistics.html.
13. "Russia Adoption Statistics," http://www.adoptionknowhow.com/russia/statistics.
14. "Russia's Lavrov Condemns Ban on U.S. Adoptions," Voice of Russia, December 18, 2012, http://sputniknews.com/voiceofrussia/2012_12_18/Russia-s-Lavrov-condemns-ban-on-US-adoptions/.
15. Steve Almasy, "Boston Marathon Bombing Victims: Promising Lives Lost," CNN, May 15, 2015, http://www.cnn.com/2013/04/16/us/boston-marathon-victims-profiles/.
16. Tom Winter, "Russia Warned U.S. About Tsarnaev, but Spelling Issue Let Him Escape," NBC News, March 25, 2014, http://www.nbcnews.com/storyline/boston-bombing-anniversary/russia-warned-u-s-about-tsarnaev-spelling-issue-let-him-n60836.
17. *Caucasian Knot* reported that 700 were killed in this fight in 2012, and 242 were killed in the first half of 2013. "Северный Кавказ: сложности интеграции (III): государственное управление, выборы, верховенство права," Кавказский Узел, October 7, 2013, http://www.kavkaz-uzel.ru/articles/231230/.
18. Tatiana Branigan and Miriam Elder, "Edward Snowden Arrives in Moscow," *Guardian,* June 23, 2013, https://www.theguardian.com/world/2013/jun/23/edward-snowden-arrives-moscow.
19. Matthew Cole and Robert Windrem, "How Much Did Snowden Take? At Least Three Times Number Reported," NBC News, August 30, 2013, http://www.nbcnews.com/news/other/how-much-did-snowden-take-least-three-times-number-reported-f8C11038702.
20. Will Englund, "Snowden Stayed at Russian Consulate While in Hong Kong, Report Says," *Wash-*

ington Post, August 26, 2013, https://www.washingtonpost.com/world/report-snowden-stayed-at-russian-consulate-while-in-hong-kong/2013/08/26/8237cf9a-0e39-11e3-a2b3-5e107edf9897_story.html?utm_term=.0a503ec21c3d.

21. "Edward Snowden: Leaks That Exposed U.S. Spy Programme," BBC News, January 17, 2014, http://www.bbc.com/news/world-us-canada-23123964.
22. Al Goodman. "The Many Mysteries of Snowden's Transit Zone," CNN, July 12, 2013, http://www.cnn.com/2013/07/11/world/europe/russia-snowden-goodman-transit/index.html.
23. Andrew Morse, "Snowden Says He Can't Get Guarantee of a 'Fair Trial,'" *Wall Street Journal*, March 6, 2015, http://www.wsj.com/articles/snowden-says-he-cant-get-guarantee-of-a-fair-trial-1425663946.
24. A U.S.-Russia summit was canceled by Eisenhower in response to the 1960 U-2 incident: http://www.history.com/topics/us-presidents/dwight-d-eisenhower/videos/eisenhower-returns-from-cancelled-summit-meeting.

22. GOING HOME

1. Michael McFaul verified, "Very Sad to Announce My Departure Later This Month. I Will Miss Russia and Its People http://t.co/OtXbRsUP5W @USEmbRu @StateDept @WhiteHouse," tweet, February 4, 2014, https://twitter.com/McFaul/status/430681242476888066.
2. Office of Inspector General, U.S. Department of State and the Broadcasting Board of Governors, *Inspection of Embassy Moscow and Constituent Posts, Russia,* September 2013, https://oig.state.gov/system/files/217736.pdf.
3. Kathy Lally, "McFaul Leaves Moscow and Two Dramatic Years in Relations Between U.S. and Russia," *Washington Post,* February 26, 2014, https://www.washingtonpost.com/world/europe/mcfaul-leaves-moscow-and-two-dramatic-years-in-relations-between-us-and-russia/2014/02/26/bb360742-9ef5-11e3-9ba6-800d1192d08b_story.html?utm_term=.16f63b3f49af.
4. "Данилин: Майкл Макфол уезжает расстроенный и разочарованный," *Деловая Газета "Взгляд,"* February 4, 2014, https://vz.ru/news/2014/2/4/671026.html.
5. Елена Чинкова, "Макфол не выдержал и уходит?" *КП,* November 8, 2013, http://www.kp.ru/daily/26156/3044707/.
6. Владимир Жириновский, "Зачем США отставка Макфола?" LDPR website, February 5, 2014, https://ldpr.ru/events/why_us_mcfauls_resignation/. Translated from Russian by Anya Shkurko.
7. Леонид Гозман, "Майкл Макфол уходит. Его ненавидели не потому, что он враг России, а потому, что друг," *Новая Газета,* February 5, 2014, https://www.novayagazeta.ru/articles/2014/02/05/58251-maykl-makfol-uhodit-ego-nenavideli-ne-potomu-chto-on-150-vrag-rossii-a-potomu-chto-drug.
8. Travis Waldron, "American Runner Wins Silver Medal in Russia, Dedicates It to LGBT Friends," Think Progress, August 4, 2013, http://thinkprogress.org/sports/2013/08/14/2465911/american-runner-in-russia-dedicates-world-championship-silver-medal-to-lgbt-friends/.
9. Information about the Gay Games can be found on the website gaygames.org.

23. ANNEXATION AND WAR IN UKRAINE

1. Bradley Klapper, "Clinton Fears Efforts to 'Re-Sovietize' in Europe," Yahoo! News, December 6, 2012, https://www.yahoo.com/news/clinton-fears-efforts-sovietize-europe-111645250--politics.html?ref=gs.
2. Vladimir Putin, "Press Statement and Answers to Journalists' Questions Following the APEC Leaders' Meeting," President of Russia official website, March 8, 2013, Eng.kremlin.ru/transcripts/6093.
3. Vladimir Putin, "Excerpts from Transcript of the Meeting of the Valdai International Discussion Club," President of Russia official website, September 19, 2013, http://en.kremlin.ru/events/president/news/19243.

4. Samuel Charap and Timothy J. Colton, *Everyone Loses: The Ukraine Crisis and the Ruinous Contest for Post-Soviet Eurasia* (New York: International Institute for Strategic Studies, 2017), 120–21.
5. "How the EU Lost Ukraine," *Spiegel,* November 25, 2013, http://www.spiegel.de/international/europe/how-the-eu-lost-to-russia-in-negotiations-over-ukraine-trade-deal-a-935476.html.
6. "Ukraine Protests After Yanukovych EU Deal Rejection," BBC News, November 30, 2013, http://www.bbc.com/news/world-europe-25162563.
7. Charap and Colton, *Everyone Loses,* 121.
8. The Russian government paid the bill by purchasing Ukrainian Eurobonds and delivering subsidized gas. Shaun Walker, "Vladimir Putin Offers Ukraine Financial Incentives to Stick with Russia," *Guardian,* December 18, 2013, https://www.theguardian.com/world/2013/dec/17/ukraine-russia-leaders-talks-kremlin-loan-deal.
9. Joshua Yaffa, "Reforming Ukraine After the Revolutions," *New Yorker,* September 5, 2016, http://www.newyorker.com/magazine/2016/09/05/reforming-ukraine-after-maidan.
10. Dmytro Gorshkov and Oleksandr Savochenko, "Ukraine Leader Signs Controversial Anti-Protest Law," Yahoo! News, January 17, 2014, https://uk.news.yahoo.com/ukraine-president-signs-anti-protest-bills-law-official-201313983.html.
11. "Ukraine Crisis: Police Storm Main Kiev 'Maidan' Protest Camp," BBC News, February 19, 2014, http://www.bbc.com/news/world-europe-26249330.
12. Andrew E. Kramer and Andrew Higgins, "Ukraine's Forces Escalate Attacks Against Protestors," *New York Times,* February 20, 2014, http://www.nytimes.com/2014/02/21/world/europe/ukraine.html?_r=0.
13. White House, Office of the Press Secretary, "Remarks by President Obama Before Restricted Bilateral Meeting," February 19, 2014, https://www.whitehouse.gov/the-press-office/2014/02/19/remarks-president-obama-restricted-bilateral-meeting.
14. Ibid.
15. Ibid.
16. "Agreement on the Settlement of Crisis in Ukraine — Full Text," *Guardian,* February 21, 2014, https://www.theguardian.com/world/2014/feb/21/agreement-on-the-settlement-of-crisis-in-ukraine-full-text.
17. "Rice: That Would be a Grave Mistake," Politico.com, February 23, 2014, http://www.politico.com/story/2014/02/kiev-ukraine-russia-susan-rice-grave-mistake-103821.
18. "Vladimir Putin Answered Journalists' Questions on the Situation in Ukraine," President of Russia official website, March 4, 2014, http://en.kremlin.ru/events/president/news/20366.
19. Ibid.
20. Ibid.
21. "Putin Acknowledges Russian Servicemen Were in Crimea," RT, April 17, 2014, https://www.rt.com/news/crimea-defense-russian-soldiers-108/.
22. "Экс-глава СБУ назвал тех, кто получил прибыль от бойни на Майдане," Vesti.ru, March 12, 2014, http://www.vesti.ru/doc.html?id=1368925&tid=105474.
23. Vladimir Putin, "Address by President of the Russian Federation," President of Russia official website, March 18, 2014, http://eng.kremlin.ru/news/6889.
24. "Is Crimea's Referendum Legal?" BBC News, March 13, 2014, http://www.bbc.com/news/world-europe-26546133.
25. Carol Morello, Will Englund, and Griff Witte, "Crimea's Parliament Votes to Join Russia," *Washington Post,* March 17, 2014, https://www.washingtonpost.com/world/crimeas-parliament-votes-to-join-russia/2014/03/17/5c3b96ca-adba-11e3-9627-c65021d6d572_story.html.
26. Paul Roderick Gregory, "Putin's 'Human Rights Council' Accidentally Posts Real Crimean Election Results," *Forbes,* May 5, 2014, http://www.forbes.com/sites/paulroderickgregory/2014/05/05/putins-human-rights-council-accidentally-posts-real-crimean-election-results-only-15-voted-for-annexation/#39f8c23010ff.
27. Vladimir Putin, "Meeting in Support of Crimea's Accession to the Russian Federation 'We Are

Together!'" President of Russia official website, March 18, 2014, http://eng.kremlin.ru/news/6892.

28. "Putin in 2008: 'Crimea Is Not Disputed Territory' and Is Part of Ukraine," *Business Insider,* April 7, 2015, http://www.businessinsider.com/putin-in-2008-crimea-is-not-disputed-territory-and-is-part-of-ukraine-2015-4.
29. Louis Charbonneau and Mirjam Donath, "U.N. General Assembly Declares Crimea Secession Vote Invalid," Reuters, March 27, 2014, http://www.reuters.com/article/us-ukraine-crisis-un-idUSBREA2Q1GA20140327.
30. Alison Smale and Michael Shear, "Russia Is Ousted from Group of 8 by U.S. and Allies," *New York Times,* March 24, 2014, http://www.nytimes.com/2014/03/25/world/europe/obama-russia-crimea.html.
31. That said, there were a few Russians on the official sanctions list that did not fit my criteria and whose inclusion made no sense to me.
32. White House, Office of the Press Secretary, "Background Briefing on Ukraine by Senior Administration Officials," March 20, 2014, http://www.whitehouse.gov/the-press-office/2014/03/20/background-briefing-ukraine-senior-administration-officials.
33. White House, Office of the Press Secretary, "Remarks by the President in Address to European Youth," March 26, 2014, https://obamawhitehouse.archives.gov/the-press-office/2014/03/26/remarks-president-address-european-youth.
34. Ministry of Foreign Affairs of the Russian Federation, "Statement by the Russian Ministry of Foreign Affairs on Retaliatory Sanctions with Regard to Several Officials and Members of the US Congress," March 20, 2014, http://www.mid.ru/BDOMP/Brp_4.nsf/arh/C2FC687EBC93876C44257CA5004D380D?OpenDocument.
35. "Russia Sanctions 9 US Officials in Response to US Sanctions on Russian Officials," CNBC, March 20, 2014, http://www.cnbc.com/2014/03/20/obama-authorizes-new-sanctions-on-russia-over-crimea.html.
36. In supporting these sanctions, Bush's national security advisor, Steve Hadley, wished in retrospect that the administration had imposed similar kinds of sanctions on Russia after the invasion of Georgia in August 2008. See David Jackson, "Bush Aide: We Should Have Sanctioned Russia over Georgia," *USA Today,* May 1, 2014, http://www.usatoday.com/story/theoval/2014/05/01/obama-george-w-bush-russia-sanctions-ukraine-georgia/8566645/.
37. Putin, as quoted in David Herszenhorn, "What Is Putin's 'New Russia'?" *New York Times,* April 18, 2014, https://www.nytimes.com/2014/04/19/world/europe/what-is-putins-new-russia.html?mcubz=2.
38. Ibid.
39. Aric Toler and Melinda Haring, "Russia Funds and Manages Conflict in Ukraine, Leaks Show," Atlantic Council, April 24, 2017, http://www.atlanticcouncil.org/blogs/ukrainealert/russia-funds-and-manages-conflict-in-ukraine-leaks-show.
40. "Ukraine Crisis: Timeline," BBC News, November 13, 2014, http://www.bbc.com/news/world-middle-east-26248275.
41. "Eastern Ukraine Insurgents Declare Referendum Victory, Seek Russia Annexation," CBS News, May 12, 2014, http://www.cbsnews.com/news/russia-walks-cautious-line-eastern-ukraine-referendums-donetsk-luhansk/.
42. "Despite Concerns About Governance, Ukrainians Want to Remain One Country," Pew Research Center, May 8, 2014, http://www.pewglobal.org/files/2014/05/Pew-Global-Attitudes-Ukraine-Russia-Report-FINAL-May-8-2014.pdf.
43. Kiev International Institute of Sociology, "The Views and Opinions of South-Eastern Residents of Ukraine: April 2014," April 20, 2014, http://www.kiis.com.ua/?lang=eng&cat=news&id=258.
44. "How Did Odessa Fire Happen?" BBC News, May 6, 2014, http://www.bbc.com/news/world-europe-27275383.
45. Somini Sengupta and Andrew E. Kramer, "Dutch Inquiry Links Russia to 298 Deaths in Explosion of Jetliner over Ukraine," *New York Times,* September 28, 2016, http://www.nytimes.com/2016/09/29/world/asia/malaysia-air-flight-mh17-russia-ukraine-missile.html.

46. Vladimir Putin, "Meeting on Economy Began with a Moment of Silence in Honour of Victims of Plane Crash over Ukrainian Territory," President of Russia official website, July 18, 2014, http://en.kremlin.ru/events/president/news/46243.
47. Julian Borger, Alec Luhn, and Richard Norton-Taylor, "EU Announces Further Sanctions on Russia After Downing of MH17," *Guardian,* July 22, 2014, https://www.theguardian.com/world/2014/jul/22/eu-plans-further-sanctions-russia-putin-mh17.
48. Michael McFaul, "Confronting Putin's Russia," *New York Times,* March 23, 2014, http://www.nytimes.com/2014/03/24/opinion/confronting-putins-russia.html.

24. THE END OF RESETS (FOR NOW)

1. Munich Security Conference, Dmitry Medvedev's speech at the panel discussion, February 13, 2016, website of the Russian Government, http://government.ru/en/news/21784/.
2. For those conversant in IR theory, you will see that I am loosely deploying Ken Waltz's three levels of analysis to frame this discussion of competing explanations. See Waltz, *Man, the State, and War: A Theoretical Analysis* (New York: Columbia University Press, 1959). In my analysis I highlight the importance of "the first image," that is, individual leaders as a driving causal variable, though also note the importance of "the second image," regime type. On more recent studies of the roles of leaders, see Chiozza and Goemans, *Leaders and International Conflict,* and Horowitz, Stam, and Ellis, *Why Leaders Fight.*
3. Paul Kennedy, *Rise and Fall of Great Powers;* John Mearsheimer, *The Tragedy of Great Power Politics;* and Ken Waltz, *Theory of International Politics* (New York: McGraw-Hill, 1979).
4. John Mearsheimer, "Why the Ukraine Crisis Is the West's Fault," *Foreign Affairs,* September/October 2014, 82, 84.
5. Stephen Kinzer, "Russia Acts Like Any Other Superpower," *Boston Globe,* May 11, 2014.
6. To be sure, sometimes weak and collapsing states negatively impact American interests, as we learned tragically on September 11, 2001. When the Soviet Union was collapsing, we also worried about the threat of "loose nukes"—a threat created by Soviet and then Russian weakness, not strength. That noted, our current conflict with Russia has not resulted from Russian weakness.
7. This is Waltz's "second image." See Waltz, *Man, the State, and War,* chapter 4.
8. Russians are not alone in advancing this explanation. Stephen Cohen has made this kind of argument most consistently, during the Clinton, Bush, and Obama administrations. See, for instance, his *Failed Crusade: America and the Tragedy of Post-Communist Russia* (New York: Norton 2000), or more recently "The Fight over a Trump-Putin Détente Begins," *Nation,* January 18, 2017, https://www.thenation.com/article/the-fight-over-a-trump-putin-detente-begins/.
9. According to the World Bank, Russia's GDP contracted 47.6 percent from 1990 to 1998; $516.8 billion in 1990 to $271 billion in 1998: http://data.worldbank.org/country/russian-federation.
10. Michael McFaul, "1789, 1917 Can Guide '90s Soviets," *San Jose Mercury News,* August 19, 1990.
11. White House, "Remarks by President Obama and President Medvedev of Russia After Bilateral Meeting," Seoul, South Korea, *The White House,* March 26, 2012, https://obamawhitehouse.archives.gov/the-press-office/2012/03/26/remarks-president-obama-and-president-medvedev-russia-after-bilateral-me.
12. Dana Bash and Adam Levy, "Boehner: Bush Would Have Punched Putin in the Nose," CNN, October 28, 2014, http://www.cnn.com/2014/10/27/politics/boehner-bush-punch-putin/.
13. Marc Thiessen, "Obama's Weakness Emboldens Putin," *Washington Post,* March 3, 2014, https://www.washingtonpost.com/opinions/marc-thiessen-obamas-weakness-emboldens-putin/2014/03/03/28def926-a2e2-11e3-84d4-e59b1709222c_story.html?utm_term=.ccda19ed14ad.
14. Rice, *No Higher Honor,* 673.
15. "Ukrainians Likely Support Move Away from NATO," Gallup.com, April, 2, 2010, http://www.gallup.com/poll/127094/Ukrainians-Likely-Support-Move-Away-NATO.aspx.
16. Two decades earlier, I argued that the shift toward more democratic politics pushed Russian foreign policy in a pro-Western direction, but worried that a reversal of democracy in Russia would have

the opposite effect. See McFaul, "The Precarious Peace: Domestic Politics in the Making of Russian Foreign Policy," *International Security* 22, no. 3 (Winter 1997/98): 5–35. The classic study on the relationship between domestic politics in Kremlin foreign policy remains Alexander Dallin, "Soviet Foreign Policy and Domestic Politics: A Framework for Analysis," *Journal of International Affairs* 23, no. 2 (1969): 250–65.

17. "Putin's Approval Rating," Levada-Center, accessed May 18, 2017, http://www.levada.ru/en/ratings/.
18. Ben Judah, *Fragile Empire: How Russia Fell In and Out of Love with Vladimir Putin* (New Haven, CT: Yale University Press, 2013).
19. World Bank, "GDP Growth (Annual %)," accessed May 18, 2017, http://data.worldbank.org/indicator/NY.GDP.MKTP.KD.ZG?end=2015&locations=RU&start=2006.
20. Tatiana Stanovaya, "Why Russia Can't Have a Public Television Network," Institute of Modern Russia, July 25, 2012, https://imrussia.org/en/analysis/politics/270-russias-newest-state-controlled-television-network.
21. Miriam Elder, "Dimitry Medvedev Proposes Electoral Reforms to Appease Russian Protesters," *Guardian,* December 22, 2011, https://www.theguardian.com/world/2011/dec/22/dmitry-medvedev-proposes-electoral-reforms.
22. Michael McFaul, "1789, 1917 Can Guide '90s Soviets," *San Jose Mercury News,* August 19, 1990.
23. Karl Marx, "The Eighteenth Brumaire of Louis Bonaparte, original version published in the first number of the monthly," *Die Revolution* (1852).
24. Seymour Martin Lipset, "Some Social Requisites of Democracy: Economic Development and Political Legitimacy," *American Political Science Review* 53, no. 1 (March 1959): 69–105.
25. I realize that I am pivoting here to the deployment of structural, not agency, arguments.
26. These are the arguments of the democratic peace theory. For a review of this literature, see McFaul, *Advancing Democracy Abroad,* chapter 2.

EPILOGUE: TRUMP AND PUTIN

1. John McCain, "Obama Has Made America Look Weak," *New York Times,* March 14, 2017, https://www.nytimes.com/2014/03/15/opinion/mccain-a-return-to-us-realism.html.
2. Cruz, as quoted in Edward-Isaac Dovere, "Politics of Blaming Obama on Ukraine," *Politico,* March 3, 2014. http://www.politico.com/story/2014/03/democrats-obama-ukraine-russia-crimea-104322.
3. Sabrina Siddiqui, "Vladimir Putin Is 'a Gangster and Thug,' says U.S. Presidential Candidate Marco Rubio," *Guardian,* October 2, 2015, https://www.theguardian.com/us-news/2015/oct/03/vladimir-putin-is-a-gangster-and-thug-says-us-presidential-candidate-marco-rubio.
4. Hillary Clinton, *What Happened* (New York: Simon & Shuster, 2017), 332.
5. Presidential Candidate Donald Trump Primary Night Speech, C-SPAN, April 26, 2016, https://www.c-span.org/video/?408719-1/donald-trump-primary-night-speech&start=1889&transcriptQuery=putin.
6. Donald Trump Campaign Rally in Hilton Head, South Carolina, C-SPAN, December 30, 2015, https://www.c-span.org/video/?402610-1/donald-trump-campaign-rally-hilton-head-south-carolina&transcriptQuery=putin&start=787.
7. Donald Trump Campaign Rally in Vandalia, Ohio, C-SPAN, March 12, 2016, https://www.c-span.org/video/?406393-1/donald-trump-campaign-rally-vandalia-ohio&transcriptQuery=putin&start=1907.
8. Alex Griswold, "Trump Defends Putin's Murder of Journalists: 'Our Country Does Plenty of Killing Also,'" *Mediaite,* December 18, 2015, http://www.mediaite.com/tv/donald-trump-defends-putins-murder-of-journalists-our-country-does-plenty-of-killing-also/.
9. Reena Flores, "Donald Trump Gives Russia's Putin an 'A' in Leadership," CBS News, September 30, 2015, https://www.cbsnews.com/news/donald-trump-gives-russias-putin-an-a-in-leadership/.
10. Tyler Pager, "Trump to Look at Recognizing Crimea as Russian Territory, Lifting Sanctions," *Politico,* July 27, 2016, http://www.politico.com/story/2016/07/trump-crimea-sanctions-russia-226292.
11. Carol Morello and Adam Taylor, "Trump Says U.S. Won't Rush to Defend NATO Countries If They

Don't Spend More on Military," *Washington Post,* July 21, 2016, https://www.washingtonpost.com/world/national-security/trump-says-us-wont-rush-to-defend-nato-countries-if-they-dont-spend-more-on-military/2016/07/21/76c48430-4f51-11e6-a7d8-13d06b37f256_story.html?tid=a_inl&utm_term=.c864abc4ab0c. Candidate Trump actually argued inaccurately that NATO allies were not paying enough to the United States. NATO does not operate that way. For elaboration, see Michael McFaul, "Mr. Trump, NATO Is an Alliance, Not a Protection Racket," *Washington Post,* July 25, 2017, https://www.washingtonpost.com/opinions/global-opinions/mr-trump-nato-is-an-alliance-not-a-protection-racket/2016/07/25/03ca2712-527d-11e6-88eb-7dda4e2f2aec_story.html?utm_term=.4f2db182f676.

12. Griswold, "Trump Defends Putin's Murder of Journalists."
13. Abby Philip, "O'Reilly Told Trump That Putin Is a Killer. Trump's Reply: 'You Think Our Country Is So Innocent?'" *Washington Post,* February 4, 2017, https://www.washingtonpost.com/news/post-politics/wp/2017/02/04/oreilly-told-trump-that-putin-is-a-killer-trumps-reply-you-think-our-countrys-so-innocent/?utm_term=.5d54f7ad3b79].
14. Hillary Clinton Remarks on Counterterrorism, C-SPAN, March 23, 2016, https://www.c-span.org/video/?407164-1/hillary-clinton-remarks-counterterrorism.
15. Office of the Director of National Intelligence, *Background to "Assessing Russian Activities and Intentions on Recent US Elections": The Analytic Process and Cyber Incident Attribution,* January 6, 2017, https://www.dni.gov/files/documents/ICA_2017_01.pdf, 1.
16. Franklin Foer, "It's Putin's World," *Atlantic,* March 2017, https://www.theatlantic.com/magazine/archive/2017/03/its-putins-world/513848/?utm_source=atltw.
17. Owen Matthews, "Alexander Dugin and Steve Bannon's Ideological Ties to Vladimir Putin's Russia," *Newsweek,* April 17, 2017, http://www.newsweek.com/steve-bannon-donald-trump-jared-kushner-vladimir-putin-russia-fbi-mafia-584962.
18. Steve Reilly and Brad Heath, "Steve Bannon's Own Words Show Sharp Break on Security Issues," *USA Today,* January 31, 2017, http://www.usatoday.com/story/news/2017/01/31/bannon-odds-islam-china-decades-us-foreign-policy-doctrine/97292068/.
19. Clinton, *What Happened,* 327.
20. ODNI, *Background to "Assessing Russian Activities and Intentions on Recent US Elections."*
21. "Plenary Session of St. Petersburg International Economic Forum," President of Russia official website, June 17, 2016, http://en.kremlin.ru/events/president/news/52178.
22. "Демократия и права человека: Трамп не сотрясает воздух пустыми лозунгами," Vesti.ru, November 13, 2016, http://www.vesti.ru/doc.html?id=2821081.
23. Владимир Ворсобин, "Член международного комитета Госдумы Виталий Милонов: Клинтон — проклятая ведьма, а Трамп — просто смешной чувак," *Комсомольская Правда,* October 22, 2016, https://www.kp.ru/daily/26598.7/3613479/.
24. "Пушков поделился мнением о кандидатах в президенты США," Lenta.ru, April 26, 2016, https://lenta.ru/news/2016/04/26/pushkov/.
25. "Киселёв: тормознутость Клинтон на руку России," Vesti.ru, September 25, 2016, http://www.vesti.ru/doc.html?id=2802850; "Клинтон с Обамой подготовились к победе Трампа," Vesti.ru, October 30, 2016, http://www.vesti.ru/doc.html?id=2816140; and "Особое Мнение, Сергей Марков," Ekho Moskvy, November 8, 2016, http://echo.msk.ru/programs/personalno/1869710-echo/.
26. Dmitri Alperovitch, *Bears in the Midst: Intrusion into the Democratic National Committee,* CrowdStrike, June 15, 2016, https://www.crowdstrike.com/blog/bears-midst-intrusion-democratic-national-committee/. Russian cyber actors penetrated the DNC email system as early as the summer of 2015, but their operations became publicly known only in June 2016.
27. In her memoir, Clinton writes, "Assange, like Putin, has held a grudge against me for a long time." Clinton, *What Happened,* 344.
28. On WikiLeaks' ties to the Russian government, see Andrei Soldatov and Irina Borogan, *The Red Web: The Kremlin's War on the Internet* (New York: Public Affairs, 2017), 312–17, 330.
29. ODNI, *Background to "Assessing Russian Activities and Intentions on Recent US Elections,"* ii.
30. "Joint Statement from the Department of Homeland Security and Office of the Director of Na-

tional Intelligence on Election Security," October 7, 2016, https://www.dhs.gov/news/2016/10/07/joint-statement-department-homeland-security-and-office-director-national.

31. ODNI, *Background to "Assessing Russian Activities and Intentions on Recent US Elections,"* ii.
32. Ibid., 4. On these other examples, see "What Does It Take to Bring Hillary Clinton to Justice?" RT, November 3, 2016, https://www.rt.com/op-edge/365204-clinton-us-investigation-justice/.
33. The RT video asserting that Chelsea Clinton's wedding was paid for by the Clinton Foundation was viewed nine million times. See ODNI, *Background to "Assessing Russian Activities and Intentions on Recent US Elections,"* 4.
34. Scott Shane, "To Sway Vote, Russia Used Army of Fake Americans," *New York Times*, September 8, 2107; and Clint Watts, "Disinformation: A Primer in Russian Active Measures and Influence Campaigns," Statement Prepared for the U.S. Senate Select Committee on Intelligence Hearing, March 30, 2017, https://www.intelligence.senate.gov/sites/default/files/documents/os-cwatts-033017.pdf.
35. Senator Mark Warner, opening statement before the Senate Select Committee on Intelligence, March 30, 2017, hearing, https://www.mediamatters.org/video/2017/03/30/sen-mark-warner-russia-used-internet-trolls-push-out-disinformation-and-fake-news-which-was-then/215870. See also Massimo Calabresi, "Inside Russia's Social Media War on America," *Time,* May 18, 2017, http://time.com/4783932/inside-russia-social-media-war-america/.
36. Clinton, *What Happened,* 362.
37. Testimony from former FBI agent Clint Watts before the Senate Select Committee on Intelligence, March 30, 2017, https://www.intelligence.senate.gov/sites/default/files/documents/os-cwatts-033017.pdf. See also Peter Stone and Greg Gordon, "Trump-Russia Investigators Probe Jared Kushner–Run Digital Operation," McClatchy DC Bureau, July 12, 2017, http://amp.mcclatchydc.com/news/nation-world/national/article160803619.html.
38. Scott Shane and Vindu Goel, "Fake Russian Facebook Accounts Bought $100,000 in Political Ads," *New York Times,* September 6, 2017, https://www.nytimes.com/2017/09/06/technology/facebook-russian-political-ads.html; and Alex Stamos, "An Update on Information Operations on Facebook," Facebook Newsroom, September 6, 2017, https://newsroom.fb.com/news/2017/09/information-operations-update/.
39. On this Russian strategy more generally, see Peter Pomerantsev, *Nothing Is True and Everything Is Possible: The Surreal Heart of the New Russia* (New York: Public Affairs, 2015).
40. Michael McFaul, "Why Putin Wants a Trump Victory (So Much He Might Even Be Trying to Help Him)," *Washington Post,* August 17, 2016, https://www.washingtonpost.com/opinions/global-opinions/why-putin-wants-a-trump-victory-so-much-he-might-even-be-trying-to-help-him/2016/08/17/897ab21c-6495-11e6-be4e-23fc4d4d12b4_story.html?utm_term=.233ca7204027.
41. Robert Faris et al., *Partisanship, Propaganda, and Disinformation: Online Media and the 2016 U.S. Presidential Election* (Cambridge, MA: Berkman Klein Center for Internet & Society at Harvard University, 2017), 88–89.
42. Clinton, *What Happened,* 342.
43. Jo Becker, Adam Goldman, and Matt Apuzzo, "Russian Dirt on Clinton? 'I Love It,' Donald Trump Jr. Said," *New York Times,* July 21, 2017, https://www.nytimes.com/2017/07/11/us/politics/trump-russia-email-clinton.html.
44. Ibid.
45. Sharon LaFraniere, David Kirkpatrick, and Kenneth Vogel, "Lobbyist at Trump Campaign Meeting Has a Web of Russian Connections," *New York Times,* August 21, 2017, https://www.nytimes.com/2017/08/21/us/rinat-akhmetshin-russia-trump-meeting.html?mcubz=3.
46. Niraj Chokshi, "Michael T. Flynn: A Timeline of His Brief Tenure," *New York Times,* February 14, 2017, https://www.nytimes.com/2017/02/14/us/politics/michael-flynn-resigns-trump-timeline.html.
47. Some of these meetings appear to be for business reasons. Others are still hard to explain. See, for instance, Matthew Rosenberg, Mark Mazzetti, and Maggie Haberman, "Investigation Turns to Kushner's Motives in Meeting with a Putin Ally," *New York Times,* May 29, 2017, https://www.nytimes.com/2017/05/29/us/politics/jared-kushner-russia-investigation.html; and Rosalind Helderman, Carol Leonnig, and Tom Hamburger, "Top Trump Organization Executive Asked Putin Aide

for Help on Business Deal," *Washington Post,* August 28, 2017, https://www.washingtonpost.com/politics/top-trump-organization-executive-reached-out-to-putin-aide-for-help-on-business-deal/2017/08/28/095aebac-8c16-11e7-84c0-02cc069f2c37_story.html?utm_term=.cf491edf7346.

48. Greg Miller, Ellen Nakashima, and Adam Entous, "Obama's Secret Struggle to Punish Russia for Putin's Election Assault," *Washington Post,* June 23, 2017, https://www.washingtonpost.com/graphics/2017/world/national-security/obama-putin-election-hacking/?utm_term=.46e76ca0a592.
49. Ibid.
50. Michael Riley and Jordan Robertson, "Russian Cyber Hacks on U.S. Electoral System Far Wider Than Previously Known," *Bloomberg,* June 13, 2017, https://www.bloomberg.com/news/articles/2017-06-13/russian-breach-of-39-states-threatens-future-u-s-elections.
51. Clinton, *What Happened,* 345.
52. #TrumpNash is an echo of the hashtag #KrymNash, or "Crimea Is Ours," which circulated widely in 2014 after Russia annexed Crimea.
53. Soldatov and Borogan, *The Red Web,* 331.
54. Zamira Rahim, "Donald Trump Has a Much Higher Approval Rating in Russia Than He Does in America, State Pollster Says," *Fortune,* January 25, 2017, http://fortune.com/2017/01/25/donald-trump-russia-approval-ratings/.
55. "Пушков: при Трампе появилась перспектива отмены санкций," Ria News, November 11, 2016, https://ria.ru/economy/20161111/1481188003.html.
56. Morris P. Fiorina, "The 2016 Presidential Election — Identities, Class, and Culture," Hoover Institution Essay on Contemporary American Politics, series no. 11 (June 2017), http://www.hoover.org/sites/default/files/research/docs/352022476-the-2016-presidential-election-identities-class-and-culture.pdf.
57. This figure comes from the Cooperative Congressional Election Study, as quoted in Danielle Kurtzleben, "Here's How Many Bernie Sanders Supporters Ultimately Voted for Trump," August 24, 2017, NPR.org, http://www.npr.org/2017/08/24/545812242/1-in-10-sanders-primary-voters-ended-up-supporting-trump-survey-finds?utm_source=twitter.com&utm_medium=social&utm_campaign=politics&utm_term=nprnews&utm_content=20170824.
58. Mirren Gidda, "Third Party Votes Could Have Cost Hillary Clinton the Presidency," *Newsweek,* November 9, 2016, http://www.newsweek.com/susan-sarandon-third-party-candidates-jill-stein-gary-johnson-hillary-clinton-519032.
59. Clinton, *What Happened,* 349.
60. Ibid., 348.
61. "James Comey, Mike Rogers Testify on Wiretapping, Russia: Live Analysis," *Wall Street Journal,* March 20, 2017, http://www.wsj.com/livecoverage/comey-house-intel-russia-hearing-trump/card/1490029657.
62. Andrew Higgins, "Maybe Private Russian Hackers Meddled in Election, Putin Says," *New York Times,* https://www.nytimes.com/2017/06/01/world/europe/vladimir-putin-donald-trump-hacking.html.
63. Vladimir Putin, "Interview to NBC," President of Russia official website, June 5, 2017, http://en.kremlin.ru/events/president/news/54688.
64. "Plenary Session of St. Petersburg International Economic Forum," President of Russia official website, June 17, 2016, http://en.kremlin.ru/events/president/news/52178.
65. Miller, Nakashima, and Entous, "Obama's Secret Struggle to Punish Russia for Putin's Election Assault."
66. Bryan Bender, "The Secret U.S. Army Study That Targets Moscow," *Politico,* April 14, 2016, http://www.politico.eu/article/the-secret-us-army-study-that-targets-moscow/.
67. Julia Manchester, "Nikki Haley: 'Everybody Knows That Russia Meddled in Our Elections,'" *Hill,* July 8, 2017, http://thehill.com/homenews/administration/341103-nikki-haley-everybody-knows-that-russia-meddled-in-our-elections.
68. Ambassador Nikki Haley, "Remarks at a UN Security Council Briefing on Ukraine," United States Mission to the United Nations, February 2, 2017, https://usun.state.gov/remarks/7668.
69. "Transcript: Secretary of State Tillerson on Trump's Meeting with Putin," NPR.org, July 7, 2017,

http://www.npr.org/2017/07/07/536035953/transcript-secretary-of-state-tillerson-on-trumps-meet ing-with-putin.

70. David A. Graham, "Trump Told the Russians That 'Nut Job' Comey's Firing Relieved 'Great Pressure,'" *Atlantic,* May 19, 2017, https://www.theatlantic.com/politics/archive/2017/05/trump-told-rus sians-that-nut-job-comeys-firing-relieved-pressure/527451/.
71. Tillerson, as quoted in "Transcript: Secretary of State Tillerson on Trump's Meeting with Putin," NPR.org, July 7, 2017, http://www.npr.org/2017/07/07/536035953/transcript-secretary-of-state-tiller son-on-trumps-meeting-with-putin.
72. Ibid.
73. Julie Hirschfeld Davis, "Trump and Putin Held a Second, Undisclosed, Private Conversation," *New York Times,* July 18, 2017, https://www.nytimes.com/2017/07/18/world/europe/trump-putin-undis closed-meeting.html.
74. Donald Trump verified, "The Democrats Had to Come Up with a Story," tweet, February 16, 2017, https://twitter.com/realdonaldtrump/status/832238070460186625?lang=en.
75. "The Trump Tweet Tracker," *Atlantic,* June 4, 2017, https://www.theatlantic.com/liveblogs/2017/06/ donald-trump-twitter/511619/.
76. BuzzFeed published the so-called Trump Dossier, containing salacious but difficult-to-verify claims about the Russian government's collection of compromising material on Trump. The complete dossier is published at: https://www.buzzfeed.com/kenbensinger/these-reports-allege-trump-has-deep-ties-to-russia?utm_term=.jugJJp1B8#.swNWWeJzg.
77. Craig Unger, "Trump's Russian Laundromat," *New Republic,* July 13, 2017, https://newrepublic.com/ article/143586/trumps-russian-laundromat-trump-tower-luxury-high-rises-dirty-money-interna tional-crime-syndicate. See also Malcolm Nance, *The Plot to Hack America* (New York: Skyhorse, 2016), chapter 4.
78. Some analysts have asserted that the Russian government had meddled in earlier elections, but the scale and scope of the 2016 activities is unprecedented.
79. Dominic Tierney, "Trump, Putin, and the Art of Appeasement," *Atlantic,* December 15, 2016, https://www.theatlantic.com/international/archive/2016/12/trump-putin-and-the-art-of-appease ment/510767/.
80. White House, Office of the Press Secretary, "Statement by President Donald J. Trump on Signing the 'Countering America's Adversaries Through Sanctions Act,'" August 2, 2017, https://www.white house.gov/the-press-office/2017/08/02/statement-president-donald-j-trump-signing-countering-americas.
81. Philip Rucker, "Trump Says He Is 'Very Thankful' to Putin for Expelling U.S. Diplomats from Russia," *Washington Post,* August 10, 2017, https://www.washingtonpost.com/news/post-politics/wp/ 2017/08/10/trump-says-he-is-very-thankful-to-putin-for-expelling-u-s-diplomats-from-russia /?utm_term=.b7dd517ec6eb.
82. Dmitry Medvedev official Facebook page, August 2, 2017, https://www.facebook.com/Dmitry.Med vedev/posts/10154587161801851.
83. "Сергей Марков: американский истеблишмент с помощью санкций ведёт войну против самого Трампа," *Parlamentskaya Gazeta,* August 1, 2017, https://www.pnp.ru/politics/sergey-markov -amerikanskiy-isteblishment-s-pomoshhyu-sankciy-vedyot-voynu-protiv-samogo-trampa.html.
84. "Vladimir Putin's Interview with Le Figaro," President of Russia official website, May 31, 2017, http:// en.kremlin.ru/events/president/news/54638.
85. Ibid.
86. "В Кремле прокомментировали сравнение Трампа с Ким Чен Ыном в программе Киселева," Interfax, April 17, 2017, http://www.interfax.ru/russia/558772.
87. "ВЦИОМ зафиксировал резкое разочарование россиян в Трампе," RBC News, August 4, 2017, http://www.rbc.ru/politics/04/08/2017/59832f3d9a794722c55aa360.
88. Angela Dewan, "U.S. Could Work with Russia on Syria No-Fly Zones, Tillerson Says," CNN.com, July 6, 2017, http://www.cnn.com/2017/07/06/politics/rex-tillerson-syria-no-fly-zone/index.html.

89. Damien Sharkov, "Putin Vows Military Response to 'Eliminate NATO Threat' If Sweden Joins U.S.-Led Alliance," *Newsweek*, June 2, 2017, http://www.newsweek.com/vladimir-putin-vows-eliminate-nato-threat-sweden-joins-619486.
90. "Moscow Calls U.S. Visa Move Attempt to Stir Up Unrest in Russia," Reuters, August 21, 2017, https://www.reuters.com/article/us-usa-trump-russia-lavrov-idUSKCN1B1146.
91. "Trump Drops to New Low, Close to 2–1 Disapproval, Quinnipiac University National Poll Finds; 71 Percent Say President Is Not Levelheaded," Quinnipiac University Poll, August 2, 2017, https://poll.qu.edu/images/polling/us/us08022017_U348kmpa.pdf/.
92. "U.S. Voters Send Trump Approval to Near Record Low, Quinnipiac University National Poll Finds; No Winner in Media War, but Voters Trust Media More," Quinnipiac University Poll, May 10, 2017, https://poll.qu.edu/national/release-detail?ReleaseID=2456.
93. Michael McFaul, "We Can't Let Trump Go Down Putin's Path," *Washington Post*, February 6, 2017, https://www.washingtonpost.com/news/democracy-post/wp/2017/02/06/we-cant-let-trump-go-down-putins-path/?utm_term=.1b417d3cc406.
94. Margaret Vice, "Publics Worldwide Unfavorable Toward Putin, Russia," Pew Research Center, August 16, 2017, http://www.pewglobal.org/2017/08/16/publics-worldwide-unfavorable-toward-putin-russia/.
95. "Confidence in Putin vs. Trump," Pew Research Center, August 15, 2017, http://www.pewglobal.org/2017/08/16/publics-worldwide-unfavorable-toward-putin-russia/pg_2017-08-16_views-of-russia_003/.

INDEX